The Aloe Genome

The book delves into the intricate world of *Aloe vera*, offering a multidimensional exploration of its biology and genetic makeup. This compilation is meticulously curated, presenting a wide array of topics crucial for understanding and enhancing *Aloe vera*'s agricultural significance. The chapters delve into cutting-edge areas like genomics, transcriptomics, proteomics, metabolomics, and nano-technological advancements specific to *Aloe vera*. It addresses molecular markers, mapping, and genome-wide characterization for breeding, in vitro propagation, and genetic transformation.

Salient Features:

1. Presents authoritative and up-to-date information, empowering with the latest knowledge from leading minds in *Aloe vera* genetics and genomics.
2. Provides exclusive access to groundbreaking insights and advancements in genomics specific to *Aloe vera*, providing readers with a unique and cutting-edge resource.
3. Delves into cutting-edge areas like genomics, transcriptomics, proteomics, metabolomics, and nano-technological advancements specific to *Aloe vera*.
4. Addresses molecular markers, mapping, and genome-wide characterization for breeding, in-vitro propagation, and genetic transformation.
5. Explores physiological responses under abiotic stress, germplasm characterization, gene discovery, and the transformative impact of genomics on plant breeding in *Aloe vera*.

Depiction and Designing of Plant Genomes
Chittaranjan Kole and Phullara Kole
Series Editor

The Cannabis Genome: A New Era in Genetics, Genomics, and Breeding
Edited by Angel Fernandez i Marti and Jaime Prohens

Designing Sorghum Genome for NextGen Agriculture
Edited by Dinesh Kumar Saini and S. V. Krishna Jagadish

The Turmeric Genome
Edited by Prasath Duraisamy, Aarthi Sounderarajan, and Chittarajan Kole

The Aloe Genome
Edited by Vinod Chhokar and Alka Jangra

For more information, please visit our website: www.routledge.com/Depiction-and-Designing-of-Plant-Genomes/book-series/DDPG

The Aloe Genome

Genetics, Genomics and Breeding

Edited by

Vinod Chhokar and Alka Jangra

CRC Press
Taylor & Francis Group
Boca Raton London New York

CRC Press is an imprint of the
Taylor & Francis Group, an **informa** business

Designed cover image: Cover design by Chittaranjan Kole; image courtesy of Getty Images

First edition published 2026
by CRC Press
2385 NW Executive Center Drive, Suite 320, Boca Raton FL 33431

and by CRC Press
4 Park Square, Milton Park, Abingdon, Oxon, OX14 4RN

CRC Press is an imprint of Taylor & Francis Group, LLC

ISBN: 978-1-032-99244-0 (hbk)
ISBN: 978-1-032-99245-7 (pbk)
ISBN: 978-1-003-60310-8 (ebk)

DOI: 10.1201/9781003603108

Typeset in Times
by Newgen Publishing UK

Dedication

This book series is dedicated to our brothers/brothers-in-law and sisters/sisters-in-law–

Bansi Badan Kole
Niranjan Kole
Bharati Nandi Nee Kole
Arati Kundu Nee Kole
Ram Ranjan Kole

Contents

Preface to the Series

Genome sequencing of the model plant Arabidopsis in 2000 followed by the crop plant rice in 2002 flagged of genome sequencing in an array of plants. As on March 05, 2024, 4604 genomes of 1482 different plant species have been sequenced, and the number has been continuing to grow exponentially. The list includes several crop plants as well as some model plants. Sequencing of the nuclear genomes was followed by sequencing of extranuclear genomes such as mitochondria and chloroplasts, and by several studies on functional genomics. However, detailed reports on them are scattered in the literature, including research articles and reviews. Therefore, a comprehensive compilation of deliberations on the sequences with enumeration of the sequences, annotation, assembly, gene families, and comparative genomics of these useful genomes and elucidation of functional genomics, including transcriptomics, metabolomics, and proteomics in these higher plants, is timely.

A number of higher plants provide the basic F5 – food, feed, fuel, fiber, and furniture – needs, and also, nutraceuticals, beverages, luxury, and aesthetic items. Genetic improvement in these plants with regard to their utilities was being done in the last century, mainly by traditional breeding. The advent of molecular breeding and genetic engineering in the last two decades of the last century strengthened the breeding strategies in the economic plants. Advances in plant genomics have ultimately revolutionized the concepts and strategies of crop improvement. Employment of high-throughput genotyping and phenotyping together with other genomics-assisted breeding tools like gene discovery, allele mining, genome-wide association studies, genomic selection, and genome-wide prediction, etc., has facilitated accelerated and precise plant breeding. Recent emergence of the genome editing technique has enormously supplemented the molecular and genomics tools, making it possible to design and tailor the genomes according to the needs. However, success stories of next-generation breeding techniques need to be compiled for easy access of the readers.

This present book series, entitled Depiction and Designing Plant Genomes, includes a number of books each dedicated to the deliberations on genomic attributes and/or genetic improvement of a higher plant, and hopefully will serve as a handy compendium for readers.

The books of this series have been edited by highly reputed scientists, and I remain thankful for their invaluable academic efforts and cooperation. We are also thankful to the publisher and specifically to its Senior Commissioning Editor (Chemistry / Agricultural Sciences), Dr. Renu Upadhyay, and her Editorial Assistant, Tisha Sharma, for their outstanding and timely help and cordial co-operation since the inception of this book series.

Finally, we must express my thanks to our children, Sourav and Devleena, for assisting us in editing the books of this series and also bearing with us for our channelizing the time that should have been kept for them towards editing this book series.

Kolkata, 3rd August 2024

Chittaranjan Kole and Phullara Kole

About the Editors

Vinod Chhokar, a distinguished Professor of Biotechnology at Guru Jambheshwar University of Science and Technology in Hisar, India, has carved a remarkable career path in the realm of Plant Sciences. His academic journey commenced with a Ph.D. in Biochemistry from CCS Haryana Agricultural University. With a solid educational foundation, he delved into the dynamic field of plant genomics and metabolomics, marking the inception of his illustrious career in 1999. His pioneering research focused on employing molecular approaches for the genetic enhancement of medicinal plants, a field that he has passionately contributed to for over two decades. During his post-doctoral research at Purdue University, West Lafayette, USA, under the prestigious BOYSCAST fellowship of the Department of Science & Technology, Govt. of India, he further honed his expertise in plant genomics. His expertise includes characterizing the genetic makeup and metabolic engineering of medicinal plants such as *Aloe vera, Asparagus racemosus, Zingiber officinale, Bryophyllum pinnatum*, and *Capsicum annuum L.* His contributions to the scientific community are reflected in his extensive publication record, boasting over 100 research papers in highly esteemed journals. A seasoned academician and researcher, he continues to shape the future of plant sciences, leaving an indelible mark on the scientific landscape. His global exposure, membership in professional organizations, and extensive mentorship of students underscore his commitment to advancing knowledge in Functional Genomics, Metabolomics, Biochemistry, and Molecular Biology.

Alka Jangra is an Assistant Professor in Agricultural Biotechnology at the College of Biotechnology, Chaudhary Charan Singh Haryana Agricultural University (CCS HAU), Hisar, India. She holds a Ph.D. in Biotechnology from Guru Jambheshwar University of Science & Technology. Her research journey is marked by a deep commitment to enhancing the genetic makeup of plants. During her doctoral studies, Alka specialized in the molecular characterization and genetic improvement of *Aloe vera*. Notably, she achieved a breakthrough by developing a highly efficient genetic transformation method for *Aloe vera* using *Agrobacterium tumefaciens*, showcasing her pioneering work in the field. During her postdoctoral work at National Agri-Food Biotechnology Institute (NABI), Mohali, under the Department of Biotechnology (Govt. of India), she played a pivotal role in the biofortification of banana and development of transgene-free genome editing systems. She has published widely in reputed international journals and contributed to several book chapters. She is the co-editor of this volume and a passionate advocate for translational plant science. Her work bridges academic discovery and applied agricultural biotechnology, addressing real-world challenges in sustainable agriculture, food security, and plant-based therapeutics. Her expertise extends to employing advanced techniques like CRISPR/Cas9 for genetic manipulation, reflecting her commitment to cutting-edge research in plant biotechnology.

Contributors

Jeevitha B.T
Department of Agricultural Biotechnology, College of Agriculture-Hassan, UAS (B) 573225, India

Vaibhav Baliyan
ICAR-Indian Agricultural Research Institute, New Delhi-110012, India

Vikas Beniwal
Department of Microbiology, Central University of Haryana, Jant-Pali, Mahendragarh-123031, Haryana, India

Bhupnesh
Department of Botany and Plant Physiology, College of Basic Science and Humanities, CCS Haryana Agricultural University, Hisar 125004, India

Vinod Chhokar
Department of Biotechnology, Guru Jambheshwar University of Science & Technology, Hisar-125001, Haryana, India

Harsh Chaurasia
Department of Genetics and Plant Breeding, College of Agriculture, Chaudhary Charan Singh Haryana Agricultural University, Hisar, Haryana-125004, India

Shweta D. C
Department of Agricultural Biotechnology, College of Agriculture-Hassan, UAS (B) 573225, India

Khushbu Dawra
ICAR- National Research Centre on Equines, Sirsa Road, Hisar-125001 (Haryana) India

Christian O. Dimkpa
Department of Analytical Chemistry, Connecticut Agricultural Experiment Station, New Haven, CT 06511, USA

Pooja Garg
Department of Biotechnology, Guru Jambheshwar University of Science & Technology, Hisar, Haryana-125001, India and ICAR- National Institute for Plant Biotechnology, New Delhi- 110012, India

Soniya Goyal
Department of Biosciences and Technology, Maharishi Markandeshwar (Deemed to be University), Mullana-133207, Ambala, India

Geetha Govind
Department of Agricultural Biotechnology, College of Agriculture-Hassan, UAS (B) 573225, India

Bharti Goyal
ICMR-National Institute of Malaria Research, New Delhi, 110077, India Academy of Scientific and Innovative Research (AcSIR), Ghaziabad, 201002, India

Vinod Goyal
Department of Botany and Plant Physiology, College of Basic Science and Humanities, CCS Haryana Agricultural University, Hisar 125004, India

Mahiti Gupta
Department of Biosciences and Technology, Maharishi Markandeshwar (Deemed to be University), Mullana-133207, Ambala, India

Sneha Gupta
Sardar Vallabhbhai Patel University of Agriculture and Technology, Meerut (U.P.) - 250110, India

Alka Jangra
Department of Agricultural Biotechnology, College of Biotechnology, Chaudhary Charan Singh Haryana Agricultural University, Hisar, Haryana-125004, India

Anamika Kashyap
ICAR- National Institute for Plant Biotechnology, New Delhi-110012, India

Trivedi Nandan Kishorchandra
Faculty of Applied Sciences and Biotechnology, Shoolini University of Biotechnology and Management Sciences, Solan (HP), 173229, India

Arthur Korte
Center for Computational and Theoretical Biology CCTB, University of Würzburg, Würzburg, Germany

Pawan Kumar
Department of Bioinformatics & Computational Biology, Chaudhary Charan Singh Haryana Agricultural University, Hisar-125004, Haryana, India

Kishor Kumar
Department of Agricultural Biotechnology, College of Biotechnology, Chaudhary Charan Singh Haryana Agricultural University, Hisar, Haryana-125004, India

Ravinder Kumar
Department of Biotechnology, Guru Jambheshwar University of Science & Technology, Hisar-125001, Haryana, India

Ranjeet Kushwaha
ICAR- National Institute for Plant Biotechnology, New Delhi- 110012, India

Shweta Mehrotra
Department of Biotechnology, Guru Jambheshwar University of Science and Technology, Hisar, Haryana, India

Monika
Department of Botany and Plant Physiology, College of Basic Science and Humanities, CCS Haryana Agricultural University, Hisar 125004, India

Ananya Mukherjee
Faculty of Applied Sciences and Biotechnology, Shoolini University of Biotechnology and Management Sciences, Solan (HP), 173229, India

Pankaj
Department of Botany and Plant Physiology, College of Basic Science and Humanities, CCS Haryana Agricultural University, Hisar 125004, India

Pooja
Department of Botany and Plant Physiology, College of Basic Science and Humanities, CCS Haryana Agricultural University, Hisar 125004, India

Sahil Prabhakar
Lambton college of applied arts and technology, Sarnia, ON N7S 6K4, Canada

Mahesh Rao
ICAR- National Institute for Plant Biotechnology, New Delhi-110012, India

Kumaraswamy R.V.
Department of Agricultural Biotechnology, College of Biotechnology CCS HAU-Hisar-125004, India

Sarita
Department of Botany and Plant Physiology, College of Basic Science and Humanities, CCS Haryana Agricultural University, Hisar 125004, India

Jyoti Sharma
Department of Biotechnology, School of Chemical and Life Sciences, Jamia Hamdard, Hamdard Nagar, New Delhi- 110064, India

Garima Sharma
Department of Biotechnology, Guru Jambheshwar University of Science & Technology, Hisar-125001, Haryana, India

Vishal Sharma
Fermentation and Phytofarming Technology Division, CSIR-Institute of Himalayan Bioresource Technology, Palampur (HP)-176061, India and Academy of Scientific and Innovative Research (AcSIR), Ghaziabad, Uttar Pradesh, India

Sonia Sihag
ICAR- National Research Centre on Equines, Sirsa Road, Hisar-125001 (Haryana) India

Dalveer Singh
Sanskriti University, Uttar Pradesh, Mathura, India-281401

Sri Devi Tallapragada
Department of Botany and Plant Physiology, College of Basic Science and Humanities, CCS Haryana Agricultural University, Hisar 125004, India

Jyoti Taunk
Department of Agricultural Biotechnology, College of Biotechnology, Chaudhary Charan Singh Haryana Agricultural University, Hisar-125004, Haryana, India

Aayushee Thakur
Department of Biotechnology, Dr Yashwant Singh Parmar University of Horticulture and Forestry, Nauni, Solan (HP). 173230, India

Shikha Tripathi
Department of Botany, Institute of Science, Banaras Hindu University, Varanasi (U.P.) – 221005, India and ICAR- National Institute for Plant Biotechnology, New Delhi-110012, India

Priya Upadhyay
ICAR- National Institute for Plant Biotechnology, Pusa Campus, New Delhi, India -110012 ICAR- Indian Agricultural Research Institute, Pusa Campus, New Delhi, India -110012

Deepika Verma
Department of Biochemistry, All India Institute of Medical Sciences (AIIMS), New Delhi, India

Shalu Yadav
Department of Biotechnology, Guru Jambheshwar University of Science and Technology, Hisar, India

Rajesh Yogi
Department of UIBT-Biotechnology, Chandigarh University, Mohali-140413, Punjab, India

1 Exploring *Aloe vera*
Botanical Insights, Medicinal Uses, and Economic Impact

Jyoti Sharma, Shikha Tripathi, Sneha Gupta, and Pooja Garg

1.1 INTRODUCTION

Aloe vera is the oldest known, and the most widely applied medicinal plant worldwide. It is a unique plant with a wealth of historical and cultural value, which is bolstered by its diverse range of applications and vast botanical history (Foster and Hunter, 2009). Its extensive pharmacological characteristics are supported by its varied chemical composition, which includes both gel and latex ingredients. These pharmacological properties have been verified by multiple scientific studies, that confirmed its efficacy in several key areas (Surjushe et al., 2008).

1.1.1 History

The aloe plant and its derivative products have been integral to medicine and healthcare as far back as the 4th century BC, when ancient Greek doctors sourced aloe from the island of Socotra in the Indian Ocean. In the ancient Nile Valley civilization of Kemet (now referred to as ancient Egypt), aloe was used for medical treatments, beauty care, and embalming. The medicinal use of aloe dates back over 4,000 years, with references found in Sumerian clay tablets from 2100 BC, and it is mentioned as a laxative in the Egyptian Papyrus Ebers from 1552 BC. Through human trade and migration, this plant gained widespread recognition and was used for treating burns and wounds across ancient civilizations (Ni and Tizard, 2004). There are at least 420 different species of aloe, belonging to the family Liliaceae. Among these species, several have been used under the common name of aloe, such as *Aloe vera* Linn., *Aloe barbadensis* Miller, *Aloe ferox* Miller, *Aloe chinensis* Baker, *Aloe. indica* Royle, and *Aloe perryi* Baker (Dagne et al., 2000). Of these, *Aloe vera* Linn., also known as *A. barbadensis* Miller, is unanimously accepted as the correct botanical source of aloe. *Aloe vera* – also referred to as Barbados aloe, *Aloe vulgaris*, *Aloe arborescens*. Cape aloe, and Socotrine or Zanzibar aloe – is the most common form used in aloe-based products (Manvitha et al., 2014). There are over 300 species of aloe, most of which are native to South Africa, Madagascar, and Arabia. These species vary in their concentrations of active ingredients. Originally from South Africa and South America, aloes are currently grown all over the world with the exception of tundra, deserts, and rain forests. In the United States, in southern Texas, aloe is grown commercially (Hossain et al., 2013).

1.1.2 Phenotype of *Aloe vera*

Aloe vera is a shrubby or arborescent perennial xerophytic succulent plant with a pea-green tint that is drought-resistant and a member of the Asphodelaceae (Liliaceae) family. The aloe plant has

DOI: 10.1201/9781003603108-1

long, thick, triangular leaves with edges tipped with spikes that can grow up to 20 inches long and five inches wide (Reynolds and Dweck, 1999). It takes four years on average to attain maturity, and its lifespan is roughly twelve years. The clear fresh parenchymal gel found in the leaf center is occasionally dried to make *Aloe vera* concentrate or diluted with water to make goods containing aloe juice. While most reference books regard *A. barbadensis* Miller as the correct name, the WHO monograph accepts *Aloe vera* Linn. as the legitimate name for this species (Das and Chattopadhay, 2004). *Aloe vera* specifically refers to the Aloe barbadensis Miller plant, which is a medicinal plant that grows in hot climates such as California, New Mexico, and the Caribbean (Saeed et al., 2004). The pericyclic tubules, which are greenish-yellow in color, are the source of the sticky latex liquid. The flowers are not utilized medicinally. The Arabic "alloeh" or Hebrew "halal," which both imply bitter, shining substance, is where the name "aloe" originates (Manvitha and Bidya, 2014).

1.1.3 Traditional Uses of *Aloe vera*

In traditional medical systems such as Ayurveda, Siddha, Unani, and homoeopathy, it plays a significant role. *Aloe vera*, or "Ghee kunwar" in Sanskrit. *Aloe vera* grows wild along the southern coast of India. It is grown in a sizable portion of several Indian states, including Tamil Nadu, Gujarat, Maharashtra, and others (Sahu et al., 2013). Aloes may grow in a variety of conditions, including desert, grassland, coastal, and even alpine regions, despite the common misconception that they can only be found in hot, dry climates (Baby and Justin, 2010).

1.1.4 Nutrient Values of *Aloe vera*

Aloe vera leaves are abundant in a variety of minerals, with different proportions of each type present. The fact that calcium makes up about 3.58% of the entire leaf emphasizes how important it is. Following at 1.22%, magnesium adds to the nutritional composition of the plant. Notably, the leaf contains significant amounts of potassium and sodium, which are necessary for cellular activity and electrolyte balance, making about 4.06% and 3.66% of the leaf, respectively (Foster et al., 2011). There are also trace levels of iron, zinc, copper, and phosphorus, with values ranging from 0.02% to 0.1%, which offer extra nutritional assistance. When ingested or applied topically for health advantages, these minerals together increase the nutritional content of *Aloe vera*, perhaps providing it with vital elements. Aloe has traditionally been used as a potent laxative to treat chronic constipation, and many pharmacopoeias still include it as a laxative. It has been mostly replaced, nevertheless, by more moderate laxatives (Mulu et al., 2015). The Aloe plant was first utilized commercially for the manufacture of a substance called Aloin, a yellow sap that was used as a laxative for a long time (Atul et al., 2011).

1.1.5 Therapeutics Uses of *Aloe vera*

Early in the 20th century, this product was mentioned in commercial, technical, and official literature and came to be associated with the name "Aloe." Later on, aloe gel, the other key constituent in aloe, started to be recognized as a valuable product in the 1950s and was utilized as a moisturizer, healing agent in cosmetics, and a base for nutritional drinks. However, it caused a great deal of misunderstanding due to the nomenclature (Grace et al., 2008). *Aloe vera* is a multipurpose plant that is mostly utilized as a medication and for food preservation. It is sold commercially in the form of tablets, sprays, ointments, lotions, beverages, jellies, and creams (Tarameshloo et al., 2012). The plant known as "healing aloe," contains enzymes, amino acids, vitamins, minerals, and other active components, (totaling more than 75), some of which may have therapeutic applications. Around the world, a variety of aloe species are used to treat ailments ranging from cancer to dermatitis (Liu et al., 2010). *Aloe vera* leaf has been shown in numerous studies to have a wide range of medicinal properties, such as antibacterial, anticancer, antioxidant, antidiabetic, antiulcer, hepatoprotective, immunomodulatory, and many more (Langmead et al., 2004). The polysaccharides included in the gel of the leaves of *Aloe vera* have been linked to numerous health benefits. It has been used for a variety of conditions, including burn wound healing, immune modulation, intestinal absorption, antidiabetic, antioxidant, hepatoprotective, anticancer, antimicrobial, antiviral, antihyperlipidemic, and antiulcer properties (Devaraj et al., 2008).

1.2 CHEMICAL COMPOSITION AND BIOLOGICAL PROPERTIES OF *ALOE VERA*

Aloe vera is well known for its gel and latex leaf lining, which include chemical components that may be biologically active. A range of polysaccharides, including acemannan and glucomannan, as well as important substances including carboxypeptidase, magnesium, zinc, calcium, glucose, cholesterol, and salicylic acid are present in the gel (Liu et al., 2006). It also includes amino acids, lignins, saponins, plant sterols, vitamins A, C, and E, and prostaglandin precursors like gamma-linolenic acid (GLA) (Hamman, 2008). However, the anthraquinone glycosides found in abundance in the latex leaf lining include barbaloin, aloin, and aloe-emodin. Aloe barbadensis contains around 200 chemicals, roughly 75 of which are biologically active (Misra and Dey, 2001). Many different types of chemicals are found in *Aloe vera* leaves, such as anthrones and their glycosides (e.g., 10-(1, 5' anhydroglucosyl)-aloeemodin- 9-anthrone, also known as aloin A and B), chromones, carbohydrates, proteins, glycoproteins, amino acids, organic acids, lipids, sugars, vitamins and minerals (Femenia et al., 1999). *Aloe vera* juice offers a minimal calorie content of ten per 8-ounce serving, making it a low-calorie beverage option. It contains no protein or fat but provides two grams of carbohydrates, including one gram of fiber. The juice is naturally sugar-free. In terms of minerals, a serving typically supplies 74 milligrams of calcium, 60 milligrams of sodium, and 110 milligrams of potassium (Ni et al., 2004). These nutritional components contribute to *Aloe vera*'s appeal as a healthful addition to diets, offering hydration and modest amounts of essential minerals. However, it is essential to note that while *Aloe vera* provides certain nutrients, its consumption should be moderate and ideally discussed with a healthcare provider, given its potential interactions with medications and individual sensitivities. Aloe gel is a popular component of many over-the-counter skin salves and is 99% water with a pH of 4.5. Glucomannan, an emollient polysaccharide, is present in the gel (Saroj et al., 2004). Its use in numerous cosmetics can be attributed to its effectiveness as a moisturizer (Babaee et al., 2012). The primary carbohydrate component of the gel, acemannan, is a water-soluble long-chain mannose polymer that has antiviral and antineoplastic properties as well as the ability to speed up wound healing and modify immunological function, notably macrophage activation and cytokine production (Im et al., 2005). Additionally, the gel contains magnesium lactate, which helps to minimize itching, salicylic acid, and other antiprostaglandin chemicals that reduce inflammation, as well as bradykininase, an anti-inflammatory (Park and Jo, 2006). Strong stimulant laxatives called anthraquinone glycosides, such as aloin, aloe-emodin, and barbaloin, are found in the latex, resin, or sap that lines leaves (Table 1.1).
The chemical composition of *Aloe vera* in detail:

- **Anthraquinones and anthrones,** such as aloe emodin, aloetic acid, anthranol, barbaloin, isobarbaloin, and emodin, along with esters of cinnamic acid, are known for their diverse pharmacological properties. Specifically, compounds like aloin and emodin exhibit analgesic, antibacterial, and antiviral activities (Choi and Chung, 2003). These natural products have been studied for their potential therapeutic applications, contributing to the understanding and development of novel treatments in medicine and healthcare.
- **Carbohydrates** encompass a wide array of compounds such as pure mannan, acetylated mannan, glucogalactomannan, galactan, and others, including pectic substances, xylan, and

TABLE 1.1
Representation of chemical constituents of *Aloe vera*

Aloe vera: Potentially Active Chemical Constituents	From the Gel	From the Latex Leaf Lining
Polysaccharides	Glucomannan, Acemannan	
Other	Carboxypeptidase, Magnesium, Zinc, Calcium, Glucose, Cholesterol, Salicylic Acid, Prostaglandin Precursors (gamma-linolenic acid [GLA]), Vitamins A, C, E, Lignins, Saponins, Plant Sterols, Amino Acids	
Anthraquinone Glycosides		Aloin, Aloe-Emodin, Barbaloin

cellulose (Femenia et al., 1999). Among these, a glycoprotein known as alprogen exhibits notable antiallergic properties, while novel anti-inflammatory compounds are also identified within this class (Chow et al., 2005). These diverse carbohydrate compounds contribute significantly to various biological activities and applications in pharmaceutical and biomedical research.

- **Chromones** such as 8-C-glucosyl-(2'-O-cinnamoyl)-7-O-methylaloediol A, 8-C-glucosyl-(S)-aloesol, and others like isoaloeresin D and neoalosin A are notable for their novel anti-inflammatory properties. These compounds have been identified for their potential therapeutic effects in managing inflammation, contributing to ongoing research into new treatments and medications (Hutter et al., 1996). Chromones represent a diverse class of natural products with promising pharmacological activities that continue to be explored for their beneficial impacts on health and well-being.
- **Enzymes** encompass a diverse group of biological catalysts essential for various physiological processes. Examples include alkaline phosphatase, amylase, bradykinase, carboxypeptidase, catalase, cyclooxygenase, lipase, oxidase, phosphoenolpyruvate carboxylase, and superoxide dismutase. Each enzyme serves specific roles: bradykinase, for instance, aids in reducing excessive inflammation when applied topically to the skin, while others play crucial roles in breaking down sugars and fats within the body (O'Brien et al., 2011). These enzymatic activities are integral to maintaining metabolic balance and supporting overall health and wellness.
- **Inorganic compounds** such as calcium, chlorine, chromium, copper, iron, magnesium, manganese, potassium, phosphorus, sodium, and zinc play crucial roles in biological systems. They are essential for the proper functioning of numerous enzyme systems involved in diverse metabolic pathways. Additionally, some of these compounds act as antioxidants, helping to neutralize harmful free radicals and thereby contributing to overall cellular health and longevity (Eshun and He, 2004). Their presence and balance are vital for maintaining physiological processes and supporting optimal bodily functions.
- **Proteins** found in *Aloe vera* gel include lectins and lectin-like substances, which contribute to its bioactive properties. *Aloe vera* also contains salicylic acid, known for its anti-inflammatory and antibacterial effects. Additionally, lignin, though inert itself, enhances the penetration of other active ingredients into the skin when included in topical formulations (Josias, 2008). Saponins, comprising approximately three percent of the gel, provide cleansing and antiseptic benefits due to their soapy nature. These constituents collectively contribute to the therapeutic and cosmetic uses of *Aloe vera*, making it a valuable natural remedy for skincare and health applications (Roy et al., 2012).
- **Vitamins** A, B12, C, E, choline, and folic acid play essential roles in maintaining overall health. Among them, vitamins A, C, and E are known for their antioxidant properties,

which help neutralize free radicals in the body. Antioxidants are crucial for protecting cells from oxidative stress, thereby supporting healthy aging and reducing the risk of chronic diseases (Patidar et al., 2012). These vitamins are obtained through a balanced diet and supplementation, contributing to various physiological functions and promoting well-being.

- **Hormones** such as auxins and gibberellins play important roles in plant growth and development, but they also exhibit therapeutic properties in humans. Auxins, for example, are involved in wound healing processes and have been noted for their anti-inflammatory actions. Gibberellins, similarly, contribute to wound repair and may have additional beneficial effects on inflammation. These natural compounds highlight the potential for plant hormones to be utilized in medicinal contexts, offering insights into novel treatments and therapies for promoting healing and reducing inflammation in various medical applications.

1.3 CULTIVATION

1.3.1 Climate

Aloe vera thrives in tropical, sub-tropical, and arid regions, even under persistent drought conditions due to its low water requirements. It flourishes in hot and dry climates but does not grow well in cold environments (Chandra and Pandey, 1989). However, *Aloe barbadensis*, which is cultivated mainly for its medicinal uses, can grow effectively under irrigated conditions (Saroj and Purohit, 2004).

1.3.2 Soil

It grows well in marginal to sub-marginal soils with low fertility and can tolerate high pH levels along with elevated sodium and potassium salts. However, it grows more quickly in medium-fertility, heavier soils. For commercial cultivation, well-drained loam to coarse sandy loam soils with moderate fertility and a pH up to 8.5 are preferred, as waterlogged soils are entirely unsuitable.

1.3.3 Land Preparation

The soil should not be disturbed too deeply, as Aloe's root system extends only 20–30 cm below the surface. Depending on soil type and agroclimatic conditions, one or two ploughings followed by leveling may be necessary. Due to Aloe's sensitivity to water stagnation, effective drainage is essential. Fields should be divided into appropriately sized plots, taking into account the slope and available irrigation sources.

1.3.4 Propagation

Propagation is achieved using root suckers or rhizome cuttings. Medium-sized root suckers are selected, carefully extracted without harming the parent plant, and then directly planted in the main field (Rajeswari et al., 2012).

1.3.5 Planting Time

Suckers should be planted in July or August during the monsoon season to enhance field survival and promote healthy plant growth. However, in irrigated conditions, planting can occur throughout the year, except during the winter months from November to February (Rajeswari et al., 2012). *Aloe vera* flowers between October and January, producing long inflorescences with numerous small, bright orange, yellow or coral flowers. The fruits develop from February to April (Manvitha and Bidya, 2014).

1.3.6 Manuring

The crop benefits significantly from the application of farmyard manure (FYM) and compost. In the first year of planting, FYM should be applied at a rate of 20 tons per hectare during land preparation, with the same amount used in subsequent years. Additionally, vermicompost at 2.5 tons per hectare can also be applied (Rajeswari et al., 2012).

1.3.7 Spacing and Planting

Suckers are planted in pits approximately 15 cm deep, created at the time of planting, and spaced 60×60 cm apart. Once the suckers are planted, the soil around the root zone should be compacted firmly, and proper drainage must be ensured to prevent water stagnation (Rajeswari et al., 2012).

1.3.8 Irrigation

Aloe is quite resilient to stress, but to achieve optimal yields, irrigation during critical growth stages is essential. It can be effectively grown under both irrigated and rainfed conditions. To ensure good yields, irrigate immediately after planting and provide two to three additional irrigations per week until the plant is established. Typically, two to three irrigations in summer and one to two in winter are sufficient for healthy growth. In semi-arid conditions, once established, Aloe can thrive even without additional irrigation (Saroj et al., 2004).

1.3.9 Intercropping

In the first year after planting, over 40% of the land may remain unused. To make the most of this space, suitable leguminous or low-competition intercrops such as cluster beans, sesame, or coriander can be planted in the gaps, which will also enhance soil conditions. However, from the second year onwards, these intercrops should be avoided as they can interfere with the growth of *Aloe vera* (Maiti and Chandra, 2002).

1.3.10 Plant Protection

Insect pests and diseases are generally not major issues for this crop. However, problems such as mealy bugs, anthracnose, termites, pad spot disease, leaf spots, and sucker rotting have been reported (Rajeswari *et al.* 2012). To prevent these issues and enhance crop quality and productivity, maintaining field sanitation is beneficial. Additionally, applying two to three foliar sprays of Dithane M-45 (0.2%) at weekly intervals can help reduce fungal diseases (Saroj et al., 2004).

1.3.11 Harvesting

An Aloe plantation begins to produce yields approximately six to eight months after planting. However, for commercial yields, it typically takes about a year to achieve, and production can continue for up to five years. After this period, replanting is necessary (Saroj et al., 2004). Typically, older outer leaves are harvested, leaving the fresh and young leaves at the top. Plants can be removed manually or with the aid of a tractor-drawn disc harrow or cultivator. Leaves showing tip necrosis should be avoided as they can be prone to microbial contamination. It is important to minimize juice loss from the cut areas. If harvesting occurs annually, the optimal period is October to November. Post-harvest, the leaves are dried in the shade, followed by sun-drying before storage. Flowers are gathered from December to January and preserved after proper drying. Harvested leaves are carefully stacked and then transported to refrigeration or processing facilities (Manvitha and Bidya, 2014).

1.3.12 Processing

Aloe vera gel is a mucilaginous substance derived from the parenchyma cells of the *Aloe vera* plant. Once exposed to air, the gel quickly oxidizes, decomposes, and loses much of its biological activity. Various researchers have outlined different methods for processing the gel, such as cold processing or heat treatment, to achieve sterilization and stabilization. Despite the differences in techniques, the underlying principle remains similar. Optimal results are achieved when the leaves are processed immediately after harvesting, as natural enzymatic reactions and bacterial growth – triggered by exposure to oxygen – can lead to the degradation of the gel matrix (Ramachandra and Rao, 2008).

The process begins with washing freshly harvested *Aloe vera* leaves in a suitable bactericide. The leaves are then processed to mechanically separate the gel matrix from the outer cortex. The addition of cellulose-dissolving compounds, such as cellulase, can help facilitate this separation. The resulting Aloe liquid is treated with activated carbon to remove color and eliminate aloin and anthraquinones, which have laxative properties, especially if the stabilized gel is intended for internal consumption. The liquid undergoes several steps of filtration, sterilization, and stabilization. The final stabilized liquid can be concentrated to reduce its water content or further processed to remove nearly all the water, resulting in a powder (Ramachandra and Rao, 2008).

In the cold processing method, all steps are carried out without using heat. Coats (1979) described using enzymes such as glucose oxidase and catalase to prevent the growth of aerobic organisms in *Aloe vera* gel, thereby sterilizing it. Additional sterilization steps in cold processing involve exposing the gel to ultraviolet light, followed by micron filtration through a 0.2 µm membrane. In heat treatment processing, sterilization is achieved by pasteurizing the aloe liquid obtained from the activated carbon treatment at high temperatures. According to Aloecorp, the biological activity of *Aloe vera* gel remains largely intact when heated at 65°C for less than 15 minutes. Longer heating periods or higher temperatures significantly reduce the activity levels. They recommend HTST (High Temperature Short Time) pasteurization, followed by rapid cooling to 5°C or below, as the best method. In these processes, stabilization can be achieved by adding preservatives and other additives. The use of sodium benzoate, potassium sorbate, citric acid, and vitamin E in combination has been reported to be effective.

1.4 ECONOMIC IMPORTANCE OF *ALOE VERA*

The *Aloe vera* plant has been renowned for centuries for its health, beauty, medicinal, and skincare benefits. Its economic importance is derived from its extensive use in agriculture, pharmaceuticals, cosmetics, and health products. *Aloe vera* promotes sustainable farming, particularly in arid areas, offering income and job opportunities. The global trade of *Aloe vera* products significantly boosts export revenues and drives economic growth and innovation across multiple industries. Here, we summarize *Aloe vera*'s applications in various sectors in Figure 1.1.

1.4.1 Cosmetic Uses

Aloe vera is extensively utilized in the cosmetic industry, where three distinct forms are used: *Aloe vera* latex, *Aloe vera* gel, and *Aloe vera* whole leaf extract. These preparations contain biological ingredients that may function individually or synergistically (Boudreau and Beland, 2006). Aloe gel is renowned for its ability to retain moisture for long periods and its soothing properties (Gage, 1996). Additionally, *Aloe vera* serves as an astringent, moisturizer, and cleanser (Manvitha and Bidya, 2014). Consequently, it is widely incorporated into various cosmetic and toiletry products, including moisturizers, cleansers, sunscreens, shampoos, conditioners, and serums.

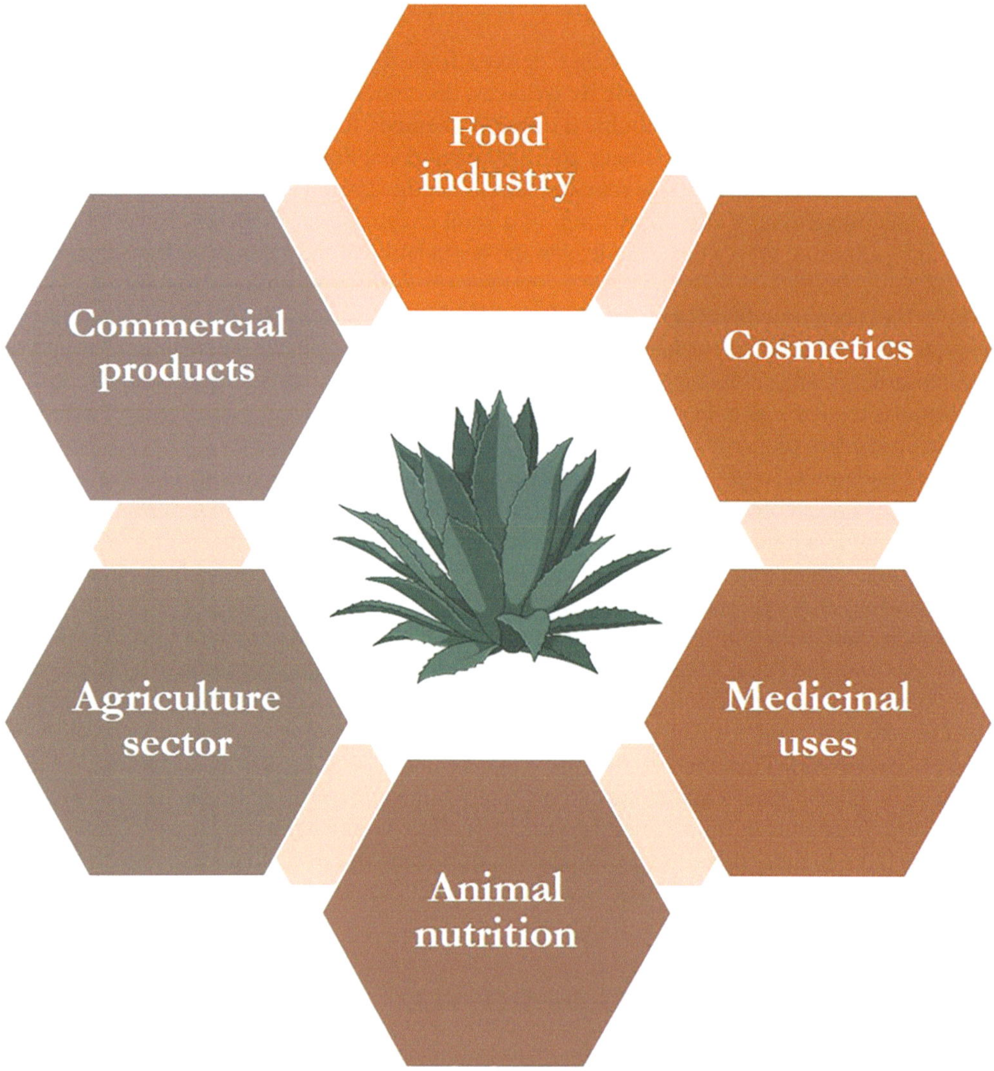

FIGURE 1.1 Representation of the economic importance of *Aloe vera*.

1.4.2 Food Uses

The food and beverage industry presents a promising opportunity for *Aloe vera*. It has been utilized in functional, such as yogurt and smoothies, and in health drinks, including tea (Eshun and He, 2004). Botanical products are widely recognized for their use as nutritional supplements that promote health and prevent diseases. According to Serrano et al. (2006), *Aloe vera* gel can serve as an edible coating to extend the quality and safety of fresh produce. For instance, aloe-coated table grapes experienced a significant delay in the loss of functional compounds, including total phenolic and ascorbic acid. *Aloe vera* also inhibits the growth of microorganisms that cause foodborne illnesses and spoilage in both humans and animals (Eshun and He, 2004). Moreover, *Aloe vera* does not alter the taste or appearance of food, making it a promising natural, safe, and environmentally friendly alternative to conventional synthetic preservatives (Serrano et al., 2006).

1.4.3 Medicinal Uses

- **Healing wounds:** The healing properties of *Aloe vera* are due to polysaccharides such as Glucomannan and the growth hormone gibberellins, which boost collagen and elastin production, resulting in reduced wrinkling and improved healing (Reynolds and Dweck, 1999; Surjushe et al., 2008). Furthermore, *Aloe vera* is effective in treating scar tissue and preventing scar formation after skin injuries, likely because of the amino acids necessary for new cell formation and enzymes that promote regeneration of the skin's deepest layers (Choi and Chung, 2003).
- **Anti-inflammatory action and immunity activity:** *Aloe vera* gel not only alleviates pain and discomfort due to its anti-inflammatory properties but also accelerates the healing process. The acetylated mannan in Aloe gel has anti-inflammatory effects similar to mannose-6-phosphate (Hamman, 2008). *Aloe vera* inhibits the cyclooxygenase pathway, reducing prostaglandin production and inflammation, making it useful for arthritis and joint problems (Reynolds and Dweck, 1999; Surjushe et al., 2008; Yu et al., 2009). Its polysaccharides activate macrophage cells to produce nitric oxide, secrete cytokines (like TNF-α, IL-1, IL-6, and INF-γ), and present cell surface markers, showcasing its immunomodulatory activities (Chow et al., 2005). Additionally, some immunomodulatory effects are linked to glycoproteins, such as lectins, found in the gel (Hamman, 2008).
- **Effects on skin exposure to UV and X-radiation:** *Aloe vera* has been noted to shield the skin from radiation damage (Roberts and Travis, 1995). It is believed that applying *Aloe vera* gel generates the antioxidant protein metallothionein, which scavenges hydroxy radicals and protects the skin from oxidative damage. Additionally, it releases the immunosuppressive Interleukin IL-10, preventing UV-induced suppression of delayed-type hypersensitivity (Byeon et al., 1988).
- **Effects on ulcers:** *Aloe vera* gel can treat gastric ulcers or prevent their formation in both animals and humans. Its antiulcer properties are due to several mechanisms, including its anti-inflammatory effects, healing properties, stimulation of mucus production, regulation of gastric secretions, and the presence of lectins (Hamman, 2008).
- **Antidiabetic activities:** Certain inorganic elements such as vanadium, manganese, and copper, along with the polysaccharides in *Aloe vera*, may play a crucial role in its antidiabetic properties (Reynolds and Dweck, 1999; Rajendran et al., 2007). This plant has been associated with reducing blood glucose levels in diabetics and lowering blood lipid or cholesterol levels by approximately 30% in patients with hyperlipidemia (Lim et al., 2003; Geremias et al., 2006).
- **Antioxidant activities:** *Aloe vera* exhibits significant antioxidant properties. Its gel contains glutathione peroxidase activity, superoxide dismutase enzymes, and a phenolic antioxidant, which likely contribute to these effects (Langmead et al., 2004; Hamman, 2008).
- **Laxative effects:** The anthraquinones in *Aloe vera* latex act as a strong laxative, enhancing intestinal peristalsis (Boudreau and Beland, 2006).
- **Antibacterial properties:** Aloe gel inhibits the microorganisms *Streptococcus pyogenes* and *Streptococcus faecalis* (Heggers, Pineless, and Robson, 1979). It is effective against both gram-positive and gram-negative bacteria that cause food poisoning or diseases in humans and animals (Habeeb et al., 2007).
- **Antifungal activity:** *Aloe vera*'s leaf pulp and liquid fraction are effective against plant pathogenic fungi (Joseph and Raj, 2010). It has also been shown to inhibit *Candida* (Feily and Namazi, 2009). Due to its antifungal properties, *Aloe vera* is used as a water conditioner in fish tanks (Shamim et al., 2004).
- **Antiviral and antitumor activity:** These effects might result from both indirect and direct actions: indirectly by stimulating the immune system and directly through anthraquinones (Surjushe et al., 2008). Consequently, clinical trials are underway to gather conclusive evidence for using *Aloe vera* in treating HIV-AIDS and cancer (Eshun and He, 2004).
- **Moisturizing and anti-aging effect:** Aloe is rich in mucopolysaccharides, which help retain moisture in the skin. It stimulates fibroblasts to produce collagen and elastin fibers,

enhancing skin elasticity and reducing wrinkles. Aloe's cohesive effects bind superficial epidermal cells, softening the skin, while its amino acids soften hardened skin cells. Zinc in the gel acts as an astringent, tightening pores. *Aloe vera* gel gloves have been shown to improve skin integrity, reduce fine wrinkles, and decrease erythema in dry skin caused by occupational exposure, highlighting its moisturizing effects (West and Zhu, 2003).

- **High cholesterol and triglyceride levels can be lowered naturally with Aloe vera:** High cholesterol is a risk factor for heart disease and strokes. Medical studies have demonstrated the internal benefits of *Aloe vera* gel. Results showed that administering *Aloe vera* gel to patients with heart disease and high cholesterol reduced their risk levels (Manvitha and Bidya, 2014).
- **Role in dentistry:** *Aloe vera* extracts have proven effective in maintaining oral hygiene, and Wynn (2005) also noted the use of *Aloe vera* gel in dental care.

1.4.4 The Use of *Aloe vera* in Animal Nutrition

In addition to its other applications, *Aloe vera* plays a significant role in promoting growth and managing the health of chickens (Bejar and Colapo, 2005; Mwale et al., 2005). For broiler chickens, adding 600 mg/kg of aloe powder, aloe water extract, aloe ethanol extract, or a combination of these to their diet can enhance production performance and immune function, with aloe water extracts showing the most significant improvements. Some studies also found that *Aloe vera* improved the acceptability of broiler meat (Wang et al., 2007).

Broilers fed *Aloe vera* exhibited significantly higher haemagglutination inhibition titres against Newcastle disease. Additionally, *Aloe vera* has been reported to help manage and control coccidiosis in chickens (Mwale et al., 2005; Chandrakesan et al., 2009). Including *Aloe vera* in the diet of laying hens led to a notable increase in egg production per hen, although there was no change in feed consumption or feed conversion ratio. Furthermore, *Aloe vera* extracts in the diet of laying hens may mitigate or treat the effects of lead intoxication in birds (Moorthy et al., 2009). Additionally, *Aloe vera* has been reported as an effective natural phytogenic growth promoter for shrimp, showing promising results.

1.4.5 Agriculture Sector

Aloe vera is highly valuable in agriculture due to its numerous benefits. It thrives in arid regions where other crops might struggle, offering a revenue source for farmers in such areas. *Aloe vera* can act as a natural pesticide (Patel, 2017) and fungicide, minimizing the need for chemical treatments and supporting soil health while reducing chemical residues on crops. Its extracts help safeguard plants from pests and diseases, boosting plant resilience and crop yield. Additionally, the vitamins, enzymes, and minerals in *Aloe vera* serve as a biofertilizer, promoting healthy plant growth and enhancing agricultural productivity.

1.4.6 Commercial Products

In addition to its health and beauty applications, *Aloe vera* is used in a wide range of commercial products due to its many benefits. It is found in food and beverages (such as yogurts and jellies), personal care items (including soaps, body washes, hand sanitizers, and deodorants), household products (such as detergents, cleaning sprays, and air fresheners), as well as textiles and other goods (such as wipes and tissues) (Adlakha et al., 2022).

1.4.7 Economic Development

Aloe vera plays a pivotal role in economic development by driving growth across various sectors. Its cultivation provides income for farmers, particularly in arid and semi-arid regions where other crops may not thrive. The growing demand for *Aloe vera* in cosmetics, pharmaceuticals, and health and wellness products generates significant revenue, creating jobs and stimulating economic activity in both rural and urban areas. Additionally, the global trade of *Aloe vera* products boosts export revenues for producing countries, enhancing their economic prospects and contributing to broader economic development. This multifaceted impact underscores *Aloe vera*'s importance as a driver of economic growth and sustainability.

1.5 FUTURE PROSPECTS

1.5.1 Emerging Trends in *Aloe vera* Research and Development

Aloe vera, known for its soothing and healing properties, has been a staple in traditional medicine and skincare for centuries. Recent advancements in research and development are unlocking new potentials for this versatile plant, propelling it into emerging fields and industries. The latest trends in *Aloe vera* research in various sectors include:

1.5.2 Genetic Engineering and Biotechnology

Recent advancements in genetic engineering and biotechnology are paving the way for new strains of *Aloe vera* with enhanced properties. Researchers are focusing on manipulating the plant's genetic makeup to increase the concentration of beneficial compounds such as polysaccharides, vitamins, and amino acids. The use of CRISPR technology could allow for precise editing of *Aloe vera* genes to enhance desirable traits, such as increased resistance to pests or higher yields of medicinal compounds (Shelake et al., 2023). Additionally, a metabolic engineering approach can optimize the pathways in *Aloe vera* that produce bioactive compounds, making the extraction process more efficient and cost-effective. These genetically modified variants could lead to more effective products for medicinal, cosmetic, and nutritional use.

1.5.3 Nanotechnology

Nanotechnology is revolutionizing *Aloe vera* applications with its impressive capabilities and groundbreaking innovations, particularly in the fields of pharmaceuticals and skincare. As a result, the application of *Aloe vera* has expanded significantly in areas such as wound healing, tissue engineering, and drug delivery through its use in hydrogels, nanoparticles, nanocomposites, nanofibers, and bio-inspired sponges (Balaji et al., 2015). Nano-encapsulation of *Aloe vera* extracts can protect its sensitive components from degradation, enhancing the bioavailability and targeted delivery of active compounds to specific tissues or cells, making treatments more effective. This technology also enables controlled release, enhancing the longevity and efficacy of *Aloe vera*-based products. In skincare, topical nanocarriers can help *Aloe vera* penetrate deeper into the skin, maximizing its hydrating and healing effects (Araujo et al., 2022).

1.5.4 Sustainable Cultivation and Processing

As sustainability becomes a critical concern globally, current research is focused on developing eco-friendly cultivation and processing methods for *Aloe vera*. Innovations in this area focus mainly on reducing the environmental impact of *Aloe vera* farming and production. Implementing organic farming practices, water-efficient irrigation systems (such as drip irrigation and rainwater harvesting), and renewable energy-powered processing plants will help in reducing the environmental footprint of *Aloe vera* production. This trend aligns with the global shift towards green and sustainable practices.

1.5.5 Clinical Research and Evidence-based Applications

While *Aloe vera* has been used traditionally for various ailments, the medical community is increasingly interested in validating its health benefits through rigorous clinical research. Conducting randomized controlled trials can provide high-quality evidence regarding the efficacy of *Aloe vera* in treating conditions such as diabetes, inflammatory diseases, and cancer (Samarh et al., 2017). Systematic reviews and meta-analyses suggest that the current shift towards evidence-based applications can facilitate the integration of *Aloe vera* into mainstream medical treatments.

1.5.6 Potential Applications in New Industries and Health Sectors

As research continues to uncover the myriad benefits of *Aloe vera*, its applications are expanding into new industries and health sectors. Below are some of the promising areas where *Aloe vera* is making a significant impact:

1.5.7 Pharmaceuticals

The pharmaceutical industry is poised to harness *Aloe vera*'s therapeutic properties in developing new medications. Its anti-inflammatory, antimicrobial, and immunomodulatory properties make it a promising candidate for a range of treatments, including wound healing, gastrointestinal issues, and autoimmune diseases (Surjushe et al., 2008; Heng et al., 2018; Kumar et al., 2019). Extracts of *Aloe vera* may be formulated into pills, creams, and injectables for chronic conditions like arthritis and inflammatory bowel disease, thereby offering natural alternatives to synthetic drugs.

1.5.8 Functional Foods and Nutraceuticals

Several researchers assert that the nutritional and functional benefits of *Aloe vera* are due to its minerals, water-soluble vitamins, and glucomannan (Vega et al., 2005; Hamman, 2008). On account of its chemical composition and structure, *Aloe vera* is considered a significant and natural source of prebiotics. Nowadays, *Aloe vera* is increasingly being incorporated into functional foods and nutraceuticals, which are foods containing prebiotics and probiotics as components with health benefits beyond basic nutrition. *Aloe vera*-based drinks, supplements, and fortified foods are gaining popularity for their potential to improve digestive health, boost immunity, and provide antioxidants (Cuvas-Limon et al., 2015). The demand for natural and health-enhancing food products is driving innovation in this sector.

1.5.9 Cosmetics and Personal Care

The cosmetic industry has long utilized *Aloe vera* for its moisturizing and healing properties. Emerging trends are now focusing on developing high-performance, natural skincare products that cater to the growing demand for clean beauty. *Aloe vera* is being used in formulations of sunscreens, anti-aging creams, and hair care products, appealing to consumers seeking effective and natural solutions (Park and Jo 2006; Kumar et al., 2019).

1.5.10 Veterinary Medicine

The benefits of *Aloe vera* are not limited to human health. The veterinary sector is exploring its use in treating animals. Being a natural antioxidant, *Aloe vera* serves as an excellent alternative to synthetic forms, effectively combating cellular and membrane damage caused by pro-oxidants (Sinha et al., 2017). *Aloe vera* based products are being developed for pet skincare, wound healing, and digestive health. This application not only benefits pets but also livestock, potentially improving the health and productivity of farm animals (Ebrahim et al., 2020).

1.5.11 Agriculture and Plant Health

Interestingly, *Aloe vera* is finding applications in agriculture as well, where its extracts are being used as natural pesticides and fertilizers. The antimicrobial properties help in protecting crops from diseases, making it an effective natural pesticide, while the nutrient content promotes healthier plant growth, enhancing yield and quality (Ndhlala et al., 2009; Rahmani et al., 2015). Additionally, the use of *Aloe vera* gel coating has generated significant interest due to its potential to extend the shelf life and storage duration of fresh horticultural produce (Hasan et al., 2021). This trend supports sustainable farming practices and reduces the reliance on chemical-based agricultural inputs. *Aloe vera* processing and production provide special agricultural problems as well as substantial economic prospects, especially when sustainable approaches are used. *Aloe vera*'s worldwide market is still expanding, which supports local economies and encourages the creation of numerous commercial goods. *Aloe vera* has bright future possibilities because of new research and possible uses in various businesses and health sectors, which might build on its already remarkable

1.6 CONCLUSION

The future of *Aloe vera* is bright, with ongoing research and development uncovering new potentials for this ancient plant. From genetic engineering to sustainable cultivation, and from pharmaceuticals to agriculture, *Aloe vera* is poised to make significant contributions across various industries. Innovations in biotechnology and nanotechnology will further enhance the efficacy and delivery of *Aloe vera*-based products, opening up new frontiers in personalized medicine and advanced skincare. As science continues to validate its benefits, *Aloe vera* is set to become an even more integral part of our health and wellness landscape, promising natural and effective solutions for the challenges of tomorrow.

REFERENCES

Adlakha K., Koul B. and Kumar A. (2022). Value-added products of Aloe species: Panacea to several maladies. *South African Journal of Botany, 147*, 1124–1135.

Araujo A. R. T. S., Rodrigues M., Mascarenhas Melo F., Peixoto D., Guerra C., Cabral C. and Paiva-Santos A. C. (2022). New-generation nanotechnology for development of cosmetics using plant extracts. In Grumezescu N. and Holban A. M. (Eds.), *Nanotechnology for the Preparation of Cosmetics Using Plant-Based Extracts,* Elsevier. 301–325. https://doi.org/10.1016/B978-0-323-85788-0.00014-2

Atul N. C., Santhosh K. C., Bhattacharjee C., Subal D. K. and Kannan K. (2011). Studies on immunomodulatory activity of *Aloe vera* (Linn). *International Journal of Applied Biology and Pharmaceutical Technology, 2*, 19–22.

Babaee N., Zabihi E., Mohseni S. and Moghadamnia A. A. (2012). Evaluation of the therapeutic effects of *Aloe vera* gel on minor recurrent aphthous stomatitis. *Dental Research Journal, 9*, 381–385.

Baby J. and Justin S. R. (2010). Pharmacognostic and phytochemical properties of *Aloe vera* linn – an overview. *International Journal of Pharmaceutical Sciences Review and Research, 4*, 106.

Balaji A., Vellayappan M. V., John A. A., Subramanian A. P., Jaganathan S. K., SelvaKumar M. and Yusof M. (2015). Biomaterials based nano-applications of *Aloe vera* and its perspective: a review. *RSC Advances, 5*(105), 86199–86213.

Bejar F. R. and Colapo R. P. (2005). Growth performance and sensory evaluation of broilers supplemented with *Aloe vera* (Aloe barbadensis) extract in drinking water. 42 PSAS National Convention. Manila, Philippines.

Boudreau M. D. and Beland F. A. (2006). An evaluation of the biological and toxicological properties of Aloe barbadensis (miller), Aloe vera. *Journal of Environmental Science and Health Part C, 24*(1), 103–154.

Byeon S. W., Pelley R. P., Ullrich S. E., Waller T. A., Bucana C. D. and Strickl F. M. (1998). Aloe barbadensis extracts reduce the production of interleukin-10 after exposure to ultraviolet radiation. *Journal of Investigative Dermatology, 110*(5), 811–817.

Chandra V. and Pandey M.C. (1989*) Cultivation of Herbal Plants* (3rd ed.). Pusa, New Delhi: Indian Agriculture Research Institute, 127–129.

Chandrakesan P., Muralidharan K., Kumar V. D., Ponnudurai G., Harikrishnan T. J. and Rani K. S. V. N. (2009). Efficacy of a herbal complex against caecalcoccidiosis in broiler chickens. *VeterinarskiArhiv*, *79*(2), 199–203.

Choi S. and Chung M. H. (2003). A review on the relationship between *Aloe vera* components and their biologic effects. *Seminars in Integrative Medicine*, 1, 53–62.

Chow J. T. N., Williamson D. A., Yates K. M. and Goux, W. J. (2005). Chemical characterization of the immunomodulating polysaccharide of *Aloe vera* L. *Carbohydrate Research*, *340*(6), 1131–1142.

Coats B. C. (1979). Hypoallergenic stabilized *Aloe vera* gel. *US Patent 4*, 178, 372.

Cuvas Limon R. B., Julio M. S., Carlos C. E. J., Mario C. H., Mussatto S. I. and Ruth B. C. (2015). *Aloe vera* and probiotics: A new alternative to symbiotic functional foods. *Annual Research & Review in Biology*, *9*(2), 1–11.

Dagne E., Bisrat D., Viljoen A. and Van Wyk B. E. (2000). Chemistry of Aloe species. *Current Organic Chemistry, 4*, 1055–1078.

Das N. and Chattopadhay R. N. (2004). Commercial cultivation of Aloe. *Natural Product Radiance, 3*, 85–87.

Devaraj S., Jialal R., Jialal I. and Rockwood R. (2008). A pilot randomized placebo controlled trial of 2 *Aloe vera* supplements in patients with pre-diabetes/metabolic syndrome. *Planta Medica*, *74*, SL77.

Eshun K. and He Q. (2004). *Aloe vera*: A valuable ingredient for the food, pharmaceutical and cosmetic industries – A review. *Critical Reviews in Food Science and Nutrition*, *44*(2), 91–96.

Ebrahim, A. A., Elnesr, S. S., Abdel-Mageed, M. A. A. and Aly, M. M. M. (2020). Nutritional significance of Aloe vera (Aloe barbadensis Miller) and its beneficial impact on poultry. *World's Poultry Science Journal*, *76*(4), 803–814. https://doi.org/10.1080/00439339.2020.1830010

Feily A. and Namazi M. R. (2009). *Aloe vera* in dermatology: A brief review. *Societaitaliana di Dermatologia e Sifilografia*, *144*(1), 85–91.

Femenia A., Sanchez E. S., Simal S. and Rosello C. (1999). Compositional features of polysaccharides from *Aloe vera* (Aloe barbadensis Miller) plant tissues. *Carbohydratre Polymers*, *39*, 109–117.

Foster M, Hunter D. and Samman S. (2011). Evaluation of the nutritional and metabolic effects of *Aloe vera*. In Benzie I. F. F. and Wachtel-Galor S., (Eds.), *Herbal Medicine: Biomolecular and Clinical Aspects*. 2nd ed. Boca Raton: CRC. 37–54.

Foster M. and Hunter D. (2009). *Aloe vera*: A review update. *HerbalGram*, *45*, 52–63.

Gage D. (1996). *Aloe vera: Nature's Soothing Healer*. Inner Traditions/Bear & Co.

Geremias R., Pedrosa R. C., Locatelli C., De Favere V. T., Coury-Pedrosa R. and Laranjeira M. C. M. (2006). Lipid lowering activity of hydrosoluble chitosan and association with *Aloe vera* L. and Brassica olearaceae L. *Phytotherapy Research*, *20*(4), 288–293.

Grace O. M., Simmonds M. S., Smith G. F. and van Wyk A. E. (2008). Therapeutic uses of Aloe L (Asphodelaceae) in southern Africa. *Journal of Ethnopharmacology*, *119*, 604–614.

Habeeb F., Shakir E., Bradbury F., Cameron P., Taravati M. R., Drummond A. J. and Ferro, V. A. (2007). Screening methods used to determine the anti-microbial properties of *Aloe vera* inner gel. *Methods*, *42*(4), 315–320.

Hamman J. H. (2008). Composition and applications of *Aloe vera* leaf gel. *Molecules*, *13*(8), 1599–1616.

Hasan M. U., Riaz R., Malik A. U., Khan A. S., Anwar R., Rehman R. N. U. and Ali, S. (2021). Potential of *Aloe vera* gel coating for storage life extension and quality conservation of fruits and vegetables: An overview. *Journal of Food Biochemistry*, *45*(4), 613–640.

Heggers J. P., Pineless G. R. and Robson M. C. (1979). Dermaide aloe aloevera gel-comparison of the antimicrobial effects. *Journal of the American Medical Technologists*, *41*(5), 293–294.

Heng H. C., Zulfakar M. H. and Ng P. Y. (2018). Pharmaceutical applications of *Aloe vera*. *Indonesian Journal of Pharmacy/Majalah Farmasi Indonesia*, *29*(3), 101–116.

Hossain M. S., Mamun Or Rashid A. N. M., Towfique N. M. and Sen M. K., (2013). A review on ethnopharmacological potential of *Aloe vera* L. *Journal of Intercultural Ethnopharmacology, 2*, 113e120.

Im, S. A., Oh, S. T., Song, S., Kim, M. R., Kim, D.-S., Woo, S., Jo, T. H., Park, Y. I. and Lee, C. K. (2005). Identification of optimal molecular size of modified Aloe polysaccharides with maximum immunomodulatory activity. *International Immunopharmacology*, *5*(2), 271–279. https://doi.org/10.1016/j.intimp.2004.09.033

Hutter J. A., Salman M., Stavinoha W. B., Satsangi N., Williams R. F., Streeper R. T. and Weintraub S. T. (1996) Antiinflammatory C-glucosyl chromone from Aloe barbadensis. *Journal of Natural Products, 59*, 541–543.

Joseph B. and Raj S. J. (2010).Pharmacognostic and phytochemical properties of *Aloe vera* in an overview. *International Journal of Pharmaceutical Sciences Review and Research, 4*(2), 106–110.

Josias H. H. (2008). Composition and applications of *Aloe vera* Leaf Gel. *Molecules, 13*, 1599–1616.

Kumar R., Singh A. K., Gupta A., Bishayee A. and Pandey A. K. (2019). Therapeutic potential of *Aloe vera*—A miracle gift of nature. *Phytomedicine, 60*, 152996.

Langmead L., Makins R. J. and Rampton D. S. (2004). Anti-inflammatory effects of *Aloe vera* gel in human colorectal mucosa in vitro. *Alimentary Pharmacology & Therapeutics, 19*(5), 521–527.

Lim B. O., Seong N. S., Choe R. W., Kim J. D., Lee H. Y., Kim S. Y., Yu B. P., Jeon T. I. and Park D. K. (2003). Efficacy of dietary *Aloe vera* supplementation on hepatic cholesterol and oxidative status in aged rats. *Journal of. Nutritional Science and Vitaminology, 49*, 292–296.

Liu C., Leung M. Y. K. and Koon J. C. M. (2006) Macrophage activation by polysaccharide biological response modifier isolated from *Aloe vera* L. var. chinensis (Haw.) Berg. *International Immunopharmacology, 18*, 1634–1641.

Liu X. D. Chen B. Y. Wu and Jiang Q. (2010). Influence of Aloe polysaccharide on proliferation and hyaluronic acid and hydroxyproline secretion of human fibroblasts in vitro. *Journal of Chinese Integrative Medicine, 8*(3), 256–262.

Maiti S. and Chandra R. (2002). *Cultivation of* Aloe vera. Gujarat: National Research Centre for Medicinal and Aromatic Plants.

Manvitha K. and Bidya B. (2014). *Aloe vera*: A wonder plant its history, cultivation and medicinal uses. *Journal of Pharmacognosy and Phytochemistry, 2*(5), 85–88.

Misra B. B. and Dey, D. (2001). Plant tissue culture: A potential source of pharmaceutically important bioactive compounds. *Critical Reviews in Biotechnology, 21*(2), 229–258.

Moorthy M., Saravanan S., Mehala C., Ravi S., Ravikumar M., Viswanathan K. and Edwin S. C. (2009). Performance of single comb white leghorn layers fed with *Aloe vera*, Curcuma longa (turmeric) and probiotic. *International Journal of Poultry Science, 8*(8), 775–778.

Mulu T., Teshale F., Gemeda S. and Sahu O. (2015). Medicated Evaluation of *Aloe vera*: Overview on characteristics and application. *World Journal of Nutrition and Health, 3*(1), 1–7.

Mwale M., Bhebhe E., Chimonyo M. and Halimani T. E. (2005).Use of herbal plants in poultry health management in the Mushagashe small-scale commercial farming area in Zimbabwe. *International Journal of Applied Research in Veterinary Medicine, 3*(2), 163–170.

Ndhlala A. R., Amoo S. O., Stafford G. I., Finnie J. F. and Van Staden J. (2009). Antimicrobial, anti-inflammatory and mutagenic investigation of the South African tree aloe (Aloe barberae). *Journal of Ethnopharmacology, 124*(3), 404–408.

Ni Y. and Tizard I. R. (2004). Analytical methodology: The gel-analysis of aloe pulp and its derivatives. *In* Reynolds T., Ed. *Aloes The Genus Aloe*. Boca Raton: CRC Press, 111–126.

Ni Y., Turner D., Yates K. M. and Tizard I. (2004) Isolation and characterization of structural components of *Aloe vera* L. leaf pulp. *International Immunopharmacol, 4*, 1745–1755.

O'Brien C., van Wyk B. E. and van Heerden F. R. (2011) Physical and chemical characteristics of Aloe ferox leaf gel. *South African Journal of Botany, 77*, 988–995.

Park Y. I. and Jo T. H. (2006). Perspective of industrial application of *Aloe vera*. In Young I. P. and Seung K. L., (Eds.), *New perspectives on Aloe*. Boston, MA: Springer US. 191–200.

Patel P. C. (2017). Pesticidal effect of *Aloe vera* extracts on some larvae of crop-damaging pests. *International Journal of Applied and Universal Research, 4*(5), 1–5.

Patidar A., Bhayadiya R. K., Nimita M., Pathan J. K. and Dubey P. K. (2012). Isolation of Aloin from *Aloe vera*, its characterization and evaluation for antioxidant activity. *International Journal of Pharmaceutical Research and Development, 4*(2), 24–28.

Rahmani A. H., Aldebasi Y. H., Srikar S., Khan A. A. and Aly S. M. (2015). *Aloe vera*: Potential candidate in health management via modulation of biological activities. *Pharmacognosy Reviews, 9*(18), 120.

Rajendran A., Narayanan V. and Gnanavel I. (2007). Study on the analysis of trace elements in *Aloe vera* and its biological importance. *Journal of Applied Sciences Research, 3*(11), 1476–1478.

Rajeswari R., Umadevi M., Rahale C. S., Pushpa R., Selvavenkadesh S., Kumar K. S. and Bhowmik, D. (2012). *Aloe vera*: The miracle plant its medicinal and traditional uses in India. *Journal of Pharmacognosy and Phytochemistry*, *1*(4), 118–124.

Ramachandra C. T. and Rao, P. S. (2008). Processing of *Aloe vera* leaf gel: A review. *American Journal of Agricultural and Biological Sciences*, *3*(2), 502–510.

Reynolds T. and Dweck A. C. (1999). *Aloe vera* leaf gel: A review update. *Journal of Ethnopharmacology*, *68*(1–3), 3–37.

Roberts D. B. and Travis E. L. (1995). Acemannan-containing wound dressing gel reduces radiation-induced skin reactions in C3H mice. *International Journal of Radiation Oncology Biology Physics*, *32*(4), 1047–1052.

Roy U., Pavel Axentiev M. S. and Diana Swisher M. A. (2012). *Aloe vera leaf: Standards of identity, analysis, and quality control.* Scotts Valley, CA: *American Herbal Pharmacopoeia*, 1–52.

Saeed M. A., Ahmad I., Yaqub U., Akbar S., Waheed A., Saleem M. and Nasir-ud-Din. (2004). *Aloe vera*: A Plant of Vital Significance. *Science Vision*; *9*, 1–13.

Sahu P. K., Giri D. D., Singh R., Pandey P., Gupta S. and Shrivastava A. K. (2013). Therapeutic and medicinal uses of *Aloe vera*: A review. *Pharmacology & Pharmacy*, *4*(08), 599–610.

Samarh S. N., Khalaf N. A. and Hajhamad M. M. (2017). Evidence based medical use of *Aloe vera* extracts, short review of literature. *International Journal of Research in Medical Sciences*, *5*(10), 41–98.

Saroj P. L., Dhandar D. G. and Singh R. S. (2004). *Indian Aloe*. Bikaner: Central Institute for Arid Horticulture.

Saroj P.L. and Purohit C.K. (2004). Indian Aloe: an alternative food with nutritional value, *SAIC News Letter*. January–March, 2004. 5 & 7

Serrano M., Valverde J. M., Guillen F., Castillo S., Martínez-Romero D. and Valero D. (2006). Use of *Aloe vera* gel coating preserves the functional properties of table grapes. *Journal of Agricultural and Food Chemistry*, *54*(11), 3882–3886.

Shamim S., Ahmed S. W. and Azhar I. (2004).Antifungal activity of Allium, Aloe, and Solanum species. *Pharmaceutical Biology*, *42*(7), 491–498.

Shelake R. M., Jadhav A. M., Bhosale P. B. and Kim J. Y. (2023). Unlocking secrets of nature's chemists: Potential of CRISPR/Cas-based tools in plant metabolic engineering for customized nutraceutical and medicinal profiles. *Plant Physiology and Biochemistry*, *203*, 108–170.

Sinha S. S., Muzamil B., Ahmad M. U. and Rehman A. Q. (2017). Ameliorative effect of *Aloe vera* supplementation in poultry feed. *Journal of Animal Research*, *7*(1), 85–90.

Surjushe A., Vasani R. and Saple D. G. (2008). *Aloe vera*: A short review. *Indian Journal of Dermatology*, *53*(4), 163–166.

Tarameshloo M., Norouzian M., Zarein-Dolab S., Dadpay M., Mohsenifar J. and Gazor R. (2012) *Aloe vera* gel and thyroid hormone cream may improve wound healing in Wistar rats. *Anatomy & Cell Biology*, *45*, 170e177.

Vega A.G., Nevenka C., Diaz L. and Lemus R. (2005). El *Aloe vera* (Aloe barbadensis Miller) como componente de alimentos funcionales. *Revista Chilena de Nutricion*, *32*, 208–214.

Wang C., Jia H., Tong J., Gao W., Sa R. and Zhang Q. (2007). Effect of Aloe powder and extract on production performance and immune function of broiler chickens. *Journal of Fujian Agriculture and Forestry University*, *12*, 6–12.

West D. P. and Zhu Y. F. (2003). Evaluation of *Aloe vera* gel gloves in the treatment of dry skin associated with occupational exposure. *American Journal of Infection Control*, *31*(1), 40–42.

Wynn R. L. (2005). *Aloe vera* gel: Update for dentistry. *General Dentistry*, *53*(1), 6–9.

Yu Z., Jin C., Xin M. and Jian Min H. (2009). Effect of *Aloe vera* polysaccharides on immunity and antioxidant activities in oral ulcer animal models. *Carbohydrate Polymers*, *75*(2), 307–311.

2 Botanical Insights

Taxonomy, Diversity, and Global Distribution of Aloe vera

Kumaraswamy R.V., Jeevitha B.T, Shweta D. C, and Geetha Govind

2.1 INTRODUCTION

Aloe (*Aloe barbadensis* L.) is a xerophytic perennial succulent plant that belongs to the family Liliaceae. The name *Aloe vera* is derived from the Arabic word "alloeh" which means "shining bitter substance"(Deshmukh et al., 2019). *Aloe vera,* initially cultivated as an ornamental plant, is now renowned for its medicinal properties worldwide (Wagan et al., 2022). It has been recognized for centuries in traditional medicine for its skin-healing properties across diverse cultures of China, India, the West Indies, and Japan (Grindlay and Reynolds, 1986). *Aloe vera* thrives well in hot arid regions of the world – the southern Mediterranean, the Middle East, North Africa (from Morocco to Egypt), West India, Southwest Asia, Central America, Mexico, Cape Verde, Madeira, and the Canary Islands (Martinez et al., 2022).

Several synonyms for *Aloe vera* have been identified, including *Aloe barbadensis Mill., Aloe indica Royle, Aloe perfoliata L. var. vera*, and *Aloe vulgaris Lam* (Riaz et al., 2021). While most aloe species are non-toxic, some can be highly poisonous. *Aloe vera* is a low-maintenance indoor plant renowned for its numerous health benefits, particularly for skin care. In addition, *Aloe vera* is renowned for its healing and moisturizing properties. Historically, it has been used to treat burns, abrasions, and dry skin, which is continued in modern times too (Jangra et al., 2022). Many cultures apply the plant's latex or inner gel directly to the skin for its soothing effects, and commercially, the gel is used in cosmetics for hydration. The use of it to improve digestion has gained immense popularity in wellness practices to improve health (Boudreau and Beland, 2006). In addition to its laxative properties, *Aloe vera* also plays a role in reducing inflammation in the gastrointestinal tract. Chronic inflammation is a hallmark of many digestive disorders, including irritable bowel syndrome (IBS) and inflammatory bowel disease (IBD). Studies have reported that *Aloe vera* helps alleviate the symptoms of these conditions by soothing inflamed tissues and promoting healing. *Aloe vera* extracts are shown to reduce inflammation in the intestines of animal models, suggesting that *Aloe vera* could be a beneficial adjunct therapy for inflammatory bowel diseases like Crohn's disease and ulcerative colitis (Kumar et al., 2020).

Aloe vera promotes wound healing through polysaccharides and gibberellins, which enhance collagen and elastin formation. It potentially reduces wrinkles and aids in scar prevention due to the amino acids and enzyme activity that stimulates skin regeneration (Reynolds and Dweck, 1999; Eshun and He, 2004; Surjushe et al., 2008). Its anti-inflammatory effects are attributed to salicylic acid, which inhibits prostaglandin production, making it beneficial for arthritis and joint issues, while polysaccharides boost immunity (Klein and Penneys, 1988; Yu et al., 2009). Additionally, *Aloe vera* aids in the recovery of burns caused by radiations (UV and X-ray), due to the therapeutic action of lectins (Maenthaisong et al., 2007). Some polysaccharides and inorganic elements

DOI: 10.1201/9781003603108-2

like vanadium, manganese, and copper present in *Aloe vera* help in reducing blood glucose levels and demonstrate antidiabetic properties. It also displays its role in decreasing cholesterol by about 30% in hyperlipidemic patients (Franz et al., 2005; Rajendran et al., 2007). In addition, it exhibits antibacterial activity against microorganisms, particularly Gram-positive bacteria linked to food poisoning, such as *Streptococcus pyogenes*, *Shigella flexneri*, and *Klebsiella sp.* (Boudreau and Beland, 2006). The antiviral and anti-tumor effects of *Aloe vera* are attributed to anthraquinones and stimulation of the immune system. Clinical trials are ongoing to explore its potential in HIV/AIDS and cancer treatments. Promising results have also been observed in age-related disease studies on pathogen-free rats.

2.2 TAXONOMY OF *ALOE VERA* AND MORPHOLOGICAL CHARACTERISTICS

The taxonomy of Aloe vera (Nadkarni, 2004) is as follows:

Kingdom: Plantae
Order: Asparagales
Division: Spermatophyte
Subdivision: Angiospermae
Class: Monocotyledoneae
Genus: Aloe
Family: Aloaceae
Species: Barbadensis Mill

The Aloe was originally classified in the Liliaceae family because of its resemblance to lilies in the form of a bulb. However, Tom Reynolds, a London-based researcher, proposed a new classification for Aloe and placed it in the botanical family Aloaceae. Aloe is a member of the large class of plants known as xerophytes. Named so because they can grow in extremely low moisture conditions and close their stomata to maintain the plant's water balance (Hoffmann et al., 2020). This ability to store water internally enables them to withstand prolonged dry weather or even droughts. Furthermore, specific chemical components have the extraordinary ability to close up any wound or damage to the epidermis almost instantly, preventing precious water from escaping. It is a genus of evergreen, shrubby, perennial plants with different colored flowers depending on the species (Prisa and Gobbino, 2021). The Aloaceae family includes approximately 350 plant varieties worldwide (López et al., 2019). In 1955, 132 species were catalogued in South Africa only. Others are spread across the African continent, Asia, the Caribbean, Spain, Central and South America, North America, and Mexico (Khan and Siddiqui, 2020). A botanical distinction can be made by examining the trunk and leaves. Aloe groups are classified as acauleas, subcauleas, and cauleas (Prisa and Gobbino, 2021). Aloe leaves have a 2 mm thick skin with elastic membranes that protect the gel, which is transparent and rich in medicinal properties (Abouelela et al., 2021). Aloe plants reproduce through pollination by insects and birds, as they are self-sterile and require fertilization from other plants (Kar and Bera, 2018). The plant also reproduces by sending out stolons and shoots from the base of the main stem, forming colonies. It has a habitat that is typical of arid and desert areas, and its height can range from a few centimeters to twenty meters, depending on the species.

Aloes have short stalks that grow close to ground level, from which their thick, tapering, green, spiny leaves emerge. Other plants in the lily family, including asparagus, tulips, turnips, garlic, onions, and leeks, are closely related to aloe. The plant is currently grown in hot, arid parts of Asia, Europe, and America, despite its native North Africa and Spain. Its tall, thick, juicy leaves have phylotaxy, which resembles a wheel. The leaves have a thorny structure on both sides, with a thorny tip. The leaves' inner content has a bitter flavor, a foul odor, and a jelly-like consistency. The leaves are composed of three layers: mucilage gel, which is the inner layer of the leaf that contains gel; sap, which is the central layer that contains latex; and rind, which is the outermost protective layer (Vogler and Ernst, 1999). Typically, each plant has 12–16 leaves, which can weigh up to three pounds when fully grown, and all the leaves of the plant are fully developed within three years. The leaves measure typically around 25–30 cm in length and 3–5 cm in width. To promote bigger leaf growth in the parent plant, the shoots of the mother plants must be cut at least twice a year. It blooms from October to January with the tall inflorescence covered in numerous tiny pink flowers.

2.3 PHYLOGENETIC RELATIONSHIPS AND EVOLUTIONARY INSIGHTS

Since the first Aloe spp. were described from Madagascar and the Mascarene Islands, a number of species have been identified till now. Assessment of Reynolds' morphological classification is necessary due to the accumulation of species and nomenclatural complexities over the past 100 years. In addition, an analysis of evolutionary relationships using molecular systematics is important (Reynolds, 1966). Ten percent or less of the species from Madagascar have been included in previous phylogenetic analyses of aloe (Manning et al., 2014; Lee et al., 2016) or Madagascar (Grace et al., 2015). Dee et al. (2018) conducted the first comprehensively sampled phylogenetic analysis of Aloe, including the species from Madagascar and the Mascarenes, filling the knowledge gap and improving the understanding of the evolutionary history that led to the remarkable variety that exists today. Out of 33 accessions of Aloe spp., 95 novel sequences were generated (29 for nuclear ITS and 33 for each of the plastid regions-matK and trnQ-rps16). Aloidendron (*A. Berger*) Klopper and Gideon F. Sm each had one species, and the core dataset had 2684 bp of nucleotide data for 43 Aloe spp. found in Madagascar and the Mascarenes. Nearly 55 Aloe spp. were included in the enlarged dataset, which includes sequences from GenBank. Compared to comparable studies of the extended dataset with more missing data, tree searches of the core dataset produced better-resolved phylogenetic trees.

The Aloe section *Lomatophyllum*, a paraphyletic group, exemplifies the hypothesized "island hopping" dispersal of aloes from Africa to Madagascar. These species are adapted to forest environments and rely on frugivorous birds for seed dispersal, though direct evidence is lacking (Rowley, 1996). Bird-mediated dispersal may explain the close evolutionary ties among species like *Aloe pembana* in shaded coastal forests on Pemba, *A. occidentalis* in Madagascar's dry forests, and *A. macra* in Réunion's forests. Another notable relationship exists between *A. suarezensis*, found on cliffs along Madagascar's northeastern coast, and two *Lomatophyllum*-group species from Mauritius: *A. tormentorii* adapted to unusually sunny habitats and *A. purpurea* thriving in deeply shaded gorges. The evolution of fleshy fruits, observed over 20 times in monocots, aligns with adaptations to shady forest habitats (Givnish et al., 2010). This study supports earlier hypotheses that fleshy fruits in Aloe are a derived adaptation to shaded environments (Perrier la Bâthie, 1926; Rowley, 1996; Rauh, 1998), as seen in the clustering of *Lomatophyllum* within Africa and Madagascar clades (A and M). No phylogenetic support was found for Groups 1–9 as defined by Reynolds (1966) in his Aloe monograph for Madagascar and the Mascarenes. Reynolds acknowledged the heterogeneity of Group 1 ("Very small to small plants") as a convenience for identification but considered the other groups cohesive. Closer examination revealed these morpho-groups to be broad assemblages, with some infraspecific taxa rendered paraphyletic.

For instance, *Aloe capitata* Baker var. *cipolinicola* H. Perrier was included in Group 9 while other subspecies of *A. capitata* were placed in Group 6. Published a year before Reynolds' death in 1967, his work may have been unfinished but remains the sole formal classification of Aloe in the region. Reynolds' groups were based on traits like growth habit, leaf arrangement, leaf morphology, inflorescences, and floral features, all of which enhance ecological adaptability, such as pollinator partitioning, substrate preference, and plant–animal interactions. The findings suggest these traits may have evolved independently multiple times, influenced by similar evolutionary pressures that preserve morphological convergence among unrelated taxa. A more recent guide by

Castillon and Castillon (2010) incorporated geographical distribution for classifying Aloe diversity, further highlighting the challenges posed by homoplasious morphological traits in dichotomous keys. The phylogenetic grouping of morphologically dissimilar species illustrates Aloe's complex evolutionary history. For example, *A. bakeri* (Group 1), a small shrubby species, clusters strongly (PP = 1) with *A. vaombe* (Group 9), a tall tree-like species. The growing recognition of Aloe species in Madagascar and the Mascarene Islands, alongside new evolutionary insights, underscores the need for an updated infrageneric classification. Further phylogenetic and morphological studies are essential to refine the evolutionary framework of this diverse and iconic succulent group.

2.4 DIVERSITY OF *ALOE VERA*

Comprehensive studies on *Aloe vera*'s genetic diversity are essential not only for its conservation but also for effective management, validation, and utilization. *Aloe vera*'s commercial significance necessitates an in-depth understanding of the genetic diversity within Aloe genotypes to support genetic improvement programs (Arolu et al., 2015). While physiological markers can measure genetic diversity, morphological markers are less reliable due to their susceptibility to climatic influences and developmental stages. Conversely, molecular markers are widely regarded as effective tools for analyzing genetic diversity and relationships among genotypes (Bisrat et al., 2000). These markers enhance the efficiency of molecular breeding by closely associating with traits of interest. Moreover, transcriptome sequencing of *Aloe vera* provides valuable insights for developing an advanced marker system and identifying genes involved in producing essential metabolites (Choudhri et al., 2018).

Various molecular markers are available to study genetic diversity in medicinal plants, with amplified fragment length polymorphism (AFLP) being a robust and highly polymorphic DNA fingerprinting method. AFLP is particularly useful for analyzing genetic diversity, population genetics, AMOVA, and plant population structures (Rozo et al., 2009). A key advantage of AFLP is that, unlike other DNA fingerprinting methods, it does not require prior genome sequence information (Blignaut et al., 2013). The AFLP process involves the complete digestion of DNA using restriction enzymes, the ligation of specific adapters designed from enzyme restriction sites to the DNA fragments, followed by PCR amplification (Sorkheh et al., 2007). The amplified products are then separated using non-denaturing polyacrylamide gel electrophoresis (Hiransuchalert et al., 2020). AFLP markers play a critical role in conserving genetic resources and aiding the development of new varieties with desirable traits. Previous studies on *Aloe vera*'s genetic diversity have largely relied on RAPD markers (Chandra and Choudhary, 2014; Das et al., 2017). However, limited research has been conducted using AFLP (Tripathi et al., 2011) or a combination of RAPD and ISSR markers (Bhaludra et al., 2014; Kumar et al., 2016), often with a small number of genotypes.

The germplasm of Aloe collected from the Indian arid zone is also subjected to diversity analysis. Nearly 74 Aloe germplasm samples were evaluated for diversity analysis using 18 morphological traits (eight quantitative and ten qualitative). The study revealed a high coefficient of variation (CV) in quantitative traits, with notable variation observed in aloin content (AC) (43.1%), leaf width (LW) (32.0%), polysaccharide content (30.7%), and inflorescence length (18.5%). The range and mean values for key biochemical traits were 0.009–10.94 and 5.77 for aloin content, and 0.16–0.61 and 0.40 mg 100 mL^{-1} for polysaccharide content, respectively. Leaf rind thickness (LRT) showed the highest variability (0.8–1.8 mm), followed by leaf length (LL) (30–57.5 mm) and LW (99.2–17.3 mm). Pearson's correlation analysis revealed both positive and negative relationships among the quantitative traits, with correlations ranging from −0.19 (AC and LL/LRT) to 0.48 (leaf thickness and LRT). Cluster analysis, performed using Euclidean distance-based unweighted pair group method with arithmetic mean (UPGMA) on six quantitative traits, produced a dendrogram that formed four major clusters at a coefficient value of 3.7, each containing two subclusters. This analysis distinguished aloe genotypes into two main groups: non-bitter and bitter types. The clustering pattern was supported by a two-dimensional scatter plot from principal components analysis (PCA), which accounted for 67.73% of cumulative variance. While the study acknowledged limitations in capturing total genetic variation, it demonstrated that morphological and biochemical traits are valuable for preliminary evaluations and provide a general framework for assessing genetic diversity among morphologically distinct Aloe genotypes (Kumar et al., 2019).

2.5 GEOGRAPHIC DISTRIBUTION AND GLOBAL DISTRIBUTION RANGE

Aloe is a drought-tolerant plant; it struggles in areas where waterlogging occurs. It therefore requires water with a low salt concentration and a pH that is sub-acidic. Additionally, crops can be found in Africa, Central America, Australia, and Mexico. They can be found in southern Spain, Portugal, Greece, and, more recently, Italy in Europe. Plants cultivated in South Florida in the Rio Grande Valley typically have leaves that weigh 600 g to 1 kg. One hectare may support between 10,000 and 12,000 plants because each plant requires an area of around one square meter (Banerjee et al., 2018). In India, it is found in Tamil Nadu, Uttar Pradesh, Rajasthan, Gujarat, Maharashtra, and Andhra Pradesh (Singh and Singh, 2021). In order to thrive in arid environments with little or irregular rainfall, this perennial succulent plant can produce water-storing tissue in its leaves (Kumar and Yadav, 2014). There are several economically important aloe species that can be used in place of *Aloe vera*, including *A. ferox Mill.*, *A. zeylanicum*, *A. Africana Mill.*, *A. arborescence Mill.*, and *A. perryi Back.*

The rising cost of allopathic medications with their well-known adverse effects has led to a recent surge in the usage of Ayurveda and medicinal plant remedies. Global trade is currently valued at around 80 million US dollars, and it is projected to rise by 35–40% in the next five years. China and India each hold a 10% market share, which might be increased by commercial cultivation, while the United States leads the market with 65%. The leaf-succulent genus Aloe L. (Asphodelaceae) is a unique feature of Madagascar's semi-arid and spiny forests. Sub-Saharan Africa, the Arabian Peninsula, and islands in the eastern Indian Ocean, such as Madagascar, the Mascarene Islands of Mauritius and Réunion, and the Seychelles, are all home to the genus. The remarkable species richness (>25% of the genus) and biodiversity (the greatest for any floristic region where Aloe is present) of Madagascar's 132 native Aloe species make it a major center of diversification for the plant. Furthermore, in Madagascar and the Mascarenes, aloe has 30 known varieties, 22 subspecies, and three hybrids (WCSP, 2015).

Four species are restricted to islands in the Mascarenes: *A. lomatophylloides* Balf.f., which is endemic to the island of Rodrigues (a territory of Mauritius); *A. purpurea* Lam. and *A. tormentorii* (Marais) L. E. Newton and G. D. Rowley are found only on Mauritius; and *A. macra* Haw. is found only on Réunion. *Aloe aldabrensis* (Marais) L. E. Newton and G. D. Rowley is endemic to the Seychelles. In Madagascar, the diversity of Aloe is richest in the central highlands, where habitats are most diverse, following a general pattern for the genus throughout its distribution (Holland, 1978; Grace et al., 2015). The region is also the centre of diversity for Aloe spp. with unusual berry-like fruits and non-winged seeds, unlike the dry dehiscent capsules containing winged, wind-dispersed seeds that are typical of Aloe (Rauh, 1995).

2.6 CULTIVATION AND COMMERCIAL PRODUCTION

Due to its eco-physiological traits, such as drought resistance, succulence, and CAM metabolism, aloe can be grown in various arid and semi-arid regions around the world. Aloe is cultivated in open fields in countries such as Kenya, Tanzania, Ethiopia, and Nigeria, with products being exported to Europe and Asia. It is also grown in southwestern Australia, China, India, Pakistan, Israel, the Caribbean islands, southern USA, Mexico, and Paraguay. In Mediterranean regions like Spain, Greece, and southern Italy, aloe is cultivated both in open fields and in cold greenhouses (Oldfield, 2004; Lucia and Lucini, 2013). Each plant produces 15 to 30 kg of leaves annually under ideal circumstances. In commercial aloe cultivation, pesticides and insecticides are typically not required because the sap's bitter flavor is enough to drive out insects and other potential predators. During flowering, aloe plants are particularly sensitive because the sap from the leaves migrates to the stem and flowers, making it easier for viruses to attack them (Banerjee et al., 2018).

2.7 CONVENTIONAL *IN-VIVO* PROPAGATION

Conventional propagation methods are insufficient to meet the growing market demand for aloe, which requires plant material that is healthy, uniform, true-to-type, and available throughout the year.

- Propagation by seed
- Suckers and rhizomes

Traditionally, *Aloe vera* is propagated using suckers (lateral shoots) and rhizome cuttings as planting material. In a conventional production system, each mother plant typically produces three to four suckers during the growing season (Gantait et al., 2014).To reduce the risk of diseases, selected suckers should be subjected to slight moisture stress under shaded conditions for about 5–10 days before transplantation. This process promotes root suberisation, enhancing their adaptability and stimulating root growth on planting (Murillo-Amador et al., 2007). The crop will be ready for harvest approximately 18 months after sowing. However, using rhizome cuttings for large-scale aloe propagation (Rajeswari et al., 2012) is impractical due to the extended time required for full plant development, which averages around two years. Additionally, the process exposes the plants to phytopathogens because of the cuts made during propagation (Gantait et al., 2014). The high risk of damaging the mother plant, producing infected seedlings, the seasonal limitations, and the restricted number of suckers, along with the slow vegetative growth and limited availability of high-quality raw material, are significant challenges for large-scale commercial production of young *Aloe vera* plants.

2.8 MEDICINAL IMPORTANCE

Aloe vera has a variety of medicinal uses, including treating burns externally and relieving constipation when consumed. There are several health benefits associated with *Aloe vera*, which is the most effective natural plant when used internally and externally. *Aloe vera* is frequently used in two parts. Because of its cathartic properties, bitter exudate is used in natural medicine. It is also frequently used as a laxative and as a bittering agent in alcoholic beverages. Despite being a potent laxative with a high barbaloin content, the dried latex should only be used for a maximum of one week and should not be used when pregnant, nursing or during menstruation. *Aloe vera*'s proposed medical application is supported by both its historical and traditional uses as well as an examination of current toxicological and pharmacological studies. The more well-known portion of the *Aloe vera* plant is the inner gel, also referred to as "pure gel."

2.8.1 Five Unique Benefits of *Aloe vera* to the Body

i. Penetration – aloe can reach deepest body tissues up to seven layers deep.
ii. Antiseptic – aloe has at least six antiseptic agents that kill bacteria, viruses, and fungi.
iii. Stimulates cell growth – aloe stimulates the birth of new healthy tissue.
iv. Settles nerves – aloe has a clearing effect on the body's nervous system.
v. Cleanses – aloe detoxifies and normalizes the body's metabolism.

2.8.2 External Uses of *Aloe vera*

Most of the vitamins and amino acids required for skin healing are found in *Aloe vera*. As a natural "band aid," the *Aloe vera* gel itself generates a glue-like substance on the skin, sealing in the nutrients and enabling them to start functioning right away while blocking out any bacteria or agents that could slow down or stop healing entirely. Additionally, the *Aloe vera* gel has a lot of water, which is necessary for the body to repair. *Aloe vera* is useful when added to topical formulations such as ointments, creams, or lotions, even though it works best when taken orally. Its moisturizing qualities are primarily responsible for protecting the wounds.

Skin care: Because it functions as an astringent, moisturizer, humidifier, and cleanser, *Aloe vera* is frequently used in dermatology. It reduces wrinkles, softens the skin, treats psoriasis, eczema, red spots, acne, herpes, mycosis, fever blisters, skin irritation, and shields the skin from pollutants. Additionally, it works wonders for sunburns, delicate skin, and the elimination and restoration of dead skin cells.

***Aloe vera* treats gum disease:** *Aloe vera* genuinely cures gum disease and gets rid of mucositis, lip fissures, and sores from oral herpes.

Reduces itching and promotes healing: *Aloe vera* juice promotes healing and reduces itching brought on by allergies and insect bites.

Reduces joint pain: *Aloe vera* improves joint and muscular pain. Using *Aloe vera* gels or sprays helps to ease joint and muscle pain brought on by arthritis.

2.8.3 *Aloe vera*'s Internal Uses

Aloe vera has been shown in numerous scientific studies to have antiviral, antibacterial, and antifungal qualities in addition to analgesic, anti-inflammatory, wound-healing, immune-modulating, and anti-tumor effects. The synergistic action of the combined nutritional components, which has a more potent effect than the separate components, is responsible for aloe's therapeutic qualities. Together, the nutrients form a potent mixture that may be used for a wide range of treatments to help treat various illnesses.

Offers relief from liver infections: *Aloe vera* juice helps the liver work better and is a great remedy for excessive alcohol consumption. Furthermore, it also keeps the liver from becoming scarred.

Treats stomach and intestinal issues: *Aloe vera* juice promotes intestinal transit, digestion, and prevents stomach ulcer formation.

Reduces inflammation: *Aloe vera* juice has 12 vital elements that reduce inflammation with relatively few adverse effects. Additionally, *Aloe vera* juice increases muscular and joint mobility.

***Aloe vera* lowers cholesterol and stabilizes blood sugar in diabetics:** Research indicates that aloe can lower blood glucose levels in animals and stimulate the pancreas to release insulin. *Aloe vera* is a natural way to reduce high cholesterol and triglyceride levels. High cholesterol is a risk factor that frequently results in heart disease and strokes. It has been demonstrated through medical research that *Aloe vera* gel has advantages when administered internally. The findings demonstrated that patients with high cholesterol and heart disease had a decreased risk when *Aloe vera* gel was given to them.

Antiviral and anti-tumor activity: *Aloe vera* helps to boost the immune system, which shields the body from illnesses linked to viruses and tumors.

2.9 CONSERVATION STATUS AND THREATS

The conservation and threats of *Aloe vera* and related species have drawn significant attention due to threats from habitat destruction, overharvesting, and climate change. Several studies have highlighted these challenges and proposed conservation strategies. Recent assessments of plant taxa indicate that Aloe species in the Horn of Africa are at a higher risk of extinction than the global average for plants (21%) (Brummitt et al., 2015), cacti (31%) (Goettsch et al., 2015), and conifers (34%) (Bachman et al., 2020) and the principal threats are the expansion and intensification of crop farming and livestock farming, gathering of plants, and unintentional effects of logging, wood harvesting, ongoing expansion of urban areas and the direct and indirect effects of fire. *Aloe peglerae* is critically endangered due to habitat loss and over-exploitation, necessitating urgent conservation efforts. A simple and efficient in vitro propagation protocol was developed, yielding up to 3906 transplantable shoots per year from a single shoot-tip explant. And the protocol supports the ex-situ conservation of this endangered species, with a 100% survival rate during acclimatization (Hlatshwayo et al., 2020). In contrast, only 16% of Aloe species are stored in seed banks, which significantly limits the potential for restoration efforts (Bachman et al., 2020). Continuous monitoring and surveying of populations will be crucial to ensure that conservation status re-assessments are thorough and to evaluate whether existing conservation gaps are being effectively addressed (Bachman et al., 2020).

2.10 FUTURE RESEARCH DIRECTIONS

Recent advancements in the study of aloe secondary metabolites suggest significant potential for developing clinical drugs to treat conditions such as malaria, viral infections, and cancer. However, research remains limited on information regarding the bioactive compounds and metabolites of aloe species other than *Aloe vera*, *Aloe barbadensis*, and *Aloe ferox*. Despite the widespread marketing of aloe-based products worldwide, ongoing concerns and controversies persist about their safety and efficacy (Ushashree et al., 2024). Addressing this gap is crucial to substantiate health claims and ensure consumer safety. Developing sustainable methods for producing aloe secondary metabolites, such as microbial synthesis or plant tissue culture, remains an area requiring further research to meet increasing demand without depleting natural resources. The structural complexity of *Aloe vera* metabolites poses challenges for chemical synthesis, requiring advanced and resource-intensive techniques. This limits progress in their synthetic replication for pharmaceutical and industrial uses (Ushashree et al., 2024). The adoption of currently imposed regulations on the usage of aloe-based products has the potential to disrupt the global supply chain of *Aloe vera*. This is particularly concerning as the commercial harvesting of aloe poses a significant threat to biodiversity, with several aloe species classified as endangered. To address these challenges, aloe companies must reconsider their business strategies by investing in research and development and focusing on sustainable production methods.

2.11 RESEARCH GAPS AND UNEXPLORED AREAS

1. Synthetic biology offers a promising solution to alleviate the dependency on natural aloe plants. De novo biosynthesis of *Aloe vera* metabolites in microbial systems using inexpensive carbon sources is an emerging area of interest. Further research is needed to optimize microbial production systems, ensuring high yields and cost efficiency.
2. The availability of a highly contiguous whole genome of *Aloe vera* provides unprecedented opportunities for genetic and metabolic engineering. However, functional annotation of genes involved in secondary metabolite biosynthesis remains incomplete, hindering the development of precision metabolic engineering strategies.

3. Beyond *Aloe vera*, many aloe species with unique metabolite profiles are under-researched and face extinction due to overharvesting and habitat loss. Comprehensive studies are needed to conserve these species while exploring their pharmacological potential.
4. Comparative metabolomics across less-studied aloe species could reveal novel bioactive compounds with therapeutic or industrial applications.
5. Evolving regulations on aloe product safety and quality standards necessitate the development of transparent and sustainable supply chain models. Research on scalable and eco-friendly production methods, such as plant tissue culture or synthetic pathways, is critical to ensure compliance without compromising supply.
6. Despite the widespread use of aloe-based products, rigorous clinical trials remain insufficient. Long-term safety assessments and efficacy studies are essential to validate health claims and address consumer concerns.
7. Aloe by-products, such as peels, are often discarded despite containing valuable bioactive compounds. Research into utilizing these by-products for insecticides, biofertilizers, or other applications could add value to the aloe supply chain while reducing waste.

Addressing these gaps through targeted research could unlock the full therapeutic potential of aloe secondary metabolites and lead to the development of novel, safe, and effective health products. And also, the industry can achieve a balance between sustainability, biodiversity conservation, and economic growth, paving the way for innovative aloe-based solutions in pharmaceuticals, cosmetics, and beyond.

2.11 SUMMARY

Aloe vera is a perennial succulent plant that belongs to the family Asphodelaceae. It has been valued for centuries for its therapeutic and medicinal properties. Originating in arid regions of Africa and the Mediterranean, its cultivation has spread worldwide. *Aloe vera* thrives in hot, arid climates and is notably drought-tolerant, with adaptations allowing it to store water in its leaves. The plant's bioactive compounds, including vitamins, minerals, polysaccharides, and amino acids, contribute to its renowned healing properties. It is used in treating burns, wounds, skin conditions, and digestive issues, and has applications in skincare and cosmetics due to its moisturizing and regenerative effects. *Aloe vera*'s internal uses include managing inflammation, enhancing immune response, lowering blood glucose, and reducing cholesterol levels, making it significant for managing conditions like diabetes, irritable bowel syndrome, and high cholesterol. Taxonomically, *Aloe vera* belongs to a genus of over 350 species characterized by thick, spiny leaves and xerophytic features. The diversity within the genus includes species with varied medicinal potentials, many of which remain under-researched. *Aloe vera*'s cultivation requires well-drained, nutrient-rich soil and careful harvesting to avoid microbial contamination and ensure leaf quality. With a global demand for its products, the *Aloe vera* industry plays a critical economic role. However, climatic and man-made challenges threaten its sustainability. Overharvesting, habitat loss, and climate change endanger *Aloe vera* and related species. Conservation efforts focus on habitat protection, ex-situ cultivation, and reducing wild harvest pressures. Innovations in synthetic biology and genetic engineering promise sustainable production of aloe metabolites, bypassing natural resource depletion. Moreover, underutilized by-products of *Aloe vera* offer potential in biofertilizer and insecticide development, contributing to waste reduction and added economic value. Despite its popularity, rigorous scientific validation of *Aloe vera*'s therapeutic claims is limited. Future research must address this through clinical trials and comparative metabolomics to unlock the potential of lesser-known aloe species. A sustainable balance between biodiversity conservation and commercial exploitation is essential. By advancing knowledge and adopting eco-friendly practices, *Aloe vera* can continue to play a transformative role in medicine, cosmetics, and global markets.

REFERENCES

Abouelela, M.E., Assaf, H.K., Abdelhamid, R.A., Elkhyat, E.S., Sayed, A.M., Oszako, T., Belbahri, L., El Zowalaty, A.E. and Abdelkader, M.S.A., El Zowalaty, A.E. and Abdelkader, M.S.A. 2021. Identification of potential SARS-CoV-2 main protease and spike protein inhibitors from the genus aloe:An in silico study for drug development. *Molecules* 26(6):1767.

Arolu, I.W., Rafii, M.Y., Hanafi, M.M., Mahmud, T.M.M. and Askani, S. 2015. Genetic divergence and evaluation of yield potential of Jatropha curcas accessions collected from Peninsular Malaysia. *J. Trop. Agric. Sci.* 38(1):127–142.

Bachman, S.P., Wilkin, P., Reader, T., Field, R., Weber, O., Nordal, I. and Demissew, S. 2020. Extinction risk and conservation gaps for Aloe (Asphodelaceae) in the Horn of Africa. *Biodivers. Conserv* 29(1):77–98.

Banerjee, N., Chatterjee, S., Chatterjee, S., Bhattacharjee, S., Bhattacharya, B. and Mukherjee, S. 2018. Health benefits of *Aloe vera*: A review. *Sci. Cult* 84(9–10):339–343.

Bhaludra, C.S.S., Hari Yadla, H.Y., Cyprian, F.S., Bethapudi, R.R., Basha, S.D. and Anupalli, R.R. 2014. Genetic diversity analysis in the genus *Aloe vera* (L.) using RAPD and ISSR markers. *Int. J. Pharmacol* 10(8):479–486.

Bisrat, D., Dagne, E., van Wyk, B.E. and Viljoen, A. 2000. Chromones and anthrones from Aloe marlothii and Aloe rupestris. *Phytochemistry* 55(8):949–952.

Blignaut, M., Ellis, A.G. and Le Roux, J.J. 2013. Towards a transferable and cost-effective plant AFLP protocol. *PloS One* 8(4):e61704.

Boudreau, M.D. and Beland, F.A. 2006. An evaluation of the biological and toxicological properties of Aloe barbadensis (miller), *Aloe vera. J. Environ. Sci. Health C* 24(1):103–154.

Brummitt, N.A., Bachman, S.P., Griffiths-Lee, J., Lutz, M., Moat, J.F., Farjon, A., Donaldson, J.S., Hilton-Taylor, C., Meagher, T.R., Albuquerque, S. and Aletrari, E. 2015. Green plants in the red: A baseline global assessment for the IUCN sampled Red List Index for plants. *PloS One* 10(8):0135152.

Castillon, J.B. and Castillon, J.P. 2010. Les aloe de Madagascar/The aloe of Madagascar. *Privately published. La Réunion.*

Chandra, D. and Choudhary, P. 2014. Diversity analysis of different accessions of Aloe barbadensis Mill.(syn. *Aloe vera* L.) collected from Rajasthan using RAPD marker system. *Bioscan* 9(1):7–10.

Choudhri, P., Rani, M., Sangwan, R.S., Kumar, R., Kumar, A. and Chhokar, V. 2018. De novo sequencing, assembly and characterisation of *Aloe vera* transcriptome and analysis of expression profiles of genes related to saponin and anthraquinone metabolism. *BMC Genom* 19:1–21.

Das, A., Nandagopal, K. and Jha, T.B. 2017. Molecular characterization of some Indian *Aloe vera* populations through RAPD and ITS markers. *Plant Biosyst* 151(4):695–703.

Dee, R., Malakasi, P., Rakotoarisoa, S.E. and Grace, O.M. 2018. A phylogenetic analysis of the genus Aloe (Asphodelaceae) in Madagascar and the Mascarene Islands. *Bot. J. Linn. Soc* 187(3):428–440.

Deshmukh, A.G., Tapre, V.V., Pawar, A.R., Deshmukh, K.M. and Parmar, J.N. 2019. Evaluation of germplasm of *Aloe vera* (Aloe barbadensis). *Int. J. Hortic. Food Sci* 1(1):20–22.

Eshun, K. and He, Q. 2004. *Aloe vera*: A valuable ingredient for the food, pharmaceutical and cosmetic industries, A review. *Crit. Rev. Food Sci. Nutr* 44:91–96.

Franz, C., Bauer, R., Carle, R., Tedesco, D., Tubaro, A. and Zitterl- Eglseer, K. 2005. Study of the assessment of plants/herbs, plant/herb extracts and their naturally or synthetically produced components as "additives" for use in animal production. www.efsa.europa.eu/en/scdocs/doc/070828.pdf

Gantait, S., Sinniah, U.R. and Das, P.K. 2014. *Aloe vera*: A review update on advancement of in vitro culture. *Acta Agric. Scand. B Soil Plant Sci* 64(1):1–12.

Givnish, T.J., Ames, M., McNeal, J.R., McKain, M.R., Steele, P.R., Depamphilis, C.W., Graham, S.W., Pires, J.C., Stevenson, D.W., Zomlefer, W.B. and Briggs, B.G. 2010. Assembling the tree of the monocotyledons: Plastome sequence phylogeny and evolution of Poales1. *Ann. Mol. Bot. Gard* 97(4):584–616.

Goettsch, B., Hilton-Taylor, C., Cruz-Piñón, G., Duffy, J.P., Frances, A., Hernández, H.M., Inger, R., Pollock, C., Schipper, J., Superina, M. and Taylor, N.P. 2015. High proportion of cactus species threatened with extinction. *Nat. Plants* 1(10):1–7.

Grace, O.M., Buerki, S., Symonds, M.R., Forest, F., van Wyk, A.E., Smith, G.F., Klopper, R.R., Bjorå, C.S., Neale, S., Demissew, S. and Simmonds, M.S. 2015. Evolutionary history and leaf succulence as explanations for medicinal use in aloes and the global popularity of *Aloe vera* . *BMC Evol. Biol* 15:1–12.

Grindlay, D. and Reynolds, T. 1986. The *Aloe vera* phenomenon: A review of the properties and modern uses of the leaf parenchyma gel. *J. Ethnopharmacol* 16: 117–151.

Hiransuchalert, R., Tongiang, B., Sae-Chua, C., Cherdsakulkij, C., Prasertlux, S., Khamnamtong, B. and Klinbunga, S. 2020. The development of species-specific AFLP-derived SCAR and SSCP markers to identify mantis shrimp species. *Mol. Biol. Rep* 47: 6807–6816.

Hlatshwayo, N.A., Amoo, S.O., Olowoyo, J.O. and Doležal, K. 2020. Efficient micropropagation protocol for the conservation of the endangered *Aloe peglerae*, an ornamental and medicinal species. *Plants* 9(4): 506.

Hoffmann, M., Kleine-Weber, H., Schroeder, S., Krüger, N., Herrler, T., Erichsen, S., ... and Pöhlmann, S. 2020. SARS-CoV-2 cell entry depends on ACE2 and TMPRSS2 and is blocked by a clinically proven protease inhibitor. *Cell* 181:271–280.

Holland, P.G. 1978. An evolutionary biogeography of the genus Aloe. *J. Biogeogr* 213–226.

Jangra, A., Sharma, G., Sihag, S. and Chhokar, V. 2022. The dark side of miracle plant-Aloe vera: A review. *Mol. Biol. Rep* 49(6):5029–5040.

Kar, S.K. and Bera, T.K. 2018. Phytochemical constituents of *Aloe vera* and their multifuctional properties: A comprehensive review. *Int. J. Pharm. Sci. Res* 9:1416–1423.

Khan, S.L. and Siddiqui, F.A. 2020. Beta-Sitosterol: As immunostimulant, antioxidant and inhibitor of SARS-CoV-2 spike glycoprotein. *Arch. Pharmacol. Ther* 2(1):12–16.

Klein, A.D. and Penneys, N.S. 1988. *Aloe vera. J. Am. Acad. Dermatol* 18:714–720.

Kumar, R., Naik, P.K., Kumar, A., Aggarwal, H., Kumar, A. and Chhokar, V. 2016. A combined approach using RAPD, ISSR and bioactive compound for the assessment of genetic diversity in *Aloe vera* (L.) Burm. F. *Indian J. Biotechnol* 15:538–549.

Kumar, S. and Yadav, J.P. 2014. Ethnobotanical and pharmacological properties of *Aloe vera*: A review. *J. Med. Plants Res* 48(8):1387–1398.

Kumar, S., Azam, M.M., Venkatesan, K., Pancholy, A. and Kulloli, R.N. 2019. Morphological and biochemical variability in aloe germplasm in hot arid region of India. *J. Herbs Spices Med. Plants* 25(2):158–171.

Kumar, S., et al. 2020. Pharmacological properties of *Aloe vera*: A review. *Phytochem. Rev* 19(3):671–690.

Lee, Y.S., Park, H.M., Kim, N.H., Waminal, N.E., Kim, Y.J., Lim, K.B., Baek, J.H., Kim, H.H. and Yang, T.J. 2016. Phylogenetic relationship of 40 species of genus Aloe L. and the origin of an allodiploid species revealed by nucleotide sequence variation in chloroplast intergenic space and cytogenetic in situ hybridization. *Genet. Resour. Crop Evol 63*:235–242.

López, Z., Femenia, A., Núñez-Jinez, G., Salazar Zúñiga, M.N., Cano, M.E., Espino, T. and Knauth, P. 2019. In vitro immunomodulatory effect of food supplement from *Aloe vera. Evid. Based Complement. Alternat. Med.* 2019(1):5961742.

Lucia, B.D. and Lucini, L. 2013. Production, quality and health benefits of aloe. Cabi Digital Library. www.soihs.it

Maenthaisong, R., Chaiyakunapruk, N. and Niruntraporn, S. 2007. The efficacy of *Aloe vera* for burnwound healing: A systematic review. *Burns* 33:713–718.

Manning, J., Boatwright, J.S., Daru, B.H., Maurin, O. and Van Der Bank, M. 2014. A molecular phylogeny and generic classification of Asphodelaceae subfamily Alooideae: A final resolution of the prickly issue of polyphyly in the alooids? *Syst. Bot* 39(1):55–74.

Martínez-Burgos, W.J., Serra, J.L., Marsiglia, F.R.M., Montoya, P., Sarmiento-Vásquez, Z., Marin, O., Gallego-Cartagena, E. and Paternina-Arboleda, C.D. 2022. *Aloe vera*: From Ancient Knowledge to the Patent and Innovation Landscape—A review. *S. Afr. J. Bot* 147:993–1006.

Murillo-Amador, B., Luna-Cisneros, M.J., Nieto-Garibay, A., Aguilar-García, M., Troyo-Diéguez, E. and García-Hernández, J.L. 2007. *Manual Para la Producción de Sábila en el Valle del Carrizal*. La Paz: Centro de Investigaciones Biológicas del Noroeste, 52.

Nadkarni, K.M. 2004. *Indian plants and drugs*. New Delhi: Srishti Book Distributors, 28–29. PMID:15129907.

Oldfield, S.A. 2004. Review of significant trade: East African aloes. In *14th Meeting of the CITES Plants Committee*, Windhoek.

Perrierde La Bâthie, H.P. 1926. *Les Lomatophyllum et les Aloe de Madagascar*. Botanique (Caen): Société linnéenne de Normandie.

Prisa, D. and Gobbino, M. 2021. Biological treatments for quality improvement and production of *Aloe vera* gel. *GSCARR* 9(1):054–063.

Rajendran, A., Narayanan, V. and Gnanavel, I. 2007. Study on the analysis of trace elements in *Aloe vera* and its biological importance. *J. Appl. Sci. Res* 3:1467–1478.

Rajeswari, R., Umadevi, M., Rahale, C.S., Pushpa, R., Selvavenkadesh, S., Kumar, K.S. and Bhowmik, D. 2012. *Aloe vera*: The miracle plant its medicinal and traditional uses in India. *J. Pharmacogn. Phytochem* 1(4):118–124.

Rauh, W. 1995. Succulent and xerophytic plants of Madagascar. Mill Valley: Strawberry Press, 1–34.

Rauh, W. 1998. Three new species of Lomatophyllum and one new Aloe from Madagascar. *Bradleya* 1998(16):92–100.

Reynolds, G.W. 1966. The aloes of tropical Africa and Madagascar. Mbabane, Swaziland: The Aloes Book Fund.

Reynolds, T. and Dweck, A.C. 1999. *Aloe vera* leaf gel: A review update. *J. Ethnopharmacol* 68:3–37.

Riaz, S., Hussain, S., Syed, S.K. and Anwar, R. 2021. Chemical characteristics and therapeutic potentials of *Aloe vera* . *RADS J. Biol. Res. Appl. Sci* 12:160–166.

Rowley, G.D. 1996. The berried aloes: Aloe section Lomatophyllum. *Excelsa* 17:59–62.

Rozo, Y., Quintero, L., Parra, M., Rodríguez, C. and Melgarejo, L.M. 2009. Analysis of genetic variability in Couepia accessions using AFLP markers. *Genet. Resour. Crop Evol* 56:77–83.

Singh, A. and Singh, P. 2021. Physical observation of Aloe vera (Aloe Barbadensis Miller) germplasm grown in Eastern UP. *Pharma Innovation* 10(7):1654–1656.

Sorkheh, K., Shiran, B., Aranzana, M.J., Mohammadi, S.A. and Martínez-Gomez, P. 2007. Application of amplified fragment length polymorphism (AFLPs) analysis to plant breeding and genetics: Procedures, applications and prospects. *J. Food Agric. Environ 5*(1):197.

Surjushe, A., Vasani, R. and Saple, D.G. 2008. *Aloe vera*: A short review. *Indian J. Dermatol* 53:163–166.

Tripathi, N., Saini, N. and Tiwari, S. 2011. Assessment of genetic diversity among *Aloe vera* accessions using amplified fragment length polymorphism. *Int. J. Med. Arom. Plants* 1:115–121.

Ushasree, M.V., Jia, Q., Do, S.G. and Lee, E.Y. 2024. New opportunities and perspectives on biosynthesis and bioactivities of secondary metabolites from *Aloe vera* . *Biotechnol. Adv* 72:108325.

Vogler, B.K. and Ernst, E. 1999. *Aloe vera*: A systematic review of its clinical effectiveness. *Br. J. Gen. Pract* 49(447):823–828.

Wagan, M.A., Miano, T.F., Nizamani, B., Shar, H.A., Wagan, M.K. and Wagan, F.A. 2022. Influence of Sunlight on Growth and Yield of *Aloe vera* Plant. *Am. J. Biosci. Bioieng* 10(4):84–88.

WCSP. 2015. *World checklist of selected plant families*. apps.kew.org/wcsp.

Yu, Z., Jin, C., Xin, M. and JianMin, H. 2009. Effect of *Aloe vera* polysaccharides on immunity and antioxidant activities in oral ulcer animal models. *Carbohydr. Polym 75*(2):307–311.

3 Genomics and Transcriptomics of *Aloe vera*

A New Frontier in Agrigenomics

Garima Sharma, Sonia Sihag, and Vinod Chhokar

3.1 INTRODUCTION

The well-being of humanity hinges on roughly 12% of Earth's estimated 300,000 to 400,000 plant species (Costion et al. 2011; Jiao et al. 2011). For more than 5000 years, plant-based herbal treatments have been essential to human health. It was discovered that in the post-Neolithic period, over 60% of plants were used for therapeutic properties (S. Chen et al. 2015). Currently, a significant number of pharmaceuticals are either directly sourced from plants or inspired by their bioactive compounds, compelling contemporary scientists to explore and develop novel drug molecules (Sharma et al. 2024). *Aloe barbadensis* Miller, frequently referred to as *Aloe vera* L., is an affiliate of the Asphodelaceae (formerly known as Liliaceae) family and has been utilized extensively in traditional medical practices around the world (Jangra et al. 2022). Contemporary omics techniques, including genomics, transcriptomics, proteomics, and metabolomics, play a vital role in unraveling the intricate networks that connect genes, proteins, and metabolites. These methods facilitate the discovery of novel drug compounds and enable the comprehensive exploration of genomes, transcriptomes, and proteomes of medicinal plants. Furthermore, they aid in the mass production of pharmaceuticals derived from plants and enhance our understanding of human cellular responses to specific chemicals or entire ethnobotanical plants used for medical applications (Mehta and Hasija 2018; Chakraborty 2018; Swamy et al. 2018; Swamy, Sinniah, and Ghasemzadeh 2018). The vast volumes of data generated by multi-omics platforms can be harnessed to predict the metabolic pathways of various therapeutic plants and pinpoint the genes responsible for synthesizing these active compounds (Pandita et al. 2021).

Plant genomes hold vital information on their origin, development, and epigenetic regulation, crucial for understanding their genetic and chemical diversity. The production of therapeutic chemicals and secondary metabolites, as well as the regulatory mechanisms underlying them, can be revealed by the meticulous routes found in high-throughput sequencing of medicinal plants. The investigation of genes and regulatory sequences with potential medical applications is made possible by genome sequencing. However, whole-genome sequencing is expensive and complicated, especially when dealing with repetitive sequences, high heterozygosity, or non-diploid genomes (Hao and Xiao 2015; Boutanaev et al. 2015; Chen et al. 2010). The advent of next-generation sequencing (NGS) has simplified comprehensive transcriptome research of therapeutic plants, leading to a better comprehension of the relationship between metabolite accumulation and gene expression patterns (Yonekura-Sakakibara, Fukushima, and Saito 2013). The transcriptomes of numerous therapeutic plants, such as *Aloe vera* (Jangra et al. 2022), *Rhodiola algida* (Crassulaceae) (Zhang et al. 2014), *Taxus mairei* (Hao et al. 2011), *Polygonum cuspidatum* Sieb. et Zucc. (Hao et al. 2012), and *Salvia sclarea* L. (Lamiaceae) (Shi Lin Chen et al. 2015) have been sequenced and are accessible

DOI: 10.1201/9781003603108-3

online through databases like the National Centre for Biotechnology Information (NCBI), Sequence Read Archive (SRA), PubMed, and Gene Expression Omnibus (GEO) database. Transcriptomics effectively retrieves genomic information from diverse non-model medicinal plants that do not have a reference genome. These studies aid in understanding essential traits that influence secondary metabolite production and in exploring the molecular mechanisms of significant medicinal processes. RNA sequencing, or RNA-seq, enables the acquisition of gene sequences from plants lacking a reference genome, facilitating integrated investigations combining transcriptomics and metabolomics for any potential therapeutic plant (Zhang et al. 2014; Hao et al. 2011; Chen et al. 2015; Yamazaki et al. 2013).

According to the Plant DNA c-value database (Zonneveld 2002), *Aloe vera* possesses a genome size estimated at 16.04 Gbp and exhibits a diploid ploidy level, featuring 14 (2n) chromosomes. Despite the extensive sequencing and analysis of genomes from various plant families such as Cannabaceae, Poaceae, Brassicaceae, Malvaceae, Cucurbitaceae, Rosaceae, Euphorbiaceae, Fabaceae, Solanaceae, Orchidaceae, and Betulaceae, the genome sequence of *Aloe vera* remains undiscovered. The lack of genetic information for *Aloe vera* is striking, especially considering the thorough genomic studies conducted on various plant species. However, a significant development occurred in 2021 with the report of a draft genome sequence of *Aloe vera* (Jaiswal et al. 2021), following an earlier milestone (Ren et al. 2020) that documented the complete chloroplast genome of Aloe. A detailed whole transcriptome sequencing examination utilizing Illumina Hi-Seq technology for both *Aloe vera* leaf and root tissues was conducted in 2018 Choudhri et al. 2018). The study generated crucial sequence data, particularly emphasizing genes associated with secondary metabolite pathways (Choudhri et al. 2018). The fusion of transcriptomics and genomics has facilitated the discovery of genes and enzymes associated with secondary metabolite production pathways. Additionally, the integration of multi-omics methodologies offers in-depth insights into the metabolic processes of medicinal plants, potentially revealing novel genes and metabolites. Encouraging the acceptance of multi-omics approaches in research projects could meaningfully enhance efficiency in conventional medicine studies (see Figure 3.1) (Alami et al. 2022).

3.2 GENOMICS IN *ALOE VERA*

Technological innovations have shaped the progress of plant genomics, with each breakthrough profoundly impacting the field's advancement. Over five decades, three generations of groundbreaking developments in sequencing technology unfolded, commencing with the introduction of first-generation sequencing in the 1970s. The evolution of sequencing technology continues at a rapid pace, persistently adapting to new challenges and changes. The achievement of whole-genome sequencing has provided a sturdy basis for significant investigation into molecular breeding of medicinal plants and has sped up the process of producing different varicties (Alami et al. 2022).

I. Genomics and phylogenetics

Employing a hybrid sequencing and assembly technique, researchers in 2021 (Jaiswal et al. 2021) combined Oxford Nanopore long-read and Illumina short-read sequences to generate the first draft of the *Aloe vera* genome. This study represents a significant milestone in the field, as no prior sequencing endeavors have targeted any member of the Asphodelaceae family. As a result, the sequencing of *Aloe vera* holds great promise for revealing crucial genetic traits within this plant family. In genome assembly, high degrees of heterozygosity pose a major problem, as de novo assemblers try to create a single haploid representation from allelic differences, which becomes more challenging as heterozygosity increases (Asalone et al. 2020). With an estimated heterozygosity of 11.3%, the *Aloe vera* genome is even more heterozygous than the wheat genome, which was assembled over more than ten years of international cooperation between several research institutes. One of the

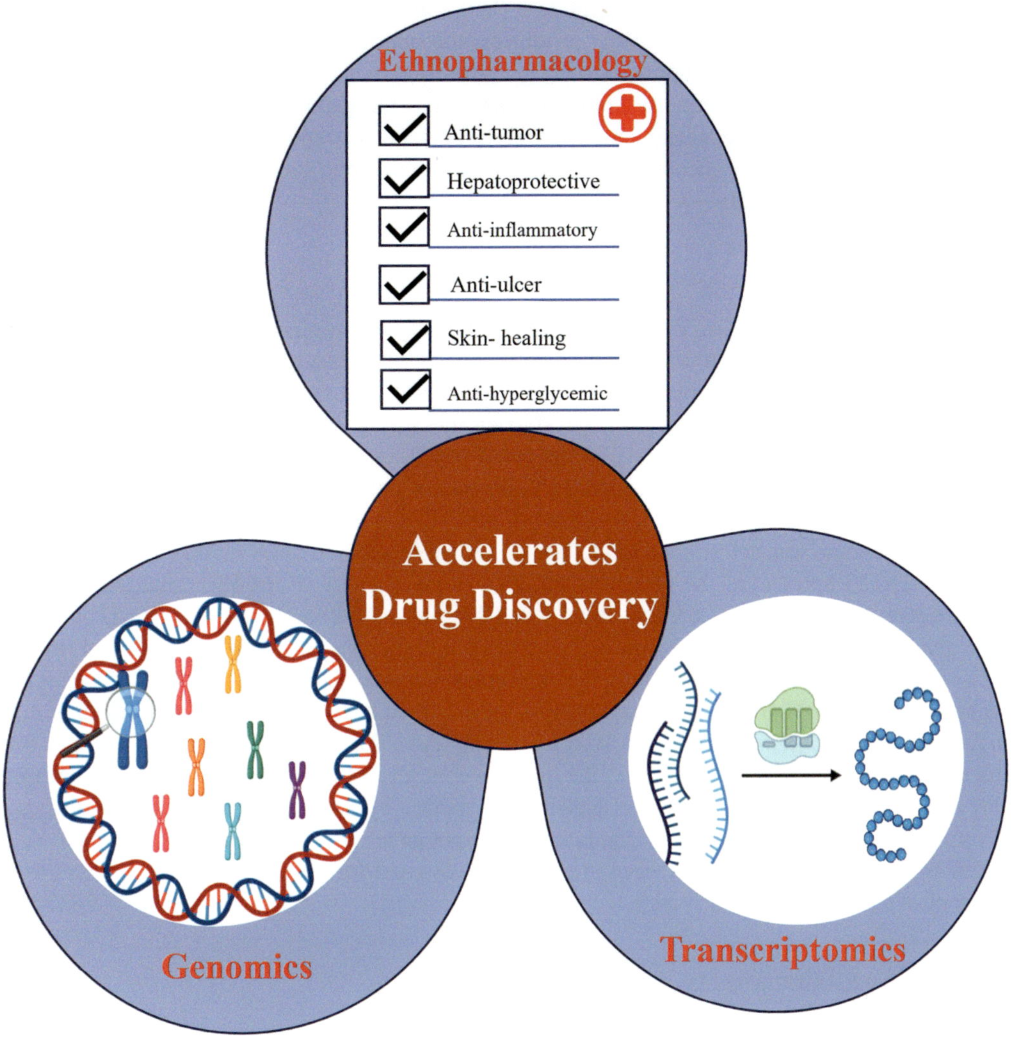

FIGURE 3.1 Integrating multi-omics and ethnopharmacology: A catalyst for rapid drug discovery.

biggest plant genomes was constructed using a hybrid strategy that used short-read sequencing (Illumina) and long-read sequencing (Nanopore) to overcome this difficulty. By creating the first genome-wide phylogeny of *Aloe vera* in comparison to every other monocot species included in the Ensembl Plant database, with *Arabidopsis thaliana* as an outgroup, (Jaiswal et al. 2021) made a substantial contribution to the field of plant phylogenetics. A small number of genomics regions were the focus of earlier research looking at *Aloe vera's* evolutionary position among monocots.

On the other hand, their study's thorough analysis made use of 1,440 loci spread throughout the genomes, offering a precise phylogenetic framework. The total lack of polytomies and unusually high bootstrap values for internal nodes further supported the correctness of the phylogeny. The study provided important new information on the evolutionary relationships among monocots, even if it was generally in line with previous phylogenies (Dunemann et al. 2014; Grace et al. 2015; Qian and Jin 2016; Silvera et al. 2014; Peng et al. 2011). They highlighted that genome-wide phylogeny was consistent with previous research as to how it represented species in the Poales order. In contrast to other researchers (Qian and Jin 2016), a unique topology for the relative locations of *Musa acuminata* (Zingiberales), *Dioscorea rotundata* (Dioscoreales), and *Aloe vera* (Asparagales) was revealed. These discrepancies most likely result from the smaller number of genomic loci employed in the previous phylogenies, as opposed to the genome-wide method used in the above-stated study, which included 1,440 one-to-one orthologs spread across the genome. These distinctions can be better understood, and the relative positions of these species can be more clearly defined when more complete monocot genomes become available and more genomic loci are added to phylogenetic analysis (Jaiswal et al. 2021).

II. Chloroplast Genome Analysis

As reported by researchers in 2020 (Ren et al. 2020), the chloroplast genome of *Aloe vera* was found to span a total length of 152,875 base pairs (bp). It exhibits a typical quadripartite structure, which is a characteristic feature of many plant chloroplast genomes. This structure is composed of two identical inverted-repeat regions (IRs), each measuring 26,596 bp in length, along with a large single-copy region (LSC) of 83,505 bp, and a small single-copy region (SCC) of 16,178 bp. Such an arrangement reflects the conserved architecture of chloroplast genomes, providing a stable framework for various genetic and functional studies. The nucleotide composition of the *Aloe vera* chloroplast genome was determined to include 47,185 bp of adenine (30.8%), 48,123 bp of thymine (31.5%), 29,326 bp of cytosine (19.2%), and 28,241 bp of guanine (18.5%), resulting in a total guanine-plus-cytosine (G+C) content of 37.7%. This balanced composition is typical of chloroplast genomes, which generally show a bias toward a higher proportion of adenine and thymine nucleotides. In terms of genetic content, the chloroplast genome encodes a total of 131 genes. These genes are categorized into three primary groups based on their functional roles. There are 85 protein-coding genes (PCGs), which are involved in various essential processes such as photosynthesis and metabolic regulation. Additionally, the genome contains 38 transfer RNA genes (tRNAs) that play a crucial role in the translation of genetic information by transferring specific amino acids to ribosomes during protein synthesis. The remaining eight genes are ribosomal RNA genes (rRNAs), which are integral components of ribosomes and are essential for the assembly and function of the chloroplast's translational machinery.

III. Aloe's Drought Tolerance Genomics

The discovery of adaptive evolution in genes linked to drought stress response was a discovery, providing a genetic explanation for *Aloe vera*'s exceptional resistance to dry environments (Jaiswal et al. 2021). *Aloe vera* is well-known for its perennial succulent leaves and the Crassulacean Acid Metabolism (CAM) pathway, which fixes carbon. It also has a number of physiological characteristics that help it to withstand drought (Jin et al. 2007). Further experimental research has shown that the plant can adjust to drought conditions by producing more sugar and expressing ubiquitin and heat-shock proteins, which improve water efficiency and preserves osmotic balance (Delatorre-Herrera et al. 2010; Hazrati et al. 2017; Huerta, Freire, and Cardemil 2013). A sizable fraction (60.81%) of the genes that showed multiple indications of adaptive evolution (MSA) were associated with functions relevant to drought stress. These genes' roles in the adaptive development of drought tolerance mechanisms in *Aloe vera* were further highlighted by patterns of co-expression and physical interactions. These discoveries offer a better knowledge of the plant's survival tactics in dry conditions and provide insight into the genetic foundation of its drought resistance (Jaiswal et al. 2021).

Furthermore, when compared to other monocot species, *Aloe vera* showed site-specific indications of adaptive evolution in a number of crucial genes related to the circadian rhythm and CAM pathway. Plants can withstand extended drought conditions thanks to the CAM pathway, which has evolved convergently across a variety of dry locations and is renowned for its high water-use efficiency

(Ming et al. 2015). This mechanism functions by separating the Calvin-Benson cycle and ambient CO_2 absorption throughout time. These activities are controlled by the circadian rhythm of the plant (Mallona, Egea-Cortines, and Weiss 2011; Yin et al. 2018). Interestingly, one particular kind of circadian rhythm development is represented by the CAM route (Silvera et al. 2014; Hartwell 2006). The adaptive evolution observed in both the CAM pathway and its circadian rhythm regulation in *Aloe vera* highlights their crucial roles in providing the species with an evolutionary advantage in drought tolerance (Jaiswal et al. 2021).

3.3 TRANSCRIPTOMICS OF *ALOE VERA*

Recent developments in next-generation sequencing (NGS) technology have accelerated research on medicinal plants, resulting in substantial cost and effort savings. These developments provide previously unheard-of possibilities for in-depth genetic investigation by enabling both direct RNA sequencing and high-throughput significant whole genome sequencing. Numerous scientific applications, such as comparative genomics, mining for small and long non-coding RNAs, finding molecular markers, and discovering novel genes, can make use of the enormous datasets produced by NGS. Furthermore, in order to meet the bioinformatics problems related to the interpretation of NGS data, commercial sequencing service providers are constantly enhancing their technologies. Because of this, many medicinal plant genome sequencing projects have been successfully finished in recent years, and several more are underway. Similarly, a vast amount of NGS-based transcriptome data has been produced for many species of medicinal plants, which has improved our knowledge of their genetic composition and possible uses. This development demonstrates how NGS has revolutionized the study of medicinal plants, creating new opportunities for the identification and development of plant-based therapeutic medicines.

The transcriptome studies have enhanced our understanding of *Aloe vera's* genetic makeup, especially in relation to its therapeutic properties and response to stress. Previously, researchers conducted de novo sequencing and assembly of the *Aloe vera* transcriptome using Illumina Hi-Seq technology. The high-quality transcriptomic database was generated from root and leaf tissues of *Aloe vera* to characterize the genes involved in the metabolism of saponin and anthraquinone compounds. *Aloe vera* root and leaf samples were subjected to transcriptome sequencing utilizing Illumina paired-end sequencing technology. The average length of the transcripts produced by de novo assembly of high-quality paired-end reads was 709 nucleotides for root tissue and 714 nucleotides for leaf tissue, totaling 161,733 transcripts. Using CD-HIT-EST to cluster non-redundant transcripts, 113,063 unigenes for root tissue and 141,310 unigenes for leaf tissue were obtained. 6,114 root and 6,527 leaf coding sequences (CDS) were functionally annotated using the KEGG pathway database to create 24 unique biological pathways.

Differential gene expression (DGE) analysis, based on FPKM values, revealed significant modulation of genes encoding enzymes involved in secondary metabolite biosynthesis. Further validation through quantitative real-time PCR (qRT-PCR) identified 16 key candidate genes associated with the biosynthesis of carotenoids, lignins, anthraquinones, and saponins. These genes exhibited pronounced upregulation in root tissues compared to leaves, as corroborated by both DGE and qRT-PCR analyses. Remarkably, the induction of methyl jasmonate elicited the highest upregulation of these genes, underscoring a pivotal link between secondary metabolite production and the plant's defense mechanisms under stress conditions. Furthermore, the identification of transcription factors like MYB, NAC, and bHLH provides opportunities to monitor their regulatory functions associated with several metabolic processes. Their finding provides significant resources for further metabolic engineering research, allowing for the regulated secondary metabolite production and extending the commercial and therapeutic applications of *Aloe vera* (Choudhri et al. 2018).

With the advent of new technologies, transcriptome sequencing has emerged as a powerful technique providing high-throughput capabilities for studying plants, especially medicinal species, without the need for preexisting genomic data. It has a wide range of applications, including the discovery of new functional genes, regulation of secondary metabolite production pathways, and detection of plant developmental processes (Huang et al. 2014). A previous study examined the total Se content and Se absorption in aloe leaves and found that 400 mg/l was the ideal concentration of Na_2SeO_4 for selenium (Se) spraying. Selenium treatment strongly impacted the accumulation of functional components, such as amino acids, phenols, flavonoids, terpenes, and indole derivatives. According to these results, adding exogenous selenium to Aloe leaves increases their selenium level while also boosting other beneficial chemicals.

Elevated quantities of bioactive compounds and antioxidants were found by metabolic profiling. Consistently, the transcriptomic analysis showed the upregulation of genes involved in antioxidant biosynthesis, sulfate transport, phytohormone signaling, transcription regulation, and phenolic metabolism. This suggests that selenium assimilation is linked to enhanced antioxidant activity and pathogen defense mechanisms. This study offers the possibility of creating selenium-enriched aloe cultivars with enhanced therapeutic potential by shedding light on the metabolic and genetic response to selenium treatment (Zou et al. 2021). The genus *Aloe* has long played a significant role in both conventional and alternative medicine because of its many bioactive secondary metabolites. The key compounds, such as aloesone, aloenin, noreugenin, aloin, saponins, and carotenoids, are known for their pharmacokinetic profiles and potential for repurposing. Multi-omics technologies are highly efficient in elucidating the enzymatic machinery and metabolic pathways associated with key metabolite production (Ushasree et al. 2024).

3.4 CONCLUSION

Aloe vera has long been an integral component of conventional and modern medicine because of its bioactive secondary metabolites. With the advent of sequencing technologies and multi-omics approaches, the understanding of the plant's genetic, transcriptomic, and metabolic complexities has significantly improved. Along with the molecular mechanisms related to stress responses, different physiological adaptations such as drought tolerance and Crassulacean acid metabolism have provided important insights into the biosynthesis of key metabolites, including aloin, saponins, carotenoids, etc. The phylogenetics, evolutionary biology, and metabolic engineering of *Aloe vera* have all significantly benefited from the discoveries obtained through the sequencing of its genome.

The studies conducted so far emphasize the limitations of earlier research that relied on fewer genomic loci while underscoring the transformative potential of genome-wide approaches in uncovering evolutionary relationships. Through comprehensive sequencing and analysis of monocot genomes, other research has established a robust foundation for refining phylogenetic frameworks, understanding genetic diversity, and investigating the unique traits of *Aloe vera* and its relatives. The thorough examination of the chloroplast genome of *Aloe vera* provides crucial new information about its genetic makeup, making it a valuable tool for comprehending the functional traits and evolutionary links of the Asphodelaceae family. In addition to expanding our understanding of the plant's special biological characteristics, this thorough genetic mapping creates new opportunities for its use in agriculture. Additionally, the CAM pathway, circadian rhythm, and drought stress response genes are among the important *Aloe vera* genes whose adaptive evolution is interpreted based on their demonstrated physiological functions in experimental investigations. *Aloe vera*'s status as a model organism for comprehending drought tolerance and environmental adaptability is strengthened by the probable phenotypic effects of this adaptive development.

The transcriptomic analysis identified the genes and pathways critical to secondary metabolite production, providing a roadmap for improving the therapeutic potential of *Aloe vera*. Integrating transcriptomics with metabolomics provides new opportunities to improve the bioactivity and therapeutic values of *Aloe vera*. Such integrative approaches not only enhance our understanding of its complex molecular structure but also pave the way for sustainable biotechnological applications.

Aloe vera's versatility and bioactive potential are highlighted by the thorough transcriptome study, especially in the formation of secondary metabolites. The identification of substantially elevated secondary metabolic genes in roots emphasizes how important these genes are for the synthesis of medicinal compounds. The correlation between methyl jasmonate-induced gene expression and plant defense systems further demonstrates the dynamic interaction between metabolism and stress responses. The discovery of transcription factors, including bHLH, NAC, and MYB, offers new opportunities to comprehend the regulatory networks governing the production of secondary metabolites. These discoveries may open the door to focused metabolic engineering, which would maximize the synthesis of useful substances for both commercial and therapeutic uses, such as anthraquinones and saponins. This genetic resource can be utilized in future research to investigate the stress-resilience characteristics of *Aloe vera*, thereby promoting its sustainable use and production.

Previous research has underscored the transformative impact of selenium (Se) treatment on the metabolic and genetic landscape of *Aloe vera*, paving the way for a promising path toward enhancing its therapeutic potential. By optimizing selenium supplementation to 400 mg/l of Na_2SeO_4, researchers achieved a significant increase in selenium accumulation within aloe leaves, while simultaneously enriching essential bioactive compounds, including amino acids, phenols, flavonoids, terpenes, and indole derivatives. This synergistic enhancement underscores selenium's pivotal role in elevating both the nutritional profile and therapeutic value of *Aloe vera*. Metabolic profiling revealed a remarkable increase in antioxidants and other bioactive components, further solidifying the therapeutic benefits of selenium-enriched aloe. Complementing these findings, transcriptomic analysis provided critical insights into the molecular underpinnings of these changes, revealing the activation of genes involved in phenolic metabolism, sulfate transport, phytohormone signaling, and antioxidant biosynthesis. These responses highlight the intricate interplay between selenium assimilation and the enhancement of bioactive molecules, as well as the bolstering of defense-related pathways. Such groundbreaking studies lay the foundation for developing selenium-enriched *Aloe vera* cultivars tailored to meet the growing demand for functional foods and medicinal products with heightened health benefits. Additionally, it opens new avenues for exploring selenium's potential in enhancing stress resilience and advancing metabolic engineering, driving innovation in sustainable agriculture and expanding commercial applications. There is a vast potential for novel advancements in biotechnology, agriculture, and medicine as we continue to unravel the genetic and molecular blueprints of *Aloe vera*. These developments highlight the value of *Aloe vera* as a model organism for studying the medicinal properties of the plants for next-generation biomedicines. Together, these discoveries shed light on the genetic foundations of *Aloe vera*'s extraordinary plasticity and highlight its significance for both practical and scientific applications. Genomics and transcriptomics studies in *Aloe vera* have revolutionized agrigenomics by providing a deeper understanding of the genetics.

REFERENCES

Alami, Mohammad Murtaza, Zhen Ouyang, Yipeng Zhang, Shaohua Shu, Guozheng Yang, Zhinan Mei, and Xuekui Wang. 2022. "The Current Developments in Medicinal Plant Genomics Enabled the Diversification of Secondary Metabolites' Biosynthesis." *International Journal of Molecular Sciences* 23 (24): 15932.

Asalone, Kathryn C, Kara M Ryan, Maryam Yamadi, Annastelle L Cohen, William G Farmer, Deborah J George, Claudia Joppert, Kaitlyn Kim, Madeeha Froze Mughal, and Rana Said. 2020. "Regional Sequence Expansion or Collapse in Heterozygous Genome Assemblies." *PLoS Computational Biology* 16 (7): e1008104.

Boutanaev, Alexander M, Tessa Moses, Jiachen Zi, David R Nelson, Sam T Mugford, Reuben J Peters, and Anne Osbourn. 2015. "Investigation of Terpene Diversification across Multiple Sequenced Plant Genomes." *Proceedings of the National Academy of Sciences* 112 (1): E81–E88.

Chakraborty, Prasanta. 2018. "Herbal Genomics as Tools for Dissecting New Metabolic Pathways of Unexplored Medicinal Plants and Drug Discovery." *Biochimie Open* 6: 9–16.

Chen, S L, Y Z Sun, Jiang Xu, H M Luo, Chao Sun, Liu He, X L Cheng, B L Zhang, and P G Xiao. 2010. "Strategies of the Study on Herb Genome Program." *Yao Xue Xue Bao= Acta Pharmaceutica Sinica* 45 (7): 807–812.

Chen, Shi Lin, Anne Osbourn, Vassiliki G Kontogianni, Li Wei Liu, and Maria J Jordán. 2015. "Temporal Transcriptome Changes Induced by Methyl Jasmonate in Salvia Sclarea." *Gene* 558 (1): 41–53.

Chen, Shilin, Jingyuan Song, Chao Sun, Jiang Xu, Yingjie Zhu, Rob Verpoorte, and Tai Ping Fan. 2015. "Herbal Genomics: Examining the Biology of Traditional Medicines." *Science* 347 (6219): S27–S29.

Choudhri, Pragati, Muniya Rani, Rajender S. Sangwan, Ravinder Kumar, Anil Kumar, and Vinod Chhokar. 2018. "De Novo Sequencing, Assembly and Characterisation of *Aloe vera* Transcriptome and Analysis of Expression Profiles of Genes Related to Saponin and Anthraquinone Metabolism." *BMC Genomics* 19 (1): 1–21. doi:10.1186/s12864-018-4819-2.

Costion, Craig, Andrew Ford, Hugh Cross, Darren Crayn, Mark Harrington, and Andrew Lowe. 2011. "Plant DNA Barcodes Can Accurately Estimate Species Richness in Poorly Known Floras." *PloS One* 6 (11): e26841.

Delatorre-Herrera, José, Ignacio Delfino, Carlos Salinas, Herman Silva, and Liliana Cardemil. 2010. "Irrigation Restriction Effects on Water Use Efficiency and Osmotic Adjustment in *Aloe vera* Plants (Aloe Barbadensis Miller)." *Agricultural Water Management* 97 (10): 1564–1570.

Dunemann, Frank, Otto Schrader, Holger Budahn, and Andreas Houben. 2014. "Characterization of Centromeric Histone H3 (CENH3) Variants in Cultivated and Wild Carrots (Daucus Sp.)." *Plos One* 9 (6): e98504.

Grace, Olwen M, Sven Buerki, Matthew R E Symonds, Félix Forest, Abraham E van Wyk, Gideon F Smith, Ronell R Klopper, Charlotte S Bjorå, Sophie Neale, and Sebsebe Demissew. 2015. "Evolutionary History and Leaf Succulence as Explanations for Medicinal Use in Aloes and the Global Popularity of Aloe vera." *BMC Evolutionary Biology* 15: 1–12.

Hao, Da Cheng, GuangBo Ge, PeiGen Xiao, YanYan Zhang, and Ling Yang. 2011. "The First Insight into the Tissue Specific Taxus Transcriptome via Illumina Second Generation Sequencing." *PLoS One* 6 (6): e21220.

Hao, Da-Cheng, and Pei-Gen Xiao. 2015. "Genomics and Evolution in Traditional Medicinal Plants: Road to a Healthier Life." *Evolutionary Bioinformatics* 11: EBO-S31326.

Hao, Da-Cheng, Pei-Gen Xiao, Yong Peng, Jingqun Dong, and Wenxia Liu. 2012. "Evaluation of the Chloroplast Barcoding Markers by Mean and Smallest Interspecific Distances." *Pakistan Journal of Botany* 44 (4): 1271–1274.

Hartwell, James. 2006. "The Circadian Clock in CAM Plants." *Annual Plant Reviews Volume 21: Endogenous Plant Rhythms*. Edited by Anthony J W Hall and Harriet McWatters. Wiley Online Library, 211–236.

Hazrati, Saeid, Zeinolabedin Tahmasebi-Sarvestani, Ali Mokhtassi-Bidgoli, Seyed Ali Mohammad Modarres-Sanavy, Hamid Mohammadi, and Silvana Nicola. 2017. "Effects of Zeolite and Water Stress on Growth, Yield and Chemical Compositions of *Aloe vera* L." *Agricultural Water Management* 181: 66–72.

Huang, X H, F Xu, H Cheng, L L, Li, and S Y Cheng. 2014. "Advances in Transcriptome Sequencing in Higher Plants." *Journal of Huanggang Normal University* 34: 28–35.

Huerta, Claudia, Matías Freire, and Liliana Cardemil. 2013. "Expression of Hsp70, Hsp100 and Ubiquitin in Aloe Barbadensis Miller under Direct Heat Stress and under Temperature Acclimation Conditions." *Plant Cell Reports* 32: 293–307.

Jaiswal, Shubham K, Shruti Mahajan, Abhisek Chakraborty, Sudhir Kumar, and Vineet K Sharma. 2021. "The Genome Sequence of *Aloe vera* Reveals Adaptive Evolution of Drought Tolerance Mechanisms." *Iscience* 24 (2): 102079.

Jangra, Alka, Garima Sharma, Sonia Sihag, and Vinod Chhokar. 2022. "The Dark Side of Miracle Plant-*Aloe vera*: A Review." Edited by Jonathan Brody and Katrin Paeschke. *Molecular Biology Reports* 49: 1–12.

Jiao, Yuannian, Norman J Wickett, Saravanaraj Ayyampalayam, André S Chanderbali, Lena Landherr, Paula E Ralph, Lynn P Tomsho, Yi Hu, Haiying Liang, and Pamela S Soltis. 2011. "Ancestral Polyploidy in Seed Plants and Angiosperms." *Nature* 473 (7345): 97–100.

Jin, Zan Min, Chang Hai Wang, Zhao Pu Liu, and Wei Jia Gong. 2007. "Physiological and Ecological Characters Studies on *Aloe vera* under Soil Salinity and Seawater Irrigation." *Process Biochemistry* 42 (4): 710–714.

Mallona, Izaskun, Marcos Egea-Cortines, and Julia Weiss. 2011. "Conserved and Divergent Rhythms of Crassulacean Acid Metabolism-Related and Core Clock Gene Expression in the Cactus Opuntia Ficus-Indica." *Plant Physiology* 156 (4): 1978–1989.

Mehta, Avni, and Yasha Hasija. 2018. "Bioinformatics Approaches for Genomics and Post Genomics Applications of Anticancer Plants." Edited by Mohd Sayeed Akhtar and Mallappa Kumara Swamy. *Anticancer Plants: Mechanisms and Molecular Interactions:* Volume 4. Springer, 283–317.

Ming, Ray, Robert VanBuren, Ching Man Wai, Haibao Tang, Michael C Schatz, John E Bowers, Eric Lyons, Ming-Li Wang, Jung Chen, and Eric Biggers. 2015. "The Pineapple Genome and the Evolution of CAM Photosynthesis." *Nature Genetics* 47 (12): 1435–1442.

Pandita, Deepu, Anu Pandita, Shabir Hussain Wani, Shaimaa A M Abdelmohsen, Haifa A Alyousef, Ashraf M M Abdelbacki, Mohamed A Al-Yafrasi, Fahed A Al-Mana, and Hosam O Elansary. 2021. "Crosstalk of Multi-Omics Platforms with Plants of Therapeutic Importance." *Cells* 10 (6): 1296.

Peng, Xi-xu, Xin-ke Tang, Ping-lan Zhou, Yao-jun Hu, Xiao-bo Deng, H E Yan, and Hai-hua Wang. 2011. "Isolation and Expression Patterns of Rice WRKY82 Transcription Factor Gene Responsive to Both Biotic and Abiotic Stresses." *Agricultural Sciences in China* 10 (6): 893–901.

Qian, Hong, and Yi Jin. 2016. "An Updated Megaphylogeny of Plants, a Tool for Generating Plant Phylogenies and an Analysis of Phylogenetic Community Structure." *Journal of Plant Ecology* 9 (2): 233–239.

Ren, Jia-Ju, Jian Wang, Kai-Ki Lee, Hui Deng, Han Xue, Nan Zhang, Jing-Chun Zhao, Tianyi Cao, Chang-Lei Cui, and Xiu-Hang Zhang. 2020. "The Complete Chloroplast Genome of *Aloe vera* from China as a Chinese Herb." *Mitochondrial DNA Part B* 5 (1): 1092–1093.

Sharma, Garima, Alka Jangra, Sonia Sihag, Siddhant Chaturvedi, Shalu Yadav, and Vinod Chhokar. 2024. "Bryophyllum Pinnatum (Lam.) Oken: Unravelling Therapeutic Potential and Navigating Toxicity." *Physiology and Molecular Biology of Plants* 30: 1–15.

Silvera, Katia, Klaus Winter, B Leticia Rodriguez, Rebecca L Albion, and John C Cushman. 2014. "Multiple Isoforms of Phospho Enol Pyruvate Carboxylase in the Orchidaceae (Subtribe Oncidiinae): Implications for the Evolution of Crassulacean Acid Metabolism." *Journal of Experimental Botany* 65 (13): 3623–3636.

Swamy, Mallappa Kumara, Sukanya Paramashivaiah, Lingayya Hiremath, Mohd Sayeed Akhtar, and Uma Rani Sinniah. 2018. "Micropropagation and Conservation of Selected Endangered Anticancer Medicinal Plants from the Western Ghats of India." Edited by Mohd Sayeed Akhtar and Mallappa Kumara Swamy. *Anticancer Plants: Natural Products and Biotechnological Implements:* Volume 2. Springer, 481–505.

Swamy, Mallappa Kumara, Uma Rani Sinniah, and Ali Ghasemzadeh. 2018. "Anticancer Potential of Rosmarinic Acid and Its Improved Production through Biotechnological Interventions and Functional Genomics." *Applied Microbiology and Biotechnology* 102: 7775–7793.

Ushasree, Mrudulakumari Vasudevan, Qi Jia, Seon Gil Do, and Eun Yeol Lee. 2024. "New Opportunities and Perspectives on Biosynthesis and Bioactivities of Secondary Metabolites from *Aloe vera*." *Biotechnology Advances* 72: 108325.

Yamazaki, Mami, Keiichi Mochida, Takashi Asano, Ryo Nakabayashi, Motoaki Chiba, Nirin Udomson, Yasuyo Yamazaki, Dayan B Goodenowe, Ushio Sankawa, and Takuhiro Yoshida. 2013. "Coupling Deep Transcriptome Analysis with Untargeted Metabolic Profiling in Ophiorrhiza Pumila to Further the Understanding of the Biosynthesis of the Anti-Cancer Alkaloid Camptothecin and Anthraquinones." *Plant and Cell Physiology* 54 (5): 686–696.

Yin, Hengfu, Hao-Bo Guo, David J Weston, Anne M Borland, Priya Ranjan, Paul E Abraham, Sara S Jawdy, James Wachira, Gerald A Tuskan, and Timothy J Tschaplinski. 2018. "Diel Rewiring and Positive Selection of Ancient Plant Proteins Enabled Evolution of CAM Photosynthesis in Agave." *BMC Genomics* 19: 1–16.

Yonekura-Sakakibara, Keiko, Atsushi Fukushima, and Kazuki Saito. 2013. "Transcriptome Data Modeling for Targeted Plant Metabolic Engineering." *Current Opinion in Biotechnology* 24 (2): 285–290.

Zhang, Faqi, Qingbo Gao, Gulzar Khan, Keming Luo, and Shilong Chen. 2014. "Comparative Transcriptome Analysis of Aboveground and Underground Tissues of Rhodiola Algida, an Important Ethno-Medicinal Herb Endemic to the Qinghai-Tibetan Plateau." *Gene* 553 (2): 90–97.

Zonneveld, Ben J M. 2002. "Genome Size Analysis of Selected Species of Aloe (Aloaceae) Reveals the Most Primitive Species and Results in Some New Combinations." *Bradleya* 2002 (20): 5–12.

Zou, Yi, Changyou Han, Fang Wang, Yanhua Tan, Saina Yang, Chuying Huang, Shengyou Xie, and Xueqin Xiao. 2021. "Integrated Metabolome and Transcriptome Analysis Reveal Complex Molecular Mechanisms Underlying Selenium Response of *Aloe vera* L." *Journal of Plant Biology* 64: 135–143.

4 Metabolomic Analysis of Aloe

Unlocking the Biochemical Secrets of a Medicinal Wonder

Bharti Goyal, Shalu Yadav, and Sahil Prabhakar

4.1 INTRODUCTION

The pursuit of understanding the intricate mysteries within the plant kingdom has long been a focus of scientific exploration. Among the diverse flora that has captivated human interest, Aloe stands out as a botanical marvel with a rich history of medicinal applications. The genus Aloe comprises over 500 species of succulent plants known for their therapeutic properties, with *Aloe vera* being the most extensively researched due to its multifaceted medicinal potential [1, 2]. *Aloe vera*, commonly referred to as the "wonder plant" or the "lily of the desert," has been revered for centuries in traditional medicine [3]. Its succulent leaves contain a gel rich in bioactive compounds, including polysaccharides, glycoproteins, enzymes, vitamins, and minerals [4]. These compounds have been shown to exhibit a wide range of biomedical applications, such as anti-inflammatory, antimicrobial, antioxidant, and potential anti-cancer properties [5]. The therapeutic reputation of *Aloe vera* spans cultures, from ancient civilizations that extolled its virtues to the contemporary era where scientific scrutiny seeks to validate its traditional claims.

Recent advancements in metabolomics, a discipline within systems biology, have become instrumental in profiling and understanding the metabolites present in biological systems, including plants [6]. Metabolomics involves the comprehensive analysis of small molecules, offering a holistic view of the biochemical landscape and providing insights into the physiological state of the plant and its responses to environmental cues [7]. For *Aloe vera*, metabolomics presents an unparalleled opportunity to delve into the intricate web of metabolites that underlie its medicinal properties. The metabolomic analysis of *Aloe vera* involves advanced analytical techniques such as mass spectrometry (MS) and nuclear magnetic resonance (NMR), coupled with sophisticated bioinformatics tools [8]. These methodologies enable the identification and quantification of a diverse array of metabolites, from primary metabolites like sugars and amino acids to secondary metabolites such as polyphenols and terpenoids. By scrutinizing the metabolic fingerprints of *Aloe vera*, researchers can unravel the biochemical pathways governing the synthesis of bioactive compounds and gain insights into the factors influencing their abundance.

Aloe vera contains a variety of metabolites with significant biomedical applications. Polysaccharides, such as acemannan, have demonstrated immunomodulatory effects [9]. Anthraquinones, including aloin and aloe-emodin, possess laxative, antimicrobial, and potential anti-cancer properties [10]. Flavonoids and other phenolic compounds provide antioxidant and anti-inflammatory benefits [11]. The therapeutic potential of *Aloe vera* is not confined to a singular compound but is instead a synergistic interplay of numerous bioactive molecules [12]. The significance of metabolomics in *Aloe vera* research extends beyond curiosity about its chemical composition; it holds profound implications for the pharmaceutical, cosmetic, and nutraceutical industries. By

 DOI: 10.1201/9781003603108-4

capturing the complexity of the metabolome, metabolomics facilitates a holistic understanding of the interactions between bioactive molecules, paving the way for developing Aloe-based products with enhanced efficacy and targeted applications.

This chapter provides a comprehensive overview of the application of metabolomics in unravelling the biochemical intricacies of Aloe. It highlights the potential for Aloe species to revolutionize healthcare and agriculture, offering novel insights into their therapeutic properties and commercial potential. Through the lens of metabolomics, *Aloe vera* transcends its identity as a botanical curiosity, emerging as a reservoir of biochemical secrets waiting to be unlocked. This exploration underscores the enduring fascination with the wonders of the natural world and the limitless potential for discovery that lies within its embrace.

4.2 METHODS AND TECHNIQUES

Metabolomics is the study of the complete set of small molecules, or metabolites, present within an organism. Metabolites include compounds like sugars, amino acids, organic acids, lipids, and secondary metabolites, and they play crucial roles in plant growth, development, and responses to environmental stimuli. Analysing the metabolome of a plant involves the use of various technologies and methodologies. Figure 4.1 shows schematics of methods to determine plant metabolites. Here's an overview of the technology used to metabolome a plant:

4.2.1 Sample Preparation

In metabolomic studies, sample preparation is crucial since it involves the extraction of metabolites using various extraction techniques, although in vivo NMR measurements that have been reported recently do not require sampling before analysis [13]. There are certain parameters of sample preparation that need to be taken care of and that can influence results and reproducibility. These include the method of sampling and harvesting time, which can greatly influence the reproducibility of the analytical sample, the diurnal and dietary composition of samples, and sample storage [14–17]. Immediately freezing of samples in liquid nitrogen is required to halt metabolic processes and prevent enzymatic degradation, but it can lead to metabolite loss [18]. Metabolites are extracted from the frozen plant tissues using appropriate solvents, which may vary depending on the metabolite class of interest (Polar/non-polar). An extensive range of metabolites can be separated with a solvent system consisting of water, methanol, and chloroform. For a broad range of metabolites, including hydrophilic, non-polar, and polar substances, this solvent system is frequently employed. In plant metabolomics, several solvent systems are used, such as methanol–water and methanol–methyl-tert-butyl-ether–water extractions, often with pure methanol. It was possible to recover nearly every kind of metabolite using a single process by developing a specialized solvent gradient extraction technique [19, 20].

4.2.2 Analytical Techniques

4.2.2.1 Mass Spectrometry

Mass Spectrometry (MS) based metabolomics relies on the principle of ionization and mass measurement to analyze metabolites within a biological sample. Using techniques such as electrospray ionization (ESI) or matrix-assisted laser desorption/ionization (MALDI), MS measures the mass-to-charge ratio of ions generated from metabolites. This approach offers high sensitivity, capable of detecting metabolites even at low concentrations. Moreover, MS provides valuable insights into the fragmentation patterns of metabolites, facilitating their identification. While MS can offer quantitative data, ensuring accuracy may necessitate the use of standards or isotopic labeling. Importantly, MS-based metabolomics is known for its high throughput, enabling the analysis of numerous

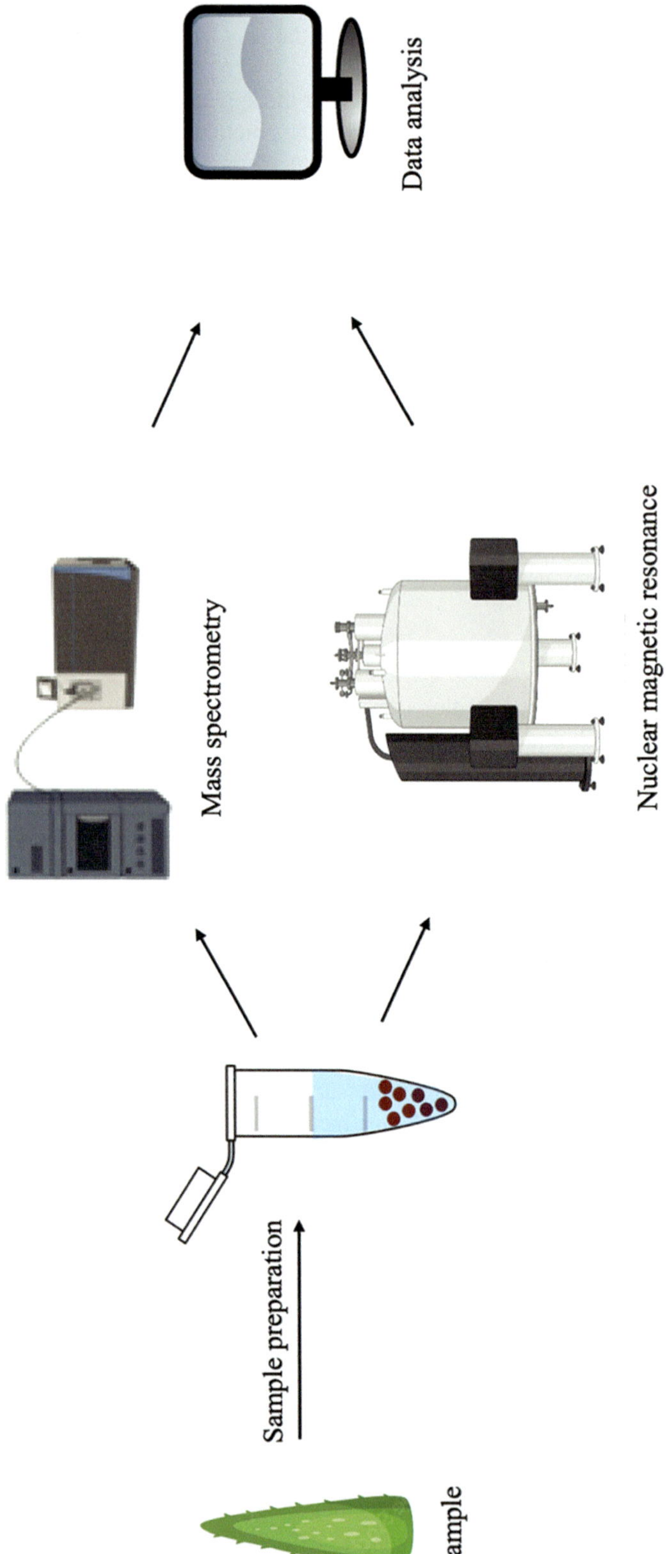

FIGURE 4.1 Schematics showing methods to determine plant metabolites.

samples within a short timeframe, thus proving invaluable for comprehensive metabolic profiling studies.

i. Gas Chromatography-Mass Spectrometry (GC-MS): GC-MS is suitable for volatile and thermally stable metabolites. It involves gas chromatography for separation and mass spectrometry for detection [21]. It is described as a gold standard technique for volatile and low molecular weight metabolites, but most of it requires two-stage chemical derivatization at room temperature to provide volatility and thermal stability prior to analysis, as sample stability is a concern [22, 23]. First, carbonyl functional groups are converted to oximes, followed by the formation of trimethylsilyl (TMS) esters with silylating reagents, which is required to eliminate undesirable slow and reversible silylation reactions with carbonyl groups, whose products can be thermally labile [24]. Metabolome coverage is largely determined by the volatility of sample components, either non-x or derivatised [25]. Many applications of GC-MS have been proposed in metabolomics, including urinary screening procedures to indicate the presence of diseases related to organic acidemias [26], measure effects of genetic or environmental modifications and stressors [24, 27–29], microbial and clinical metabolomics [30].
ii. Liquid Chromatography-Mass Spectrometry (LC-MS): LC-MS is commonly used for metabolomics due to its high sensitivity and ability to separate and identify a wide range of metabolites [21]. It involves the use of liquid chromatography to separate metabolites, followed by electron-spray ionisation (ESI) for detection and quantification. It differs from GC-MS in several ways, including lower analysis temperatures, and sample volatility and derivatization are not required; this simplifies sample preparation [31–33].
iii. Direct Infusion-mass spectrometry: Some MS-based methods involve direct infusion of samples into the mass spectrometer, which is particularly useful and high throughput for rapid profiling. In this, crude samples were directly injected into an electrospray mass spectrometer, resulting in one mass spectrum per sample, which is representative of the composition of the sample [21].

4.2.2.2 Nuclear Magnetic Resonance (NMR)

NMR spectroscopy can provide information about the chemical structure and concentration of metabolites. It is a rapid, non-destructive, high-throughput screening method and requires minimal sampling [21]. NMR spectroscopy exploits the magnetic properties of certain atomic nuclei. When placed in a strong magnetic field, nuclei with an odd number of protons or neutrons (such as ^1H, ^13C, ^15N, ^31P) have a property called spin. These nuclei behave like tiny magnets and can align with or against the applied magnetic field, creating different energy levels. In NMR, a sample is placed in a strong magnetic field, and a radiofrequency (RF) pulse is applied. This RF pulse provides energy that can flip the spin of the nuclei from a lower energy state to a higher energy state. When the RF pulse is turned off, the nuclei relax back to their original states, releasing energy in the form of RF signals. The RF signals emitted by the nuclei are detected by the NMR instrument and transformed into a spectrum using Fourier Transform (FT) techniques. The resulting NMR spectrum displays peaks corresponding to different chemical shifts, coupling patterns, and intensities, which are used to identify and quantify metabolites. NMR spectroscopy offers high specificity, reproducibility, and the ability to quantify metabolites without the need for standards [34–36]. It is widely used in plant metabolomics, functional genomics, and systems biology to understand plant metabolism, identify biomarkers, and improve crop traits [37–39]. Databases such as Metlin, HMDB, and KEGG provide reference metabolite information for identification. Specialized software packages like XCMS, MZmine, and MetaboAnalyst help in data processing and statistical analysis [40].

MS and NMR are often used complementarily in metabolomics studies to obtain a more comprehensive understanding of the metabolome. The two techniques may have different strengths in detecting specific classes of metabolites, and their integration can enhance overall metabolome coverage. Integrated data analysis approaches are employed to combine information from both MS and NMR datasets, providing a more holistic view of the metabolome. The choice between MS-based and NMR-based metabolomics depends on factors such as the research question, the types of metabolites of interest, sensitivity requirements, and the need for quantitative or structural information. Often, researchers use both techniques in tandem to capitalize on their respective strengths.

4.3 METABOLOMIC CONSTITUENTS OF ALOE

The growing demand for Aloe species (*Xanthorrhoeaceae* family) in food supplements, cosmetics, and pharmaceutical development has sparked a heightened focus on exploring the phytochemical composition of these plants in recent times. Surprisingly, only a limited number of Aloe species have been intentionally cultivated and made available in the market, possibly due to the relatively scant knowledge about their potent bioactive constituents. Furthermore, numerous species in this genus are in danger of extinction because of habitat loss and overuse for the trade in medicinal and decorative plants [41]. An alternative to prevent extension due to habitat loss is to use in vitro tissue culturing of aloe species. In a recent study on *Aloe arborescens,* which is one of the most significant and highly valued medicinal species in the Aloe genus worldwide, it was found that the condensed tannins, flavonoids, phenolics, and total iridoids were found in varying concentrations in the regenerated shoots when it is treated with the different cytokinin concentrations in vitro [42]. Their study proves that the selection of the kind and amount of exogenously supplied cytokines during tissue culture can increase the amount of bioactive metabolites [43].

The Aloe species offers a unique source of bioactive compounds and naturally occurring antioxidants that can be extracted from its leaves, roots, and flowers. This genus is known to contain more than 130 phytoconstituents of various classes, including glycoproteins, phenolic compounds, flavonoids, phytosterols, coumarins, alkaloids, pyrones, anthrones, naphthalenes, indoles, and anthraquinones [44]. Some of the notable bioactive compounds found in Aloe species are discussed below, and their chemical structures are depicted in Figure 4.2.

4.3.1 Anthraquinones

Anthraquinones are a group of naturally occurring compounds found in various plants, including those belonging to the Aloe genus within the Xanthorrhoeaceae family (formerly known as Aloaceae) [4]. Aloe species, known for their medicinal and therapeutic properties, contain several bioactive compounds, among which anthraquinones play a significant role [45].
Some of the common anthraquinones found in Aloe species include:

1. Aloin: Aloin is one of the best ingredients in aloe but levels are highly variable, depends on species and strain, as well as growing conditions. It is a diastereoisomeric mixture of aloin A (barbaloin) and aloin B (isobarbaloin), an anthrone C-glucoside component of aloes [46]. It is a bitter-tasting, yellow-brown colored compound found in the outer rind of the aloe plant and constitutes up to 30% of the aloe plant's dried leaf exudates. It is used as a stimulant-laxative, treating constipation by inducing bowel movements. It has a role as a laxative and a tyrosinase inhibitor [47].
2. Aloe-emodin: Aloe emodin is a dihydroxyanthraquinone. It has a role as an antineoplastic agent and a plant metabolite. It is functionally related to a chrysazin. This anthraquinone is known for its anti-inflammatory, antioxidant, and anti-cancer properties. It is found in various Aloe species and has been studied for its potential therapeutic applications [19], [48].
3. Chrysophanol: It is a trihydroxyanthraquinone present in various Aloe species, plays a significant role in the plant's medicinal properties. Characterized by its structure, which includes a methyl group at the C-3 position of chrysazin, chrysophanol exhibits a range of biological

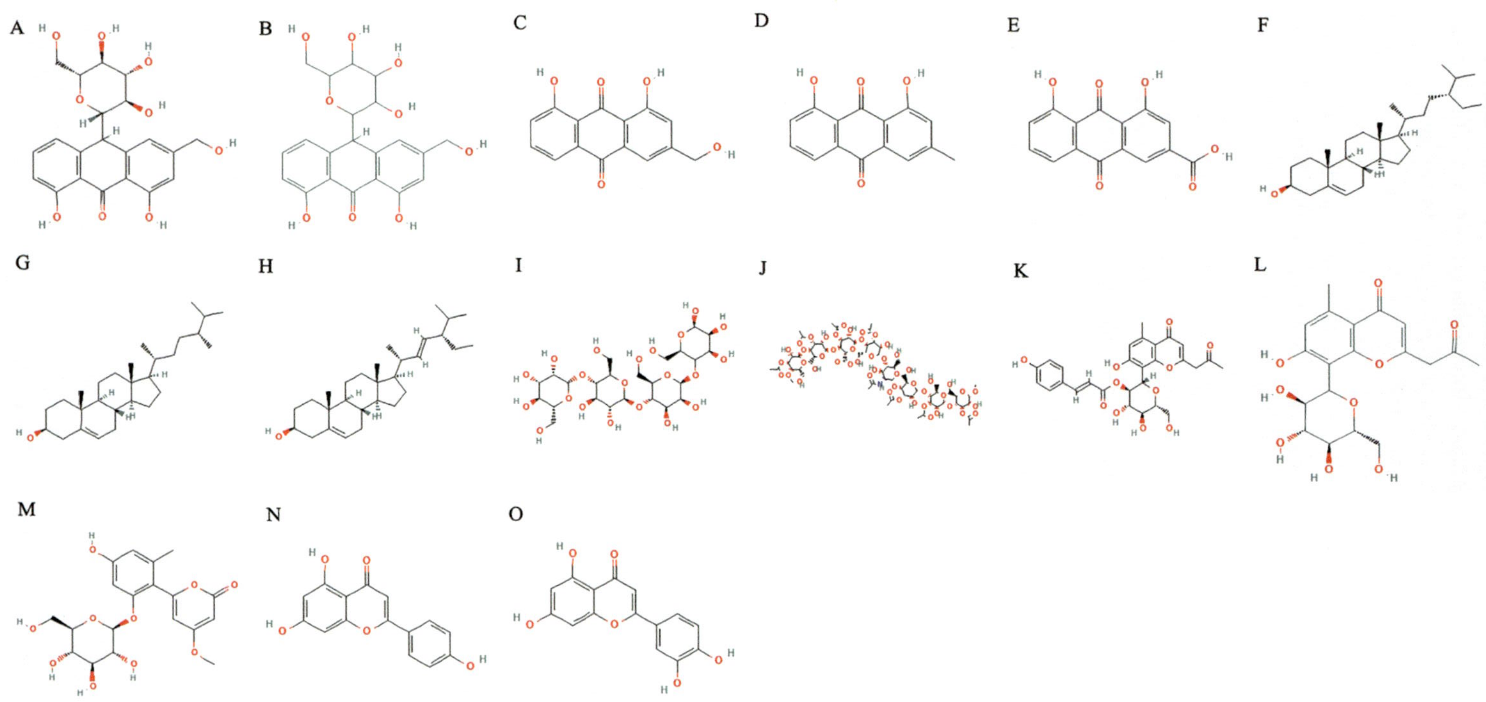

FIGURE 4.2 Chemical structures of essential Aloe metabolites.

Notes: Anthraquinones (A-E) Aloin, Barbaloin, Aloe-emodin, Chrysophanol, Rhein, Phytosterols (F-H) β-Sitosterol, Campesterol, Stigmasterol, Polysaccharides (I-J) Glucomannans, Acemannan, Flavanoids (K-O) Aloeresin A, Aloeresin B, Aloenin, Apigenin, Luteolin.

activities. Notably, it demonstrates antibacterial, antifungal, and anti-inflammatory effects, contributing to the therapeutic benefits attributed to Aloe. The compound's ability to combat microbial infections and modulate inflammation underscores its importance in traditional and modern medicine [49, 50]. Chrysophanol's functional relationship with chrysazin further emphasizes its potential as a bioactive constituent in Aloe-based treatments.

4. Rhein: Rhein is a dihydroxyanthraquinone characterized by its yellow needle-like crystals or yellow-brown powder form. While it is not as prevalent as other anthraquinones in Aloe species, rhein has been detected in certain varieties. It is noted for its antioxidant and anti-inflammatory properties and has garnered interest for its potential therapeutic applications [51, 52].

These anthraquinones contribute to the diverse pharmacological activities exhibited by Aloe species, including their use in treating wounds, skin conditions, gastrointestinal disorders, and more. However, it is essential to note that while these compounds offer various potential health benefits, their efficacy and safety should be further investigated through scientific research and clinical studies. Additionally, the concentration of anthraquinones can vary depending on factors such as species, growing conditions, and processing methods, so standardized preparations are important for consistent therapeutic outcomes. There are species of aloe that contain anthraquinones, which are laxatives. It is crucial to remember that using some anthraquinones in excessive amounts may have unintended consequences.

4.3.2 Polysaccharides

Polysaccharides are complex carbohydrates made up of long chains of monosaccharide units bonded together. In Aloe species, the polysaccharides primarily include glucomannans and acemannan.

1. Glucomannans: Glucomannans are found in the gel-like substance of *Aloe vera* leaves. It is a heteroglycan consisting of beta-(1->4)-linked D-glucose (G) and D-mannose (M) in a proportion of 5:8 and the repeating unit follows pattern: GGMMGMMMMMGGM. It has a role as a nutraceutical. Glucomannans are known for their ability to retain water, which contributes to the hydrating and moisturizing properties of *Aloe vera* gel when applied topically. They also possess immunomodulatory properties, potentially enhancing the body's immune response [48].
2. Acemannan: Acemannan is a specific type of glucomannan found in *Aloe vera*. It is a complex polysaccharide consisting mainly of acetylated mannose. Acemannan has been studied for its various therapeutic effects, including immunostimulatory, anti-inflammatory, wound-healing, and antiviral properties. It is believed to modulate the immune system, potentially enhancing its ability to fight infections and promote tissue repair. Overall, polysaccharides in Aloe species contribute significantly to their medicinal properties, making them valuable ingredients in various healthcare and cosmetic products [4].

4.3.3 Phytosterols

Phytosterols are plant-derived compounds that closely resemble cholesterol and are known for their potential health benefits, especially in managing cholesterol levels. *Aloe vera*, a prominent member of the Aloe genus, has been studied for its phytosterol content, which includes compounds such as campesterol, β-sitosterol, and stigmasterol, among others. These phytosterols are primarily located in the gel and latex of *Aloe vera* leaves. Campesterol is a plant sterol that, like other phytosterols, may help manage cholesterol levels by inhibiting cholesterol absorption in the digestive system [53]. β-Sitosterol, a member of the phytosterol family, is characterized by a β-hydroxy group at the three-position of the stigmast-5-ene structure. It functions as a sterol methyltransferase inhibitor and exhibits antioxidant properties. β-Sitosterol has been studied for its cholesterol-lowering effects and potential role in reducing heart disease risk [54]. Stigmasterol is another important phytosterol present in *Aloe vera*, known for its potential anti-inflammatory and cholesterol-lowering effects [55]. Lupeol, while not a phytosterol but a triterpene, also contributes to *Aloe vera*'s bioactive profile, demonstrating various biological activities including anti-inflammatory and antioxidant effects [56]. Research indicates that the phytosterols in *Aloe vera* may contribute to its diverse biological activities, such as anti-inflammatory, antioxidant, and cholesterol-lowering effects. However, the specific composition and concentration of these phytosterols can vary depending on factors like species, growing conditions, and extraction methods [57]. Understanding these variations is crucial for optimizing the health benefits of *Aloe vera* and its derived products.

4.3.4 Flavonoids

Flavonoids, a diverse group of polyphenolic compounds, are prevalent in plants and known for their broad range of biological activities, including antioxidant, anti-inflammatory, anti-cancer, and antimicrobial effects [58, 59]. In Aloe species, these flavonoids play a significant role in their medicinal properties and therapeutic effects. Aloeresin A and B are found in the resin of Aloe species such as *Aloe vera* and *Aloe castanea*. Aloeresin A and B exhibit various biological activities, including antioxidant and anti-inflammatory properties [60]. Aloenin, a glycoside present in *Aloe africana* and *Aloe nyeriensis*, is noted for its anti-inflammatory effects. It is also found in the gel of *Aloe vera* [59]. Aloenin A and B are dihydroisocoumarins, isolated from Aloe species, and are recognized for their anti-inflammatory and wound-healing properties [61]. Apigenin, a trihydroxyflavone substituted with hydroxy groups at positions 4', 5, and 7, is found in several Aloe species. Apigenin is known for its role in inducing autophagy in leukemia cells and has antioxidant and anti-inflammatory activities [62]. Luteolin is a tetrahydroxyflavone with hydroxy groups at positions 3', 4', 5, and 7. Luteolin demonstrates significant antioxidant and anti-inflammatory effects. It also exhibits potential for apoptosis induction and chemoprevention, protecting cells from oxidative damage and inhibiting tumor proliferation [63]. Quercetin is a pentahydroxyflavone with hydroxy groups at positions 3, 3', 4', 5, and 7. Quercetin is abundant in various edible plants, including some Aloe species. It is known for its broad range of activities, including antioxidant, antibacterial, and anti-cancer effects. Kaempferol is a tetrahydroxyflavone that has hydroxy groups located at positions 3, 5, 7, and 4'. Kaempferol acts as an antioxidant and is under investigation for its potential anti-cancer properties. It also functions as an antibacterial agent and plant metabolite [64]. Other flavonoids include Naringenin, present in some Aloe species. Naringenin is a flavanone with notable antioxidant and anti-inflammatory effects. Dihydroisorhamnetin is a flavonoid derivative, dihydroisorhamnetin and is involved in various biological activities, including antioxidant properties. Isovitexin is recognized for its antioxidant and anti-inflammatory effects and is found in particular Aloe species [65]. These flavonoids, along with other bioactive compounds in Aloe species, contribute to their therapeutic effects and health benefits.

4.4 BIOMEDICAL POTENTIAL OF ALOE METABOLITES

Metabolites present in the aloe species show various kinds of activities depicted in Figure 4.3.

The various activities of some metabolites are listed below:

- Antioxidant activity: Antioxidants are substances that prevent the oxidation of other compounds. *Aloe vera*'s antioxidant activity contributes to its protective effects against oxidative stress-related conditions. By scavenging free radicals and modulating oxidative pathways, *Aloe vera* may play a role in preventing or managing conditions associated with

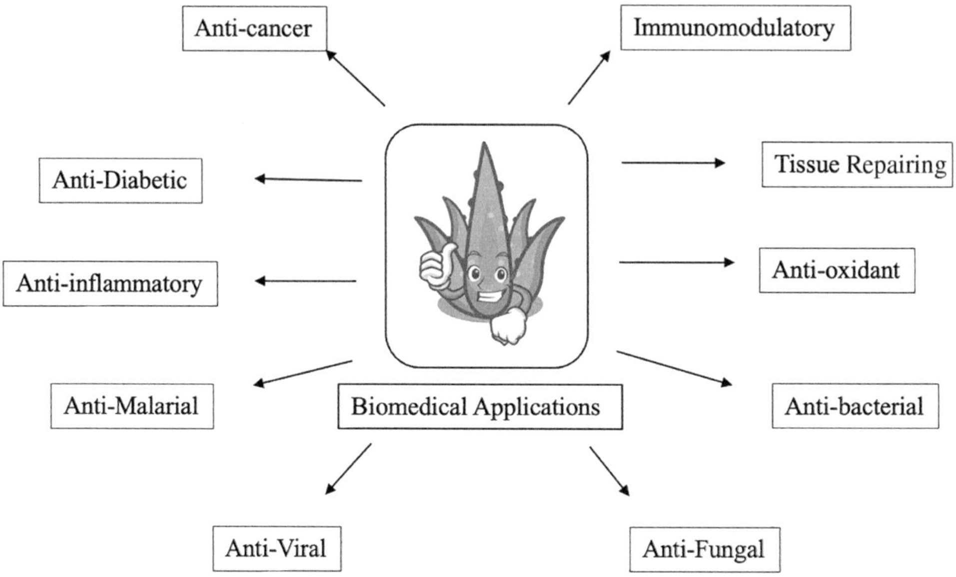

FIGURE 4.3 Biomedical applications of Aloe.

oxidative damage, including neurodegenerative disorders and cardiovascular diseases. Free radical scavenging activity was measured through DPPH and ABTS assays, while the FRAP assay measures the content of electron-donating species with some redox potential [66]. Free radical scavenging activity of the methanol, acetone, and ethanol extracts of Aloe ferox Mill at a concentration of 0.5 mg/mL showed higher inhibition against ABTS, hydrogen peroxide, and nitric oxide radicals [67]. Aloesin, aloeresin A, 2″-O-trans-p-coumaroylaloenin, caffeoylquinic acid isomers, luteolin-C-glucoside-O-pentoside, and isoorientin were identified as metabolites responsible for the radical scavenging properties [68]. At 1 mg/ml, Aloe perryi methanolic extract showed 94% radical scavenging activity comparable to the activity of ascorbic acid [69].

- Anti-diabetic: α-glucosidase, one of the main therapeutic targets, breaks down glycosidic bonds in complex carbohydrates to release absorbable monosaccharides; therefore, inhibiting α-glucosidase can slow the release of glucose, delay absorption, and help in managing postprandial hyperglycemia [70]. Phytosterols isolated from *Aloe vera* gel were shown to have anti-hyperglycemic effects – lophenol, 24-methyl-lophenol, 24-ethyl-lophenol, cycloartanol, and 24-methylene-cycloartanol. After administration for 28 days, fasting blood glucose levels decreased to approximately 64%, 28%, 47%, 51%, and 55% of control levels, respectively [71]. Aloin has been reported as a potent inhibitor of α-glucosidase enzyme [72]; however, aloin, *Aloe perryi* resin, *and Aloe vera* gel showed IC50 of 0.34, 0.67, and 1.09 μg/mL, comparable to acarbose (0.54 μg/mL) as a positive control. On administering 2mg/kg of *Aloe perryi* extract, diabetic albino rats showed a 22.5% reduction of the fasting glucose level after 54 hours [73]. *Aloe vera* and *Aloe barbadensis* induced 40 and 44% reduction in glucose levels upon oral administration in diabetic rats [45]. Flavonoids, such as apigenin and luteolin, present in various aloe species act as an antioxidant agent by protecting beta cells from oxidation and increasing islet cells proliferation, resulting in stimulating insulin production [74].
- Wound healing and tissue repair: One of the most well-established biomedical applications of *Aloe vera* is in wound healing. The gel's ability to accelerate the wound-healing process, reduce inflammation, and minimize scarring has been validated through both traditional use and scientific studies [75]. Key molecular mechanisms, such as the stimulation of fibroblast activity

and collagen synthesis, contribute to its efficacy in tissue repair [76]. *Aloe vera* (or yellow aloe) leaves include a mucilage gel that consists of some glycoproteins and polysaccharides, which help prevent inflammation and pain and stimulate skin growth and healing. The mucilage of this plant can be used for the treatment of internal and external wounds [75, 77]. Glucomannan stimulates and proliferates fibroblast growth factor, improves collagen production and secretion [78]. Oral dosing of *Aloe vera* mucilage in rats with diabetes has shown increased expression of the gene of vascular endothelial growth factor (VEGF) and TGF β-1 in the area of the wound in the skin of rats, resulting in stimulated fibroblasts to reconstruct the extracellular matrix at the wound site [79].

- Gastrointestinal health: Traditionally consumed for gastrointestinal health, *Aloe vera* has been studied for its potential in conditions such as irritable bowel syndrome (IBS) and inflammatory bowel disease (IBD). The anti-inflammatory and immunomodulatory effects of *Aloe vera* may contribute to its beneficial effects on gut health [75].
- Immunomodulation: Research suggests that *Aloe vera* exhibits immunomodulatory effects, influencing both innate and adaptive immune responses. This property has implications for conditions involving immune dysregulation, making *Aloe vera* a subject of interest in autoimmune disorders and immunodeficiency states [45, 80, 81].
- Antimicrobial activity: The rapidly emerging rate of antibiotic resistance necessitates alternative solutions rather than the development of new drugs. The antimicrobial properties of *Aloe vera* have been investigated against a range of pathogens, including bacteria, viruses, and fungi. This makes it a potential candidate for topical applications in wound care and the development of natural antimicrobial agents. The ethanol extract gave a better minimum inhibitory concentration (MIC) (0.125, 0.125, and 0.40mg/ml) on *Escherichia coli, Staphylococcus aureus, and Candida albicans,* respectively [82].
- Anti-inflammatory effects: Chronic inflammation underlies many health conditions, from arthritis to cardiovascular diseases. *Aloe vera's* anti-inflammatory effects, mediated by various bioactive compounds, position it as a natural anti-inflammatory agent with potential applications in managing inflammatory disorders. Studies showed that the aqueous extract of *Aloe vera* produced dose-dependent and significant inhibition of carrageenan-induced paw oedema in rats at 400 mg/kg, comparable to that of the standard drug indomethacin, while at 600 mg/kg b.w. of *Aloe vera* exhibited the most significant anti-inflammatory activity in [83]. Similarly, aqueous extract of *Aloe barbadensis* revealed that 25, 50, and 100 mg/kg of the extract reduced the formalin-induced oedema significantly ($P<0.05$) at the beginning of three hours when compared to the control group [84].
- Cancer Research: Emerging research suggests that certain compounds in *Aloe vera* may exhibit anti-cancer properties. Aloe-emodin is one of the highly researched anthroquinones in cancer cell lines. In bladder cancer cells (T24), aloe-emodin induced time-and dose-dependent apoptosis [85]. The cell death induction was accompanied by perturbation of mitochondrial membrane potential and reduced levels of cyclin-dependent kinase (CDK) 1, cyclin B1, and BCL-2 after treatment with aloe-emodin. Cervical cancer cells (HeLa) were treated with aloe-emodin, which caused cell cycle arrest in the G2/M phase. The cells showed a decrease in cyclin A and CDK2, which reduces the cell's ability to proliferate, and suppression of protein kinase Ca (PKCa) and c-MYC, signifying that proliferation and differentiation were suppressed. Increases in cyclin B1, CDK1, and alkaline phosphatase (ALP) activity were observed along with inhibition of proliferating cell nuclear antigen (PCNA), showing decreased proliferation [86]. Monoblastic leukemia (U937) cells treated with aloe-emodin, resulting in reduced proliferation rate while reactive oxygen species (ROS) and NO production, phagocytosis, and intracellular acidity increased [87], the significance of which is currently elusive. In another study, aloe-emodin treatment resulted in time-and dose-dependent irreversible cell death of human lung non-small cell carcinoma (H460) [88]. Aloe-emodin decreased BCL-2, which

abrogated the inhibition of pro-apoptotic proteins (such as BAK and BAX) and increased gene expression of p38 and caspase-3 activity, exacerbating apoptosis. The effects of aloin on cell viability were studied using Jurkat T cells, revealing a dose-dependent reduction in cell size, compromised membrane integrity, and loss of mitochondrial membrane potential, indicating a mitochondrial-dependent pathway for aloin-induced apoptosis. Additionally, aloin altered the cell cycle [89]. In HaCat cells exposed to UVB, aloin pretreatment downregulated NO generation and iNOS mRNA synthesis, inhibited NF-kappa B P65 activation, and increased cell proliferation, suggesting protective effects against UVB-induced damage [90]. A low molecular weight fraction of Aloe barbadensis gel (LMWF) demonstrated similar cytotoxic potency to known toxic substances in chemiluminescence assays [91]. Among five purified compounds from *Aloe vera* tested on K562 leukemia cells and their multidrug-resistant variant, only aloe emodin showed significant antitumor effects, especially in the MDR cell line [92]. While this area is still in the early stages of exploration, preliminary findings highlight the need for further investigation into the potential role of *Aloe vera* in cancer prevention and adjuvant therapy.

- Others: Aloe gel increases the absorption of vitamin C and E by slowing down the absorption process, which results in a significant increment in the concentration of plasma [81]. HIV therapy shows that there is a significant increase in the level of CD4 cells when patients were given the combination of *Aloe vera* gruel and antiretroviral drugs as compared to the patients given only antiretroviral drugs [80]. On treatment of oral submucous fibrosis, it is found that the patient suffering from OSMF shows significant improvement when given the gel extracts of *Aloe vera,* and mouth washing with aloe extract helps in treating Oral mucositis patients [93, 94]. Aqueous extract of *Aloe vera* increases levels of thiols in albino rats when lowered due to hepatotoxicity of paracetamol [95]. Extracts from *Aloe barbadensis* stimulate bile secretion and restore lipid peroxidation to protect the liver from hepatotoxin [96]. *Aloe vera* gel has a reduced level of plasma triglycerides and LDL in PCOS rats, resulting in decreased overall androgen production and increased flux of the steroidogenesis pathway toward estrogen synthesis [97].

4.4 FUTURE DIRECTIONS

The exploration of the metabolome of *Aloe vera* is an evolving field that holds promise for diverse applications in medicine, agriculture, cosmetics, and beyond. As researchers continue to delve into the biochemical intricacies of this remarkable plant, several future directions and potential areas of focus emerge: unravelling novel bioactive compounds using metabolomics to identify and characterize new molecules with therapeutic potential; engineering metabolic pathways to enhance the production of specific metabolites with desired therapeutic properties; integrating metabolomics with genomics and transcriptomics for a comprehensive view of the molecular mechanisms governing metabolite synthesis; investigating the metabolome of *Aloe vera* under different environmental stresses to understand its adaptive responses; studying the pharmacokinetics and bioavailability of *Aloe vera* compounds in humans and animals; employing metabolomics for quality control of *Aloe vera*-based products to ensure consumer safety and product efficacy; tailoring *Aloe vera*-based interventions to individual metabolomic profiles for personalized medicine and nutraceuticals; exploring *Aloe vera* metabolomics in ecological interactions for agroecosystem management; integrating metabolomics data with other omics disciplines and systems biology approaches for a systems-level understanding of *Aloe vera* metabolism; and prioritizing ethical considerations and sustainable practices in *Aloe vera* research to ensure long-term viability. In conclusion, the future of metabolomics in *Aloe vera* research combines technological advancements, interdisciplinary collaborations, and a commitment to sustainable and ethical practices, promising new therapeutic avenues while respecting the balance between scientific exploration and the preservation of nature.

4.5 CONCLUSION

In summary, this chapter provides a comprehensive overview of the application of metabolomics in unravelling the biochemical intricacies of Aloe. It highlights the potential for Aloe species to revolutionize healthcare and agriculture, offering novel insights into its therapeutic properties and commercial potential. This chapter serves as a valuable resource for researchers, scientists, and practitioners interested in Aloe and metabolomics. In conclusion, the metabolomics analysis of *Aloe vera* stands at the crossroads of tradition and innovation, weaving together threads of ancient wisdom and modern scientific inquiry. The pages that follow delve into the biochemical intricacies of this medicinal wonder, exploring the metabolites that define its therapeutic legacy. Through the lens of metabolomics, *Aloe vera* transcends its identity as a botanical curiosity, emerging as a reservoir of biochemical secrets waiting to be unlocked – a testament to the enduring fascination with the wonders of the natural world and the limitless potential for discovery that lies within its embrace.

REFERENCES

[1] "*Aloe vera* leaf gel: A review update." [Online]. Available: www.sciencedirect.com/science/article/abs/pii/S0378874199000859
[2] S. R. Cousins and E. T. F. Witkowski, "African aloe ecology: A review," *J Arid Environ*, vol. 85, pp. 1–17, 2012, Available: www.sciencedirect.com/science/article/abs/pii/S0140196312001280
[3] K. Eshun and Q. He, "*Aloe vera*: A valuable ingredient for the food, pharmaceutical and cosmetic industries – A review," *Crit Rev Food Sci Nutr*, vol. 44, pp. 91–96, 2004. doi: 10.1080/10408690490424694.
[4] J. H. Hamman, "Composition and applications of *Aloe vera* leaf gel," *Molecules*, 13, pp. 1599–1616, Aug. 2008. doi: 10.3390/molecules13081599.
[5] M. D. Boudreau and F. A. Beland, "An evaluation of the biological and toxicological properties of Aloe barbadensis (Miller), *Aloe vera," J Environ Sci Health Part C*, vol. 24, pp. 103–154, Jan. 2006. doi: 10.1080/10590500600614303.
[6] K. Saito and F. Matsuda, "Metabolomics for functional genomics, systems biology, and biotechnology," *Annu Rev Plant Biol*, vol. 61, pp. 463–489, Jun. 2010, doi: 10.1146/annurev.arplant.043008.092035.
[7] O. Fiehn, "Metabolomics — the link between genotypes and phenotypes," in *Functional Genomics*, Springer Netherlands, 2002, pp. 155–171. doi: 10.1007/978-94-010-0448-0_11.
[8] K. Dettmer, P. A. Aronov, and B. D. Hammock, "Mass spectrometry-based metabolomics," *Mass Spectrom Rev*, 26, pp. 51–78, Jan. 2007. doi: 10.1002/mas.20108.
[9] S. A. Im, S. T. Oh, S. Song, M. R. Kim, D. S. Kim, S. S. Woo, T. H. Jo, Y. I. Park, and C. K Lee, "Identification of optimal molecular size of modified Aloe polysaccharides with maximum immunomodulatory activity," *Int Immunopharmacol*, vol. 5, 271–279, 2005.
[10] E. Dagne, D. Bisrat, A. Viljoen, and B.-E. Van Wyk, "Chemistry of Aloe species," *Curr Org Chem*, vol. 4, no. 10, pp. 1055–1078, Mar. 2005, doi: 10.2174/1385272003375932.
[11] A. Kumar, A. Mahajan, and Z. Begum, "Phytochemical screening and in vitro study of free radical scavenging activity of flavonoids of *Aloe vera*," *Res J Pharm Technol*, vol. 13, no. 2, pp. 593–598, Feb. 2020, doi: 10.5958/0974-360X.2020.00112.2.
[12] A. Catalano, J. Ceramella, D. Iacopetta, M. Marra, F. Conforti, F. R. Lupi, D. Gabriele, F. Borges, and M. S. Sinicropi, "Aloe vera—An extensive review focused on recent studies," *Foods*, vol. 13, no. 13, p. 2155, Jul. 2024, doi: 10.3390/foods13132155.
[13] I. Tkáč, P. G. Henry, P. Andersen, C. D. Keene, W. C. Low, and R. Gruetter, "Highly resolved in vivo 1H NMR spectroscopy of the mouse brain at 9.4 T," *Magn Reson Med*, vol. 52, no. 3, pp. 478–484, 2004, doi: 10.1002/mrm.20184.
[14] E. M. Lenz, J. Bright, I. D. Wilson, S. R. Morgan, and A. F. P. Nash, "A H NMR-based metabonomic study of urine and plasma samples obtained from healthy human subjects." [Online]. Available: www.sciencedirect.com/science/article/abs/pii/S0731708503004102
[15] K. S. Solanky *et al.*, "Application of biofluid H nuclear magnetic resonance-based metabonomic techniques for the analysis of the biochemical effects of dietary isoflavones on human plasma profile," *Anal Biochem*, vol. 323, pp. 197–204.

[16] O. Al-Jowder, E. K. Kemsley, R. H. Wilson, and N. K. Sharma, "Mid-infrared spectroscopy and authenticity problems in selected meats: A feasibility study References (11) Differentiation of meat from food animals by enzyme assay." [Online]. Available: www.sciencedirect.com/science/article/abs/pii/S0308814696002890

[17] F. Ma, and L. Cheng, "The sun-exposed peel of apple fruit has higher xanthophyll cycle-dependent thermal dissipation and antioxidants of the ascorbate–glutathione pathway than the shaded peel," *Plant Sci,* vol. 165, 819–827, 2003.

[18] C. Wittmann, J. O. Krömer, P. Kiefer, T. Binz, and E. Heinzle, "Impact of the cold shock phenomenon on quantification of intracellular metabolites in bacteria." [Online]. Available: www.sciencedirect.com/science/article/abs/pii/S0003269704000740

[19] U. Roessner, C. Wagner, J. Kopka, R. N. Trethewey, and L. Willmitzer, "Simultaneous analysis of metabolites in potato tuber by gas chromatography-mass spectrometry," *Plant J*, vol. 23, no. 1, pp. 131–142, Jul. 2000, doi: 10.1046/j.1365-313X.2000.00774.x.

[20] W. Weckwerth, K. Wenzel, and O. Fiehn, "Process for the integrated extraction, identification and quantification of metabolites, proteins and RNA to reveal their co-regulation in biochemical networks," *Proteomics*, vol. 4, no. 1, pp. 78–83, Jan. 2004, doi: 10.1002/pmic.200200500.

[21] W. B. Dunn and D. I. Ellis, "Metabolomics: Current analytical platforms and methodologies," *TrAC – Trends in Analytical Chemistry*, vol. 24, no. 4, pp. 285–294, 2005, doi: 10.1016/j.trac.2004.11.021.

[22] C. Deng, X. Zhang, and N. Li, "Investigation of volatile biomarkers in lung cancer blood using solid-phase microextraction and capillary gas chromatography-mass spectrometry," *J Chromatogr B*, vol. 808, pp. 269–277, 2004.

[23] R. M. M. Perera, P. J. Marriott, and I. E. Galbally, "Headspace solid-phase microextraction – Comprehensive two-dimensional gas chromatography of wound induced plant volatile organic compound emissions," *Analyst*, vol. 127, no. 12, pp. 1601–1607, Dec. 2002, doi: 10.1039/b208577a.

[24] U. Roessner, C. Wagner, J. Kopka, R. N. Trethewey, and L. Willmitzer, "Simultaneous analysis of metabolites in potato tuber by gas chromatography-mass spectrometry," *Plant Journal*, vol. 23, no. 1, pp. 131–142, Jul. 2000, doi: 10.1046/j.1365-313X.2000.00774.x.

[25] C. Wagner, M. Seow, and J. Kopka, "Construction and application of a mass spectral and retention time index database generated from plant GC/EI-TOF-MS metabolite profiles." [Online]. Available: www.sciencedirect.com/science/article/abs/pii/S0031942202007033

[26] K. Tanaka and D. G. Hine, "Compilation of gas chromatographic retention indices of 163 metabolically important organic acids, and their use in detection of patients with organic acidurias." [Online]. Available: www.sciencedirect.com/science/article/abs/pii/S0021967300819901

[27] U. Roessner-Tunali *et al.*, "Metabolic profiling of transgenic tomato plants overexpressing hexokinase reveals that the influence of hexose phosphorylation diminishes during fruit development," *Plant Physiol*, vol. 133, no. 1, pp. 84–99, Sep. 2003, doi: 10.1104/pp.103.023572.

[28] O. Fiehn, J. Kopka, P. Dörmann, T. Altmann, R. N. Trethewey, and L. Willmitzer, "Metabolite profiling for plant functional genomics," *Nat Biotechnol*, vol. 18, no. 11, pp. 1157–1161, Nov. 2000, doi: 10.1038/81137.

[29] W. B. Dunn and D. I. Ellis, "Metabolomics: Current analytical platforms and methodologies," *TrAC Trends Anal Chem,* vol. 24, pp. 285–294, 2005,

[30] C. Grote and J. Pawliszyn, "Solid-phase microextraction for the analysis of human breath," *Anal Chem*, vol. 69, no. 4, pp. 587–596, 1997, doi: 10.1021/ac960749l.

[31] W. J. Leavens, S. J. Lane, R. M. Carr, A. M. Lockie, and I. Waterhouse, "Derivatization for liquid chromatography/electrospray mass spectrometry: Synthesis of tris(trimethoxyphenyl)phosphonium compounds and their derivatives of amine and carboxylic acids," *Rapid Commun Mass Spectrom*, vol. 16, no. 5, pp. 433–441, 2002, doi: 10.1002/rcm.559.

[32] R. Bakhtiar, L. Ramos, and F. L. S. Tse, "High-throughput mass spectrometric analysis of xenobiotics in biological fluids," *J Liq Chromatogr Relat Technol*, vol. 25, no. 4, pp. 507–540, 2002, doi: 10.1081/JLC-120008809.

[33] M. Brönstrup, "High-throughput mass spectrometry for compound characterization in drug discovery," in *Modern Mass Spectrometry*, C. A. Schalley, Ed. Berlin: Springer, 2003, pp. 283–302, doi: 10.1007/3-540-36113-8_9.

[34] J. C. Edwards, "Principles of NMR." Danbury CT: Process NMR Associates LLC, 2009.

[35] Marion D. "An introduction to biological NMR spectroscopy," *Mol Cell Proteomics*, vol. 12, no. 11, pp. 3006–3025, Nov. 2013, doi: 10.1074/mcp.O113.030239.
[36] B. Diehl, "Principles in NMR spectroscopy," in *NMR Spectroscopy in Pharmaceutical Analysis*, U. Holzgrabe, I. Wawer, and B. Diehl, Eds. Amsterdam: Elsevier, 2008, pp. 1–41, doi: 10.1016/B978-044451893-6.50003-5.
[37] J. G. Bundy *et al.*, "Earthworm species of the genus Eisenia can be phenotypically differentiated by metabolic profiling," *FEBS Lett*, vol. 521, no. 1–3, pp. 115–120, Jun. 2002, doi: 10.1016/S0014-5793(02)02854-5.
[38] J. L. Ward, C. Harris, J. Lewis, and M. H. Beale, "Assessment of H NMR spectroscopy and multivariate analysis as a technique for metabolite fingerprinting of Arabidopsis thaliana," *Phytochemistry*, vol. 62, no. 6, pp. 949–959, Mar. 2003.
[39] H.-K. Choi, Y. H. Choi, M. Verberne, A. W. M. Lefeber, C. Erkelens, and R. Verpoorte, "Metabolic fingerprinting of wild type and transgenic tobacco plants by H NMR and multivariate analysis technique," *Phytochemistry*, vol. 65, pp. 857–864.
[40] J. J. Chylla, R. A. Ulrich, E. L. Markley, and John L, "Databases and software for NMR-based metabolomics: 41 databases and software for NMR-Based metabolomics: Ingenta connect," 2013. [Online]. Available: www.ingentaconnect.com/content/ben/cmb/2013/00000001/00000001/art00004
[41] M. Cardarelli, Y. Rouphael, M. Pellizzoni, G. Colla, and L. Lucini, "Profile of bioactive secondary metabolites and antioxidant capacity of leaf exudates from eighteen Aloe species," *Ind Crops Prod*, vol. 108, pp. 44–51, Dec. 2017, doi: 10.1016/j.indcrop.2017.06.017.
[42] A. N. B. Singab, H. M. El-Hefnawy, A. Esmat, H. A. Gad, and J. A. Nazeam, "A systemic review on Aloe arborescens pharmacological profile: Biological activities and Pilot Clinical Trials," *Phytother Res*, vol. 29, 1858–1867, 2015, doi: 10.1002/ptr.5483.
[43] S. O. Amoo, A. O. Aremu, and J. Van Staden, "Unraveling the medicinal potential of South African Aloe species," *J Ethnopharmacol,* vol. 153, pp. 19-41, Apr. 11, 2014, doi: 10.1016/j.jep.2014.01.036.
[44] C. Egbuna *et al.*, "Correction to: Aloe Species as Valuable Sources of Functional Bioactives," in *Functional Foods and Nutraceuticals*, C. Egbuna and G. Dable-Tupas, Eds. Cham: Springer International Publishing, 2022, pp. C1–C1, doi: 10.1007/978-3-030-42319-3_28.
[45] A. M. El Sayed, S. M. Ezzat, M. N. Khalil, and S. S. El-Hawary, "Chemical composition and evaluation of possible alpha glucosidase inhibitory activity of eight Aloe species," *J Med Plants Res*, vol. 10, no. 13, pp. 167–178, 2016, doi: 10.5897/JMPR.
[46] F. Zonta, P. Bogoni, P. Masotti, and G. Micali, "High-performance liquid chromatographic profiles of aloe constituents and determination of aloin in beverages, with reference to the EEC regulation for flavouring substances," *J Chromatogr A*, vol. 718, pp. 99–106, 1995.
[47] W. G. Wamer, P. Vath, and D. E. Falvey, "In vitro studies on the photobiological properties of aloe emodin and aloin A," *Free Radic Biol Med*, vol. 34, pp. 233–242, 2003.
[48] Y. Ni, D. Turner, K. M. Yates, and I. Tizard, "Isolation and characterization of structural components of *Aloe vera* L. leaf pulp," *Int Immunopharmacol*, vol. 4, pp. 1745–1755, 2004.
[49] C. Liu and Y. Huang, "Chinese herbal medicine on cardiovascular diseases and the mechanisms of action," *Front Pharmacol,* vol. 7, p. 469, 2016, doi: 10.3389/fphar.2016.00469.
[50] L. Xie, H. Tang, J. Song, J. Long, L. Zhang, and X. Li, "Chrysophanol: A review of its pharmacology, toxicity and pharmacokinetics," *J Pharm Pharmacol*, vol. 71, pp. 1475–1487, 2019.
[51] W. Zhou, G.-A. Bounda, F. Yu, and A. Feng, "Pharmacological potential action of rhein and its diverse signal transduction: A systematic review," *World J Pharm Res*, vol. 3, pp. 3599–3626, 2014.
[52] Y. X. Zhou, W. Xia, W. Yue, C. Peng, K. Rahman, and H. Zhang, "Rhein: A Review of Pharmacological Activities," *Evid Based Complement Alternat Med*, vol. 2015, p. 578107, 2015, doi: 10.1155/2015/578107.
[53] C. M. Greene, D. Waters, R. M. Clark, J. H. Contois, and M. L. Fernandez, "Plasma LDL and HDL characteristics and carotenoid content are positively influenced by egg consumption in an elderly population," *Nutr Metab*, vol. 3, no. 1, p. 6, Jan. 2006.
[54] H. Gylling, P. Simonen, L. Kaipiainen, and I. Wester, "Methodological aspects of phytosterol measurements in biological samples," *Curr Med Chem*, vol. 26, no. 37, pp. 6776–6785, Dec. 2018, doi: 10.2174/0929867325666180713160330.
[55] P. K. Srivastava and A. K. Pandey, "Natural products and derivatives: Biological and pharmacological activities," *Cell Arch*, vol. 15, pp. 1–38, 2015.

[56] M. B. C. Gallo, and J. M. Sarachine, "Biological activities of lupeol," *Int J Biomed Pharm Sci*, vol. 3, pp. 46–66.

[57] T. Pullaiah, editor. *Phytochemical Composition and Pharmacy of Medicinal Plants*, Vol. 2. Florida: CRC Press.

[58] C. A. Rice-Evans, N. J. Miller, and G. Paganga, "Antioxidant properties of phenolic compounds," *Trends Plant Sci,* vol. 2, 152–159, 1997, doi: 10.1016/S1360-1385(97)01018-2.

[59] A. J. Harborne, *Phytochemical Methods a Guide to Modern Techniques of Plant Analysis*. Springer science & business media, 1998 Apr 30.

[60] M. Ahmed and F. Hussain, "Chemical composition and biochemical activity of Aloe vera (Aloe barbadensis Miller) leaves," *Int. J. Chem. Biochem. Sci.*, vol. 3, no. 5, pp. 29–33, 2013.

[61] H. I. Abd-Alla, M. Shaaban, K. A. Shaaban, N. S. Abu-Gabal, N. M. M. Shalaby, and H. Laatsch, "New bioactive compounds from Aloe hijazensis," *Nat Prod Res*, vol. 23, no. 11, pp. 1035–1049, 2009, doi: 10.1080/14786410802242851.

[62] B. Salehi, A. Venditti, M. Sharifi-Rad, D. Kręgiel, J. Sharifi-Rad, A. Durazzo, M. Lucarini, A. Santini, E. B. Souto, E. Novellino, and H. Antolak, "The therapeutic potential of apigenin," *Int. J. Mol. Sci.*, vol. 20, no. 6, p. 1305, Mar. 2019, doi: 10.3390/ijms20061305.

[63] M. Lopez-Lazaro, "Distribution and Biological Activities of the Flavonoid Luteolin," *Mini-Reviews in Medicinal Chemistry*, vol. 9, no. 1, pp. 31–59, Dec. 2008, doi: 10.2174/138955709787001712.

[64] M. Y. Park, H. J. Kwon, and M. K. Sung, "Evaluation of aloin and aloe-emodin as anti-inflammatory agents in aloe by using murine macrophages," *Biosci Biotechnol Biochem*, vol. 73, no. 4, pp. 828–832, 2009, doi: 10.1271/bbb.80714.

[65] A. M. Viljoen, B. E. Van Wyk, and F. R. Van Heerden, "Distribution and chemotaxonomic significance of flavonoids in Aloe (Asphodelaceae)," *Plant Systematics and Evolution*, vol. 211, no. 1–2, pp. 31–42, 1998, doi: 10.1007/BF00984910.

[66] R. L. Prior, "Oxygen radical absorbance capacity (ORAC): New horizons in relating dietary antioxidants/bioactives and health benefits." [Online]. Available: www.sciencedirect.com/science/article/pii/S1756464614003971

[67] O. A. Wintola and A. J. Afolayan, "Phytochemical constituents and antioxidant activities of the whole leaf extract of Aloe ferox Mill.," *Pharmacogn Mag*, vol. 7, no. 28, pp. 325–333, Oct. 2011, doi: 10.4103/0973-1296.90414.

[68] C. Breaud *et al.*, "LC-MS Based Phytochemical Profiling towards the Identification of Antioxidant Markers in Some Endemic Aloe Species from Mascarene Islands," *Antioxidants*, vol. 12, no. 1, Jan. 2023, doi: 10.3390/antiox12010050.

[69] R. A. Mothana, U. Lindequist, R. Gruenert, and P. J. Bednarski, "Studies of the in vitro anticancer, antimicrobial and antioxidant potentials of selected Yemeni medicinal plants from the island Soqotra," *BMC Complement Altern Med*, vol. 9, Mar. 2009, doi: 10.1186/1472-6882-9-7.

[70] S. Kumar, S. Narwal, V. Kumar, and O. Prakash, "α-glucosidase inhibitors from plants: A natural approach to treat diabetes," Jan. 2011. doi: 10.4103/0973-7847.79096.

[71] M. Tanaka *et al.*, "Identification of five Phytosterols from Aloe vera Gel as anti-diabetic compounds," *Biol Pharm Bull,* vol. 29, pp. 1418–1422, 2006.

[72] F. Ghamari, S. M. Ghaffari, M. Salami, F. Moosavi-Movahedi, F. Farivar, A. Johari, A. A. Saboury, J. M. Chobert, T. Haertlé, and A. A. Moosavi-Movahedi, "Synergic study of α-glucosidase inhibitory action of aloin and its antioxidant activity with and without camel β-casein and its peptides," *Protein Pept. Lett.*, vol. 20, no. 5, pp. 607–612, May 2013. DOI: 10.2174/0929866511320050015

[73] T. S. Aldayel, M. H. Grace, M. A. Lila, M. A. Yahya, U. M. Omar, and G. Alshammary, "LC-MS characterization of bioactive metabolites from two Yemeni Aloe spp. with antioxidant and antidiabetic properties." [Online]. Available: www.sciencedirect.com/science/article/pii/S1878535220300 40X#b0050

[74] N. Deora and K. Venkatraman, "*Aloe vera* in diabetic dyslipidemia: Improving blood glucose and lipoprotein levels in pre-clinical and clinical studies," *J Ayurveda Integr Med*, vol. 13, p. 100675, 2002.

[75] K. Eshun and Q. He, "*Aloe vera*: A valuable ingredient for the food, pharmaceutical and cosmetic industries – A review," *Crit Rev Food Science Nutr*, vol. 44, pp. 91–96, 2004, doi: 10.1080/10408690490424694.

[76] S. A. Hashemi, S. A. Madani, and S. Abediankenari, "The review on properties of *Aloe vera* in healing of cutaneous wounds," *BioMed Res Int*, vol. 2105, p. 714216, 2015, doi: 10.1155/2015/714216.

[77] L. R. S. Moreira and E. X. F. Filho, "An overview of mannan structure and mannan-degrading enzyme systems," *Applied Microbiol Biotechnol*, vol. 79, pp. 165–178, May 2008. doi: 10.1007/s00253-008-1423-4.

[78] M. D. Boudreau and F. A. Beland, "An evaluation of the biological and toxicological properties of Aloe barbadensis (Miller), *Aloe vera*," *Journal of Environmental Science and Health Part C*, vol. 24, pp. 103–154, Jan. 2006. doi: 10.1080/10590500600614303.

[79] A. Atiba, "The effect of *Aloe vera* oral administration on cutaneous wound healing in Type 2 diabetic rats," *J Vet Med Sci*, vol. 73, pp. 583–589, 2010, doi: 10.1292/jvms.10-0438(https).

[80] O. S. Olatunya, A. M. Olatunya, H. C. Anyabolu, E. A. Adejuyigbe, and O. A. Oyelami, "Preliminary trial of *Aloe vera* gruel on HIV infection," *J Alter Compleme Med*, vol. 18, no. 9, pp. 850–853, Sep. 2012, doi: 10.1089/acm.2010.0735.

[81] J. A. Vinson, H. Al Kharrat, and L. Andreoli, "Effect of *Aloe vera* preparations on the human bioavailability of vitamins C and E." [Online]. Available: www.sciencedirect.com/science/article/abs/pii/S0944711305000656?via%3Dihub

[82] M. C. Stanley, O. E. Ifeanyi, and O. Godson Eziokwu, "Antimicrobial effects of *Aloe vera* on some human pathogens," 2014. [Online]. Available: www.ijcmas.com

[83] A. Devaraj and T. Karpagam, "Evaluation of anti-inflammatory activity and analgesic effect of *Aloe vera* leaf extract in rats," 2011. [Online]. Available: www.irjponline.com

[84] U. G. Egesie, K. E. Chima, and N. Z. Galam, "Anti-inflammatory and analgesic effects of aqueous extract of *Aloe vera* (*Aloe barbadensis*) in rats," *Afr J Biomed Res*, vol. 14, pp. 209–212, 2011.

[85] J.-G. Lin, G.-W. Chen, T.-M. Li, S.-T. Chouh, T.-W. Tan, and J.-G. Chung, "Aloe-Emodin Induces Apoptosis in T24 Human Bladder Cancer Cells Through the p53 Dependent Apoptotic Pathway," *Journal of Urology*, vol. 175, no. 1, pp. 343–347, Jan. 2006, doi: 10.1016/S0022-5347(05)00005-4.

[86] J. M. Guo, B. X. Xiao, Q. Liu, S. Zhang, D. H. Liu, and Z. H. Gong, "Anticancer effect of aloe-emodin on cervical cancer cells involves G 2/M arrest and induction of differentiation," *Acta Pharmacol Sin*, vol. 28, no. 12, pp. 1991–1995, Dec. 2007, doi: 10.1111/j.1745-7254.2007.00707.x.

[87] C. Tabolacci *et al.*, "Aloe-emodin as antiproliferative and differentiating agent on human U937 monoblastic leukemia cells," *Life Sci*, vol. 89, no. 21–22, pp. 812–820, Nov. 2011, doi: 10.1016/j.lfs.2011.09.008.

[88] F. T. Yeh, C. H. Wu, and H. Z. Lee, "Signaling pathway for aloe-emodin-induced apoptosis in human H460 lung nonsmall carcinoma cell," *Int J Cancer*, vol. 106, no. 1, pp. 26–33, Aug. 2003, doi: 10.1002/ijc.11185.

[89] E. J. Buenz, "Aloin induces apoptosis in Jurkat cells," *Toxicol Vitro*, vol. 22, pp. 422–429, 2008.

[90] X.-Z. Song, A.-E. Xu, Z.-G. Bi, "Effect of aloin on inducible nitric oxide synthase and nuclear factor Kappa B synthesis of hacat cells irradiated by Ultraviolet B." [Online]. Available: www.pifukezazhi.com/EN/abstract/abstract509.shtml

[91] H. Avila, J. Rivero, F. Herrera, and G. Fraile, "Cytotoxicity of a low molecular weight fraction from *Aloe vera* (Aloe barbadensis Miller) gel," *Toxicon,* vol. 35, pp. 1423–1430, 1997.

[92] S. Grimaudo, M. Tolomeo, R. Gancitano, N. Dalessandro, and E. Aiello, "Effects of highly purified anthraquinoid compounds from *Aloe vera* on sensitive and multidrug resistant leukemia cells," *Oncol Rep*, vol. 4, pp. 341–343, 1997.

[93] S. Alam, I. Ali, K. Y. Giri, S. Gokkulakrishnan, S. S. Natu, M. Faisal, A. Agarwal, and H. Sharma, "Efficacy of Aloe vera gel as an adjuvant treatment of oral submucous fibrosis," *Oral Surg Oral Med Oral Pathol Oral Radiol*, vol. 116, no. 6, pp. 717–724, 2013, doi: 10.1016/j.oooo.2013.08.003.

[94] M. Sahebjamee *et al.*, "Comparative efficacy of *Aloe vera* and benzydamine mouthwashes on radiation-induced oral mucositis: A triple-blind, randomised, controlled clinical trial," *Oral Health Prev Dent*, vol. 13, no. 4, pp. 309–315, 2015, doi: 10.3290/j.ohpd.a33091.

[95] V. Nayak *et al.*, "Connecting you to content on EBSCOhost Title Hepatoprotective activity of *Aloe vera* Gel against Paracetamol Induced Hepatotoxicity in albino rats," 2024. [Online]. Available: https://openurl.ebsco.com/EPDB%3Agcd%3A12%3A28476444/detailv2?sid=ebsco%3Aplink%3Ascholar&id=ebsco%3Agcd%3A64935701&crl=c

[96] B. K. Chandan *et al.*, "Hepatoprotective potential of Aloe barbadensis Mill. against carbon tetrachloride induced hepatotoxicity," *J Ethnopharmacol*, vol. 111, pp. 560–566, 2007.
[97] B. N. Desai, R. H. Maharjan, and L. P. Nampoothiri, "Aloe barbadensis Mill. formulation restores lipid profile to normal in a letrozole-induced polycystic ovarian syndrome rat model," *Pharmacognosy Res*, vol. 4, no. 2, pp. 109–115, Apr. 2012, doi: 10.4103/0974-8490.94736.

5 Proteomics Research in Aloe

Vinod Goyal, Shweta Mehrotra, Monika, Pankaj, Pooja, Sri Devi Tallapragada, Vaibhav Baliyan, Deepika Verma, and Arthur Korte

5.1 INTRODUCTION

5.1.1 Overview and Importance of Proteomics Research in Aloe

Traditional knowledge about the therapeutic powers of has been transmitted across generations, forming the basis for traditional medicine practices globally. This practice has a long history, spanning thousands of years. In contrast, modern medicine is a more recent development, only around a century old (Babu *et al.*, 2021). Combining the old ways with modern approaches could offer more potent treatments than relying solely on either traditional or modern methods. Certainly, exploring the components in plants from traditional medicine has unveiled new molecules. Some of these have found their place in modern pharmacological formulations, and scientists have figured out how they work (Kumar *et al.*, 2024). *Aloe vera* thrives in arid regions and is primarily cultivated in India, with prominent growth areas including Rajasthan, Tamil Nadu, Andhra Pradesh, and Karnataka. *Aloe vera*, a succulent and resilient perennial plant, belongs to the Asphodelaceae family. It holds significance in traditional medicine, known as "ghrita kumari" in India and "luhui" in Chinese practices. Among the 420 Aloe species, *Aloe barbedensis* stands out as one of the most potent bioactive plants, boasting around 75 nutrients and 200 bioactive compounds (Kaur & Bains, 2023). The leaves of the *Aloe vera* plant are the primary focus for medicinal purposes, and they consist of three distinct parts: latex, gel, and rind (Figure 5.1). These parts contain a variety of phytochemicals with therapeutic properties. Latex and gel are key components of an *Aloe vera* leaf; they differ significantly in both chemical composition and therapeutic applications (Baruah *et al.*, 2016). Latex is sourced from pericyclic cells, whereas the gel, a clear liquid, originates from parenchyma cells. The gel is the innermost, transparent, mucilaginous pulp that contains parenchyma cells, serving as a reservoir for various phytochemicals (Rajeswari *et al.*, 2012). These include polysaccharides, amino acids, enzymes, glucomannans, lipids, sterols, vitamins, flavonoids, alkaloids, anthraquinones, anthrones, coumarins, chromones, and pyrones. *Aloe vera* gel comprises polysaccharides (55%), sugars (17%), minerals (16%), proteins (7%), lipids (4%), and phenolic substances (1%) on a dry matter basis. The latex of *Aloe vera* is the thick, yellow sap found beneath the rind, containing vascular bundles with a high concentration of phenolic compounds. This includes flavonoids, phenolic acids, anthraquinones, anthrones, coumarins, chromones, and pyrones (Jangra *et al.*, 2022).

In traditional Hindu medicine, *Aloe vera* is employed to address gastrointestinal and skin issues. Moreover, its anti-inflammatory properties and ability to promote skin regeneration make it a quick remedy for minor burns and wounds. *Aloe vera* is additionally utilized to alleviate coughs, manage diabetes, alleviate stomachaches, combat skin inflammation and infections, and act as an antihelminthic agent (Jadhav *et al.*, 2020). These uses highlight the diverse pharmacological activities associated with *Aloe vera* (Figure 5.2). *Aloe vera* gel is recognized for its therapeutic applications in

DOI: 10.1201/9781003603108-5

FIGURE 5.1 An overview of *Aloe vera* plant. Cross section of *Aloe vera* plant showing *Aloe vera* gel, latex and rind.

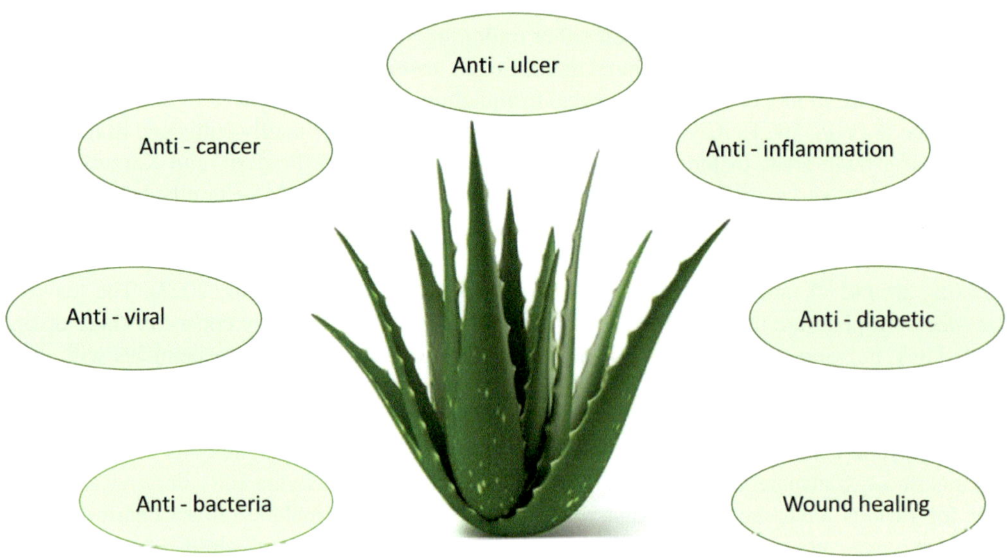

FIGURE 5.2 Medicinal benefits of *Aloe vera* plant.

wound healing, as it possesses the ability to penetrate and soothe tissue while inhibiting microbial growth and inflammation (Martínez-Burgos *et al.*, 2022). Owing to its myriad health benefits, aloe gel has also been incorporated into various food products such as beverages, milk, ice cream, and yogurt. The therapeutic properties of *Aloe vera* are predominantly attributed to its polysaccharides. Additionally, other components, including anthraquinones, anthrones, alkaloids, flavonoids, amino acids, enzymes, and vitamins, contribute to its healing properties (Ushasree *et al.*, 2024). These include anti-inflammatory, immunomodulatory, skin regenerative, antitumor, antimicrobial, antihelminthic, antidiabetic, antiulcer, and hepatoprotective effects, among others. *Aloe vera* has also been embraced as a "natural" component in the modern Food and Drug Industry. Due to its diverse effects, there has been a concerted effort to pinpoint the bioactive substances responsible for its properties (Kumar *et al.*, 2024).

A recent trend involves the exploration of plant-derived proteins, and *Aloe vera* has not been overlooked. Proteins are vital biological compounds composed of amino acids, serving crucial roles in various life processes such as structure, metabolism, transportation, immunity, signaling, and regulation. The term "proteome", introduced by Australian Ph.D student Marc Wilkins in 1994, encompasses all the proteins an organism can produce, with each species possessing its distinct proteome. Proteomics investigates how proteins interact, their functions, compositions, and structures, along with their cellular activities (Aslam *et al.*, 2016). Unlike genomics, which focuses on genes, proteomics offers a more in-depth understanding of the structure and function of an organism. However, it is more complex than genomics since protein expression changes based on time and environmental conditions (Burger & Chouchani, 2024).

Proteomics, the large-scale study of proteins, plays a pivotal role in unraveling the complex molecular landscape of *Aloe vera*, a succulent plant renowned for its medicinal properties. Understanding the proteome of Aloe is crucial for unlocking the mechanisms underlying its therapeutic effects, enhancing agricultural practices, and optimizing the utilization of Aloe in various industries.

5.1.2 Proteomic Techniques Applied to Aloe Research

a) Two-Dimensional Gel Electrophoresis (2D-GE): This technique separates proteins based on charge and size, providing a comprehensive view of the Aloe proteome. Researchers utilize 2D-GE to identify and quantify proteins, unraveling the dynamic changes in protein expression under different conditions.
b) Mass Spectrometry (MS): Mass spectrometry is an indispensable tool in Aloe proteomics. It enables the identification of proteins by analyzing their mass-to-charge ratios. Techniques such as liquid chromatography-mass spectrometry (LC-MS/MS) offer high sensitivity and accuracy, facilitating the identification of *Aloe vera* proteins.
c) Protein Profiling and Quantification: Proteomic studies involve profiling the abundance of proteins in different Aloe tissues or in response to environmental stimuli. Quantitative proteomics techniques, such as isobaric tags for relative and absolute quantitation (iTRAQ) or tandem mass tags (TMT), enable the comparison of protein expression levels under diverse conditions.

Proteomics is essential for *Aloe vera* research, providing valuable insights into how proteins function, interact, and are structured within the plant cells. This method helps researchers explore the various roles proteins play in the healing properties of *Aloe vera*, such as aiding in wound recovery, adjusting the immune response, fighting against microbes, and acting as antioxidants (Ramírez *et al.*, 2024). Using proteomic techniques, scientists can pinpoint and describe specific proteins such as lectins, glycoproteins, peptides, and enzymes found in *Aloe vera* extracts. This allows them to understand how the pharmacological activities of plants work and discover potential beneficial compounds. Moreover, proteomics helps compare the patterns of protein expression during different environmental conditions and growth stages of the *Aloe vera* plant. This analysis assists in understanding how external factors, such as stress or growth conditions, regulate protein expression. This knowledge is vital for improving *Aloe vera* cultivation and processing methods to maximize its therapeutic benefits (Santos *et al.*, 2021). Proteomics in aloe has large and varied importance, as mentioned below.

a) Identification of Therapeutic Proteins: Proteomics enables the identification and characterization of proteins responsible for these therapeutic benefits, including anti-inflammatory, antioxidant, and wound-healing effects. Understanding the precise proteins involved allows for targeted applications in pharmaceuticals and nutraceuticals.

b) Quality Control in Aloe Products: Proteomics research contributes to ensuring the quality and authenticity of Aloe-based products. By identifying specific protein markers, researchers can establish quality control measures to verify the presence of authentic *Aloe vera* in products, preventing adulteration and ensuring consumer safety.
c) Insights into Stress Responses: Proteomic studies shed light on the response of *Aloe vera* to environmental stresses such as drought, salinity, or pathogens, for developing stress-tolerant Aloe varieties and optimizing cultivation practices for enhanced yield and quality.
d) Agricultural Applications: Proteomics research aids in understanding the molecular mechanisms involved in the growth of *Aloe vera*, development, and response to agronomic conditions. This knowledge can inform agricultural practices, leading to improved cultivation techniques and increased crop resilience.
e) Optimizing Processing and Storage: Proteomics is instrumental in optimizing the processing and storage of Aloe-based products. By studying changes in protein profiles during processing or storage, researchers can enhance product stability and shelf life, ensuring that consumers receive the maximum benefits from Aloe-derived products.
f) Unraveling Aloe-Beneficial Compounds: Proteomics complements other omics approaches, such as genomics and metabolomics, in unraveling the complex interactions between genes, proteins, and metabolites in *Aloe vera*. This systems biology approach provides a holistic understanding of the biology of the plant and its potential applications.
g) Facilitating Biotechnological Interventions: Proteomic insights provide a foundation for biotechnological interventions in *Aloe vera*. Genetic modification and molecular breeding strategies guided by proteomic data can lead to the development of Aloe varieties with enhanced therapeutic properties, increased yield, or improved stress tolerance.

Recently, researchers exploring *Aloe vera* made noteworthy discoveries. They found three interesting lectins: a 35-kDa lectin with lymphoproliferative activity (Koike *et al.*, 1995) and two immunomodulatory lectins, aloptine A (7.5-kDa) and aloptine B (24-kDa) (Im *et al.*, 2010). Additionally, three glycoproteins were identified: aloectine A, recognized for its anti-tumoral and antiulcer effects, a 29-kDa substance that boosts the growth of human dermal cells (Das *et al.*, 2011), and G1G1M1DI2, a 5.5-kDa molecule that stimulates mitosis and cellular migration (Choi *et al.*, 2001). Moreover, they discovered an octapeptide with anti-tumoral and antioxidant properties (El-Shemy *et al.*, 2010), and a 14-kDa protein displaying antifungal activity against *Candida paraprilosis, C. krusei, and C. albicans* (Cabello-Ruiz *et al.*, 2019). A previous study also emphasized the antimicrobial activity of the protein fraction from lyophilized *Aloe vera* gel against *C. albicans* and *Staphylococcus aureus* (Ruiz *et al.*, 2015). The components in the protein extract were identified using quantitative time-of-flight mass spectrometry (qTOF-MS) combined with bidimensional nanochromatography. In conclusion, proteomics research in *Aloe vera* is instrumental in unraveling the molecular intricacies of Aloe. From identifying therapeutic proteins to optimizing agricultural practices and ensuring product quality, proteomics plays a crucial role in harnessing the full potential of *Aloe vera* across various industries. As technology advances, proteomics will continue to play a central role in shaping the future applications and understanding of *Aloe vera* in medicine, agriculture, and beyond.

5.2 FUNDAMENTALS OF PROTEOMICS

Proteomics is the study of the entire complement of proteins present in a biological system, such as cells, tissues, or organisms, at a given moment, including their structure, functions, interactions, modifications, connections, and occurrences in biological systems. It delves into the intricate world of proteins, which serve as essential components in cells. Proteins are crucial for all living beings, performing various tasks such as DNA synthesis and replication. The genome acts as a blueprint, containing the genetic information of an organism and guiding RNA transcription in the transcriptome. This transcribed information then directs protein synthesis, forming the proteome. The metabolome, consisting of small molecules such as carbohydrates, lipids, and amino acids, is influenced by the dynamic interplay among the genome, transcriptome, and proteome, shaping an organism's functional landscape and reflecting its physiological state (Figure 5.3). The entire collection of proteins expressed by a genome, tissue, cell, or organism at a given time is known as the proteome.

In contrast to the relatively static nature of genes, proteins undergo continuous synthesis, modification, and degradation in response to cellular demands and external stimuli. Thus, compared to analyzing the genome alone, studying the proteome offers a more comprehensive view of what is happening inside a cell. The study of the proteome is known as proteomics. A variety of techniques are used to study proteins under proteomics. There are three primary categories into which proteomics can be divided:

- Large-scale identification of proteins and their post-translational changes using protein micro-characterization
- "Differential display" proteomics, which differentiates the levels of proteins and has potential applications in the medical field.
- Research on interactions among proteins using techniques such as mass spectrometry and electrophoresis.

The fundamentals of proteomics involve several key steps:

a) Sample Preparation: This initial step involves extracting proteins from the biological sample, which, in the case of *Aloe vera*, could be various tissues or extracts. Proper sample preparation is crucial to preserving the integrity of proteins for downstream analyses.
b) Protein Separation: Proteins are often complex mixtures, and separation techniques are employed to isolate individual proteins. Gel-based techniques, including two-dimensional gel electrophoresis (2D-GE) or liquid-based techniques such as liquid chromatography (LC), are commonly used for this purpose.

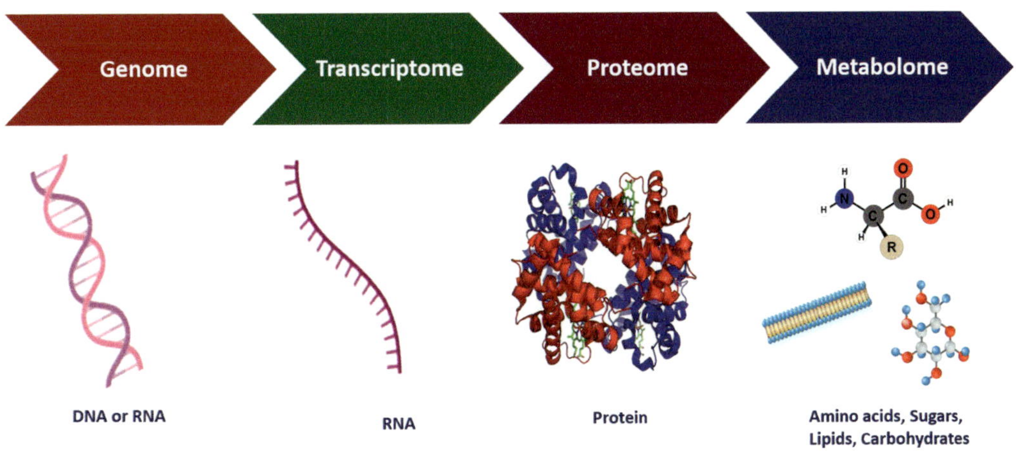

FIGURE 5.3 General diagram illustrating the connections between the transcriptome, proteome and metabolome.

c) Mass Spectrometry (MS): Mass spectrometry is a central tool in proteomics for protein identification and characterization. It involves ionizing proteins and measuring the mass-to-charge ratio of resulting ions. High-resolution mass spectrometers, such as those used in Liquid Chromatography LC-MS/MS and Tandem MS/MS, provide accurate identification of peptides and proteins.
d) Database Searching: Identified peptides are compared to protein databases using bioinformatics tools. This process involves matching experimental mass spectra with theoretical spectra derived from known protein sequences. Database searching enables the identification of proteins present in the sample.
e) Post-Translational Modification (PTM) Analysis: Proteomics allows the identification and characterization of post-translational modifications in Aloe proteins. PTM analysis provides insights into the functional diversity of proteins and their regulatory roles.
f) Quantitative Proteomics: Techniques such as isobaric tags for relative and absolute quantitation (iTRAQ) enable the quantification of protein abundance in Aloe. Quantitative proteomics provides valuable information on changes in protein expression under different conditions.

In conclusion, the fundamentals of proteomics in Aloe involve a systematic approach to identify and characterize proteins. From sample preparation to mass spectrometry analysis and bioinformatics, each step contributes to unraveling the complexities of the protein landscape of Aloe to enhance our understanding of the therapeutic properties of *Aloe vera* and opens avenues for its diverse applications in medicine, agriculture, and cosmetics.

5.2.1 Protein Identification and Characterization

Gene expression and activity are governed by proteins. Proteomics will significantly advance our understanding of how genes function in the post-genomic era. Since proteins contribute to many of the structural components of cells and aid in keeping them together and structured, their characterization is crucial. A large number of hormones are proteins. The technique of examining a single protein by means of separation and detection is known as protein characterization (Walport *et al.*, 2021). The distinguishing features of the structure and function of protein, such as its molecular weight, purity, activity, and so forth, aid in identifying the special protein. Characterizing proteins is important for two main areas of structural proteomics. The first step involves assessing the quality of the protein following its production. The ability of crystals to diffract light well is one of the primary barriers to understanding their structure. The two most popular techniques for identifying proteins are mass spectrometry and Edman Degradation (ED). ED, the more traditional technique, involves gradually removing N-terminal amino acids using chemical techniques. Peptide sequencing takes many hours, and the sensitivity ranges from two to five pmol for pure peptides (Deutzmann, 2004). It is the identification of protein characteristics, including domains, oligomeric state, post-translational modifications, and interactions with ligands. MS is an automated technique that is incredibly quick. It is the preferred technique for precise analysis and applications on a large scale (proteomics) (Yates *et al.*, 2009).

Proteomics is an effective method for identifying the functions of the proteins in *Aloe vera*. This includes the separation of proteins into smaller chains of amino acids called peptides, followed by their analysis using methods such as mass spectrometry (Simões *et al.*, 2012). Usually, integrated techniques such as one- or two-dimensional gel electrophoresis, liquid chromatography, and mass spectrometry are employed. Scientists can identify the original proteins by comparing the peptide sequences to databases. Proteomics is also capable of measuring the concentration of every protein. This makes it easier to comprehend how the relative abundance of certain proteins may vary depending on the situation, such as growth stage or stress. Compared to more conventional focused approaches that concentrate on certain proteins, proteomics offers a more objective and thorough understanding of the *Aloe vera* proteome (Cabello Ruiz *et al.*, 2015). It can identify post-translational modifications (PTMs), which have the potential to drastically change the function of proteins. These changes, for example, may involve glycosylation or phosphorylation. Through the use of proteomics on *Aloe vera*, scientists can uncover novel therapeutic proteins, comprehend how environmental conditions impact protein expression, and obtain important insights into the biological processes of the plant. Protein assays must be continually improved in terms of accuracy, sensitivity, and durability in order to determine the safety as well as effectiveness of biological products. Additionally, novel assays targeting specific issues must be developed.

5.3 METHODS IN PROTEOMICS ANALYSIS

Proteomics is a rapidly expanding field that involves the study of proteins and their functions in biological systems (Kumar & Mann, 2009). A variety of methods are used to accurately analyze and interpret the vast amount of proteomic data. Some of the key methods used are spectrometry techniques such as LC-MS, MALDI-MS, and MS/MS (Figure 5.4).

Proteomics in *Aloe vera* involves sophisticated techniques for protein identification and characterization to gain insights into the complex protein composition of Aloe, and contributes to a deeper understanding of its therapeutic properties and potential applications in various industries. The methods involved are mentioned below.

a) Two-Dimensional Gel Electrophoresis (2D-GE): In Aloe proteomics, 2D-GE is often employed for protein separation. This technique separates proteins based on their isoelectric point (pI) and molecular weight. Spots on the gel represent individual proteins, and these can be excised for further analysis.
b) Liquid Chromatography-Mass Spectrometry (LC-MS/MS): LC-MS/MS is a powerful method for protein identification in *Aloe vera*. After enzymatic digestion of proteins into peptides, liquid chromatography separates these peptides before they enter the mass spectrometer. The mass spectrometer then identifies and quantifies the peptides, which are mapped back to the original proteins.
c) Tandem Mass Spectrometry (MS/MS): In MS/MS, peptides are selected from the initial mass spectrometry analysis and fragmented. The resulting spectra provide sequence information, aiding in the identification of the peptides and the corresponding proteins.
d) De Novo Sequencing: In cases where a complete protein sequence is not available in databases, de novo sequencing is employed. This approach involves reconstructing the amino acid sequence directly from the mass spectra. It is particularly useful for identifying novel or less-characterized proteins in *Aloe vera*.
e) Post-Translational Modification (PTM) Analysis: Proteomics allows the characterization of post-translational modifications in Aloe proteins. PTMs play a crucial role in protein function, and identifying them provides insights into the regulatory mechanisms within Aloe cells.
f) Quantitative Proteomics: Beyond identification, proteomics also facilitates the quantification of protein abundance. Isobaric tags for relative and absolute quantitation (iTRAQ) and tandem mass tags (TMT) are commonly used techniques for quantitative proteomics in *Aloe vera*.

These techniques allow for the identification and quantification of proteins, as well as the analysis of protein interactions and modifications (Bennike *et al.*, 2014). Proteomics, the large-scale study of proteins, provides insights into the structure, function, and regulation of proteins in biological systems. The field has evolved to address various biological questions, including the effects of environmental changes on protein expression, the identification of protein modifications in response to stress, and the elucidation of cellular pathways associated with disease states. A study conducted on the carbon assimilation in *Eucalyptus urophylla* grown under high atmospheric CO_2 concentrations

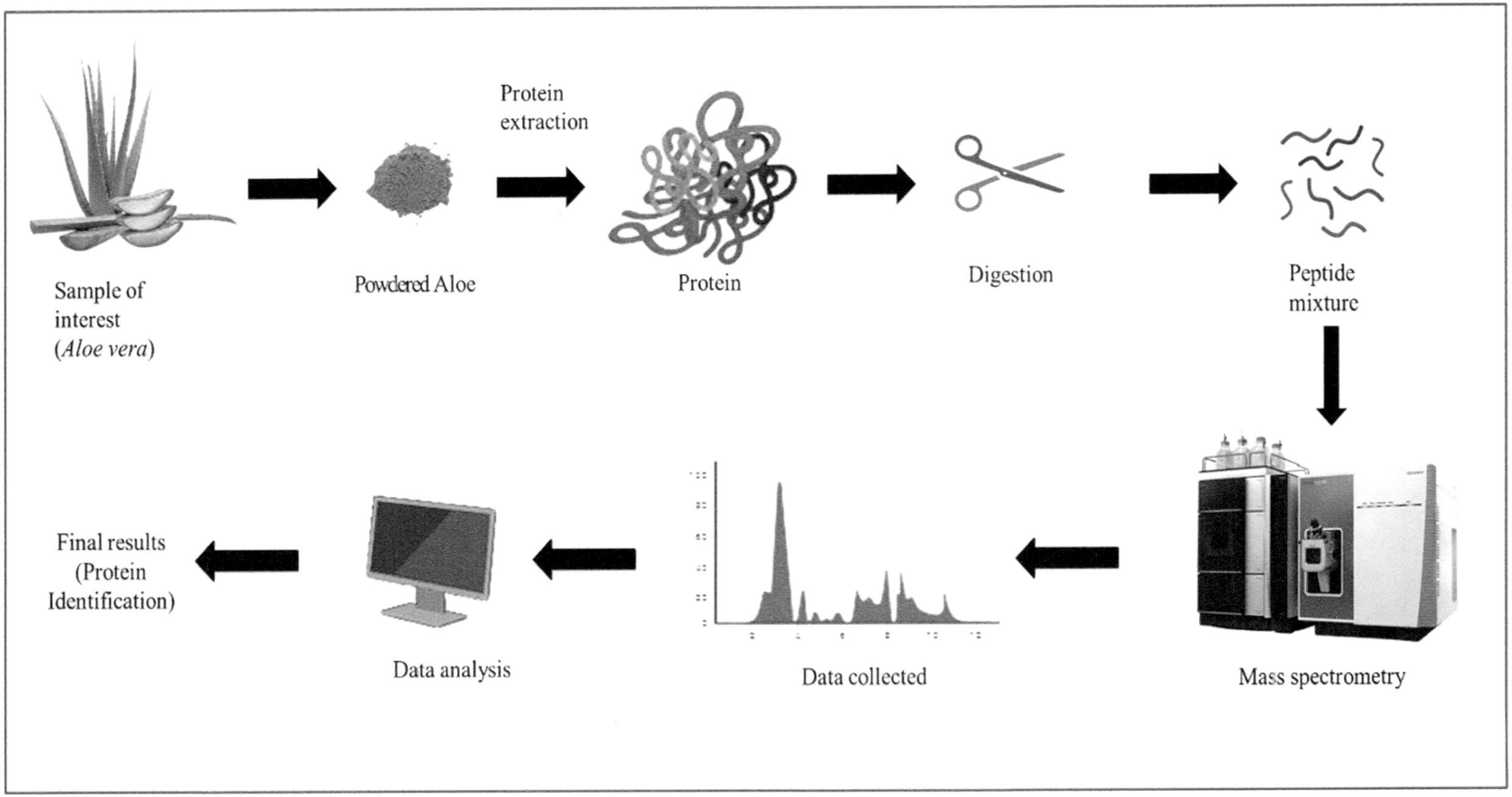

FIGURE 5.4 Steps of proteomic analysis using mass-spectrometry.

highlighted the impact of environmental conditions on plant proteomes (Santos & Balbuena, 2017). Similarly, another research on *Aspergillus fumigatus* revealed oxidative modifications of proteins by increased levels of intracellular reactive oxygen species during hypoxia adaptation (Shekhova *et al.*, 2019). This demonstrates the role of proteomics in understanding stress responses and environmental changes at the molecular level.

Quantitative proteomics allows system-wide protein identification and quantification, and allows for both targeted and non-targeted proteomics applications. It is a robust method utilized for both exploratory and focused proteomic investigations to discern the overall proteomic changes in a cell, tissue, or organism. The majority of quantitative proteomic studies involve protein or peptide labeling with isotopes, allowing for their differentiation through mass spectrometry. Other relative quantitation techniques (such as SILAC, ICAT, ICPL, and isobaric tags) are also being employed to compare the abundance of proteins or peptides across samples. Known concentrations of isotopically labeled synthetic peptides are added to unlabeled samples for absolute quantitation of target peptides using selected reaction monitoring (SRM). Label-free approaches are also accessible for both relative and absolute quantitation as a cheaper alternative. Despite its complexity, quantitative proteomics plays a crucial role in enhancing our comprehension of global protein expression and modifications that underlie the molecular mechanisms of biological processes and disease conditions. The subsequent explanation will provide a comprehensive description of the methodologies previously addressed.

5.3.1 Liquid Chromatography-Mass Spectrometry (LC-MS)

LC-MS is a powerful technique widely used in proteomics research, allowing researchers to perform comprehensive proteome profiling and the detection of post-translational modifications. It also provides deep and highly sensitive proteome coverage without the need for prefractionation (Thakur *et al.*, 2011). LC-MS involves enzymatic digestion of proteins into peptides followed by liquid chromatography tandem mass spectrometry, which helps overcome complexity and dynamic range challenges. Additionally, nano-flow LC-MS/MS is commonly used in proteome research due to its excellent sensitivity, although it may sacrifice robustness (Bian *et al.*, 2020). New approaches are also being developed to reduce analyte losses during sample preparation, aligning biological starting material with the detection limits of LC-MS analysis (Cong *et al.*, 2020). LC-MS-based proteomics allows for the analysis of thousands of proteomes in a robust, reproducible, and quantitative manner (Bian *et al.*, 2020). It is also used for analyzing protein/peptide mixtures in shotgun proteomics. It has been pivotal as it allows the study of a large number of peptide elements in samples in an automated and high-throughput manner (Mueller *et al.*, 2008).

In conclusion, LC-MS plays a crucial role in proteomics research by enabling deep proteome coverage, high sensitivity, and robust quantitative analysis of proteins and peptides. Researchers continue to refine and optimize LC-MS methodologies to enhance throughput, sensitivity, and accuracy in proteomic studies.

5.3.2 Matrix-Assisted Laser Sesorption/Ionization (MALDI-MS)

Matrix-Assisted Laser Desorption/Ionization (MALDI) mass spectrometry (MS) provides powerful capabilities for structural identification, protein, and peptide analysis. This technique involves a laser striking a matrix of small molecules to induce the molecules of the analyte to go into the gaseous phase without fragmentation. MALDI in conjunction with MS analysis has been employed to generate complementary data sets to liquid chromatography-electrospray ionization (LC-ESI) MS/MS (Bodnar *et al.*, 2003).

Nano-based techniques such as nano-MALDI-MS have been projected as platforms with the potential to transition MALDI-MS from research labs to clinical applications, highlighting the evolving landscape of MALDI technology (Gopal *et al.*, 2015). The integration of MALDI-MS with other techniques, such as liquid chromatography (LC) and microfluidics, has enabled enhanced proteome coverage and sample preparation efficiency (Maus *et al.*, 2018). Furthermore, MALDI-TOF MS has been instrumental in identifying dysregulated proteins in breast cancer and other cancer subtypes, highlighting its role in biomarker discovery and cancer research (Neagu *et al.*, 2023).

5.3.3 Tandem Mass Spectrometry (MS/MS)

Tandem Mass Spectrometry (MS/MS) is a crucial tool in plant-based proteomics research, allowing for in-depth analyses of plant proteomes and their responses to differential conditions. In this method, ions of interest are quadrupole filtered on the basis of their *m/z* and are fragmented by several dissociation methods such as collision-induced dissociation (CID), electron-transfer dissociation (ETD), and electron-capture dissociation (ECD). Separation of these fragments occurs on the basis *m/z* ratios in the second round of MS. It is generally used to sequence proteins and oligonucleotides due to the ability of the fragments to predict sequences of peptides or nucleotides.

MS-based proteomics has significantly improved the precision and scope of plant studies (Chen *et al.*, 2021). Large-scale proteome quantifications using spectral counting allows researchers to explore the intricate details of plant proteomes (Majeran *et al.*, 2012). In plant proteomics, tandem mass spectrometry is commonly used alongside liquid chromatography (LC) for protein identification and quantification (Wang *et al.*, 2018). This approach has been instrumental in studying plant responses to biotic stress, exploring the diversity of plant proteomes, and investigating the effects of nanoparticles on crop plants (Liu *et al.*, 2019). By leveraging MS/MS, researchers can accurately identify low-abundance proteins, protein-protein interactions, and cellular signaling networks (Hossain *et al.*, 2020). Its application in plant proteomics has been essential for studying secondary plant defense mechanisms, metabolic adaptations, and signal transduction pathways in response to environmental stresses, such as drought (Simova-Stoilova *et al.*, 2015).

5.3.4 Quantitative Proteomics

Quantitative proteomics is another important method in proteomics analysis. It aims to measure changes in protein abundance across different conditions. In Aloe, this can provide insights into its response to environmental stress, growth stages, or other factors. Quantitative proteomics involves comparing protein expression levels between different samples to determine relative quantitation or absolute quantitation. This can be achieved through various strategies, such as label-based quantification using stable isotope labeling or label-free quantification based on mass spectrometry of intact proteolytic peptides (Bantscheff *et al.*, 2007). Two-dimensional gel electrophoresis, two-hybrid analysis, and protein microarrays cannot analyze a large number of proteins. It does not provide the depth of informative proteome analysis as compared to mass spectrometry. Mass spectrometry techniques in proteomics can be used as high-throughput discovery-based methods to identify and analyze a wide range of proteins in a sample. Furthermore, the integration of bioinformatics and computational tools has become indispensable for the analysis of large-scale proteomic data. These tools aid in the interpretation of complex data sets, the identification of protein pathways and the prediction of protein structures and functions.

5.3.5 Relative and Absolute Quantitation

Quantitative proteomic analysis involves comparative methodologies to assess peptide levels within samples. Relative quantitation involves the separate analysis of samples via mass spectrometry and subsequent comparison of their spectra to determine peptide abundance relative to each other. This methodology is characteristic of label-free quantitation strategies (Nyalwidhe *et al.*, 2017). By contrast, absolute proteomic quantitation utilizes isotopic peptides for precise quantification.

This method involves introducing known concentrations of synthetic, heavy isotopologues of target peptides into an experimental sample, followed by liquid chromatography-tandem mass spectrometry (LC-MS/MS) analysis. Unlike relative quantitation, where peptide abundance is compared directly, absolute quantitation involves comparing the abundance of the target peptide in the experimental sample to that of the heavy peptide. This information is then used to back-calculate the initial concentration of the standard through referencing a predetermined standard curve, resulting in the absolute quantitation of the target peptide (Ankney *et al.*, 2018).

5.3.6 Label- free and Label-based Quantitative Proteomics

Label-based quantitative proteomics encompasses various techniques, i.e., metabolic labeling, isobaric tags, and isotopic tags, which play a crucial role in protein quantification. Stable Isotope Labeling by Amino Acids in Cell Culture (SILAC) is a metabolic labeling method that involves integrating stable isotopes into proteins during cellular growth for accurate protein expression level measurement (Fabrik *et al.*, 2014). Isobaric chemical labeling, utilizing tags such as Isobaric Tags for Relative and Absolute Quantitation (iTRAQ) or Tandem Mass Tags (TMT), has also gained popularity due to its ability to analyze multiple samples simultaneously (Bąchor *et al.*, 2019). iTRAQ allows for the simultaneous quantification of proteins from multiple samples. Proteins are labeled with isobaric tags, and differences in reporter ion intensities in the mass spectrometer provide quantitative information. TMT is a labeling technique for multiplexed protein quantification. Proteins from different samples are labeled with distinct isotopic tags, enabling simultaneous analysis and quantification. Isotopic tags such as (15) N metabolic labeling have proven instrumental in identifying disease-related changes and has significance in disease research and biomarker discovery (Filiou *et al.*, 2015). These label-based approaches offer high accuracy and precision in quantification, which are essential for comparative proteomic studies (Filiou *et al.*, 2012). The use of stable isotopes in metabolic labeling not only ensures precise quantification but also eliminates random errors that can arise from chemical labeling methods (Jiang *et al.*, 2023). In summary, label-based quantitative proteomics methods are indispensable for advancing our understanding of complex biological systems by enabling precise and comprehensive protein quantification. These techniques provide researchers with powerful tools to investigate protein expression dynamics, identify biomarkers, and unravel the mechanisms underlying diseases, making them vital in modern proteomic research.

Label-free quantitative proteomics has become a powerful and versatile tool in the field of proteomics. Label-free methods directly compare peptide or protein abundances based on spectral counting or chromatographic peak intensities without the use of chemical tags. This approach is advantageous for its simplicity and reduced sample manipulation. This approach provides a cost-effective alternative to labeled quantitation methods, offering reliable quantitative information without the need for expensive labeling reagents (Neilson *et al.*, 2011). Label-free samples are separately collected, prepared, and analyzed by LC-MS/MS or LC-MS, unlike contrasting quantitation techniques. With advancements in mass spectrometry, label-free quantitation has gained recognition due to enhanced accuracy and reliability in peptide mapping across samples (Li *et al.*, 2012). The simplicity of sample preparation in label-free techniques has contributed to their growing interest in quantitative proteomics applications (Woo *et al.*, 2020). This method has been crucial in biomarker identification, disease etiology studies, and understanding complex biological systems (Nahnsen *et al.*, 2013).

The Protein Abundance Index (PAI) is a metric that quantifies protein abundance in large-scale proteomic analyses. It is calculated using the number of peptides detected and the number of theoretically observed tryptic peptides for each protein. The exponentially modified PAI (emPAI) is an enhanced version of PAI that takes into account the exponential relationship between peptide detection and protein abundance. This modification allows for a more accurate estimation of protein abundance. Another metric, Absolute Protein Expression (APEX), utilizes spectral counts and correction factors to ensure that protein abundance is directly proportional to the number of peptides observed. APEX provides a reliable measure of absolute protein abundance in proteomic studies.

In conclusion, label-free quantitative proteomics is a valuable approach in modern proteomics research, offering researchers a reliable, cost-effective, and versatile method for quantitative analysis across diverse biological samples and research areas. Label-free quantitation is excellent for measuring immense changes in protein expression, but its efficiency is limited in the case of minute changes in protein expression. Quantitative proteomics in Aloe provides a dynamic understanding of the response of the plant to various conditions and aids in identifying key proteins involved in therapeutic properties, stress tolerance, or developmental processes. The integration of mass spectrometry techniques and quantitative proteomics enhances our ability to explore the intricate protein landscape of *Aloe vera*, contributing to advancements in agriculture, medicine, and cosmetic applications.

5.4 APPLICATIONS OF PROTEOMICS

Proteomics holds immense potential across various fields, contributing to advancements in biomedical research, agriculture, and biotechnology. In medicine, proteomics plays a crucial role in understanding diseases, identifying potential biomarkers, and developing targeted therapies. It enables researchers to analyze the entire protein complement of cells, tissues, or organisms, providing insights into complex biological processes. In agriculture, proteomics aids in crop improvement by studying plant proteins involved in growth, stress response, and disease resistance. This knowledge can be leveraged to develop crops with enhanced yields, nutritional content, and resilience to environmental challenges. Furthermore, proteomics contributes to food safety by detecting and monitoring contaminants in food products. Overall, proteomics is a versatile and indispensable tool that continues to revolutionize our understanding of biological systems, paving the way for innovative applications and advancements in diverse scientific disciplines. Various applications of proteomics are explained below:

5.4.1 Agriculture/Crop Improvement

Globally, approximately 925 million people are currently experiencing undernutrition, and this number is expected to reach two billion by 2050, as per United Nations projections. Despite being a core objective in the millennium goals, malnutrition remains widespread, necessitating an increase in food production to support the rapidly growing population. Unfortunately, the amount of cultivable land is decreasing, and soil health is deteriorating. These challenges, coupled with the impacts of climate change, are causing declines in food production (Mustafa & Komatsu, 2021). Proteomics has emerged as a powerful tool in identifying a diverse array of proteins crucial for sustaining crop health, enhancing resistance to stressors, and elevating product quality. By characterizing specific genes, researchers gain the capability to track how crops respond to adverse conditions and can even predict the nutritional content of crops (Moreno-Hernández *et al.*, 2020). This involves observing the expression levels of targeted genes, which may vary under specific conditions, and these variations can now be precisely identified. The progressive accumulation of data from proteomic analyses enables a deeper comprehension of the genes and pathways responsible for stress response and tolerance in crops (Yan *et al.*, 2022). Furthermore, advancements in proteomic analysis techniques have occurred over time, with the introduction of new and improved tools refining the quality of protein extraction, separation, quantification, and identification processes. Notably, innovations in high-throughput analysis have enabled the collection of larger datasets simultaneously, minimizing errors in measuring protein levels (Yang *et al.*, 2021).

5.4.2 Proteomics and Stress Responses

Stress poses a significant challenge to the growth and productivity of agricultural crops. Difficult conditions can result in delayed seed germination, hindered plant growth, and reduced crop yield. Proteins that play a crucial role in an organ's main function tend to gather specifically in that organ, tissue, or organelle (Jan *et al.*, 2023). Proteomics in Aloe contributes to understanding the response of the plant to environmental stressors. Identification of stress-responsive proteins enables the development of Aloe varieties with enhanced tolerance to conditions such as drought, salinity, or pathogens, thereby improving crop resilience. Komatsu & Hossain (2013) emphasized the importance of conducting organ-specific proteomic analyses to pinpoint proteins commonly accumulated in organs facing various environmental stresses. The proteomics approach is utilized to study how plants respond to stresses and the intricate biochemical processes involved (Gong *et al.*, 2015). Plant stress proteomics can identify potential candidate genes for enhancing plant resistance to stresses (Barkla, 2016). Various signaling pathways are activated in response to stresses, creating a complex regulatory network involving transcription factors, ion balance, antioxidants, hormones, kinase cascades, reactive oxygen species (ROS), and osmolyte production (Yin *et al.*, 2015). However, plant cells respond differently to abiotic stresses in different organs. Analyzing proteomics at the organ level, along with studying subcellular organelles from leaves to roots, can offer more detailed insights into cellular mechanisms that regulate stress responses and signal transmission in various organelles, potentially enhancing crop stress tolerance (Haq *et al.*, 2024). Proteomic analysis aids in identifying proteins associated with the response to pests and diseases of Aloe. This information is crucial for developing strategies to enhance resistance and reduce the dependence on chemical interventions in agriculture: Quantitative proteomics helps elucidate the proteins involved in nutrient uptake and utilization in *Aloe vera*. This knowledge can inform agricultural practices to optimize nutrient management, resulting in enhanced crop yield and quality.

5.4.3 Proteomics-based Fungicides

This potential hinges on the assumption that most drug targets are proteins, and proteomics can identify candidate proteins participating in specific biological processes. The scientific research community is actively incorporating various alterations in the design of chemical fungicides, leveraging compiled genomic and proteomic data (Marghany *et al.*, 2024). Additionally, bioinformatics plays a role in predicting proteins with fungicidal properties. The development of biosynthetic fungicides, with a focus on disease-associated targets, has emerged as a novel direction in fungicide design. In-depth studies on the proteins and genes involved in fungal infections are crucial for certain projects (Cai *et al.*, 2019). One way to control plant diseases caused by fungi is by using modified natural compounds. These compounds have the potential to specifically target and inhibit the proteins responsible for the infection cycle. If these compounds are biodegradable and highly specific, they can minimize their impact on the environment. Additionally, they have the advantage of not easily entering the food chain. Gathering information about the proteins of fungal plant pathogens through proteomic studies could motivate the development of new and eco-friendly fungicides (de Ávila *et al.*, 2023).

5.4.4 Role of Proteomics in Plant Breeding

Proteomics provides plant biologists and breeders with new insights into gene identification through DNA analysis. In plant breeding, Marker-Assisted Selection (MAS) is a widely employed technique that utilizes DNA markers to choose specific genes or quantitative trait loci (QTLs) (Das *et al.*, 2015). This approach is extensively used to enhance the selection of desired genes and improve the development of superior breeding lines. In crop breeding, the goal is to improve overall crop quality, reduce yield loss, enhance disease resistance, and protect crops from various biotic and abiotic stresses (Ghatak *et al.*, 2017). The use of proteomics in evaluating crop plants, especially in crop breeding, has significantly expanded over the last decade. Proteomics offers advantages over other "omics" methods because it provides a holistic view of cellular proteins, contributing to a more precise assessment of crop plants. Recent advances in proteomic technologies have expanded our understanding of plant responses to abiotic stresses at the genetic and molecular levels. However, the primary focus of this review is to provide a critical assessment of recent approaches that offer potential for improving plant tolerance to various abiotic stresses while minimizing crop losses (Vanderschuren *et al.*, 2013).

5.4.5 Medicine

Proteomics plays a crucial role in advancing drug development by offering a comprehensive map of protein interactions linked to disease pathways. In recent years, the use of MS in proteomics research has grown across various scientific fields (Doll *et al.*, 2019). Drug discovery is a complicated and expensive process, but emerging technologies, such as proteomics, can make it easier and faster. It is believed that discovering a new drug can cost around $70 million. Different MS-based platforms, such as clinical, functional, and chemical proteomics, play a crucial role in today's drug discovery (Jain, 2016). These proteomic approaches go beyond simply identifying drug targets in the complex drug discovery process. They also allow researchers to investigate aspects such as how a drug interacts with its target (selectivity and specificity), its effectiveness, resistance, and toxicity, and to uncover the mechanisms behind how a new drug works. Clinical proteomics helps uncover and comprehend the roles of proteins in different health problems (Jain, 2016). This aids in speeding up the early detection, prognosis, and finding new targets for treatment. Using proteomics tools in medicine can enhance our understanding of how diseases develop and help identify potential new targets for drug development. Proteomic studies contribute to the identification of bioactive compounds in Aloe that may have pharmaceutical potential. This knowledge aids in drug discovery efforts, potentially leading to the development of novel medications. The use of proteomics is gaining popularity in cancer research. Proteomics technologies help identify potential biomarkers and patterns of protein expression. These findings are valuable for assessing tumor prognosis, predicting outcomes, classifying tumors, and identifying potential responders to specific therapies. Proteomics is also instrumental in understanding the underlying biology of cancer by revealing changes in signaling pathways within tumor cells. This knowledge facilitates the development of targeted therapies for various pathways involved in cancer. As a result, extensive cancer proteome databases have been established, integrating vast datasets with molecular biology and clinical information to advance our understanding of cancer (Amrani *et al.*, 2021). Proteomics facilitates the identification of proteins that vary across different Aloe varieties or in response to environmental factors, thus supporting personalized medicine approaches, tailoring treatments based on the specific therapeutic protein profiles of *Aloe vera* extracts. With the rapid expansion of proteomics technology and an overwhelming increase in scientific articles utilizing proteomics approaches, it would not be surprising to see proteomics applications in various fields of investigation. These could include the food industry, toxicology, development, neuroscience, and apoptosis. The versatile nature of proteomics is contributing to its widespread use and potential impact across diverse scientific domains (Su *et al.*, 2021).

Multiple proteomic studies have demonstrated the positive effects of *Aloe vera* on various issues. Recently, *Aloe vera* has been found to influence proteins associated with mitochondrial function, the vascular system, and the glycolysis/pentose pathway in diabetes. These findings offer additional insights and enhance our understanding of the potential molecular mechanisms by which *Aloe vera* could contribute to the development of treatment and preventive strategies for kidney damage associated with diabetes mellitus (DM) (Santos *et al.*, 2021) .

5.4.6 Cosmetics

Twenty years ago, scientists did not anticipate the effectiveness of synthetic or natural peptides in cosmetics. Through proteomics studies, researchers identified specific peptide sequences targeting vital skin layers. Since 2000, the use of peptides in beauty products has grown, underscoring the need for a deep understanding of these molecules and their interactions with the skin (Newton *et al.*, 2019). Proteomics studies highlight the value of peptides, particularly those with anti-aging, antioxidant, and skin brightening effects, in cosmetics. Peptides can now be modified for enhanced performance, such as improved solubility and skin penetration. (He & Jia, 2022). Proteomics in Aloe contributes to understanding the proteins responsible for skin-enhancing properties. Bioactive proteins identified through proteomic analysis guide the formulation of cosmetic products aimed at promoting skin health, hydration, and anti-aging effects. Proteomic insights into *Aloe vera* proteins benefit the development of hair care products. Understanding the proteins involved in hair nourishment and strength allows for the creation of formulations that address specific hair care needs. Questions persist about improving compound penetration into the skin, presenting opportunities for novel peptide discoveries through proteomics research in cosmetics. The extensive in vitro studies on skin protection and wound healing involving *Aloe vera* and its active compounds, such as aloesin, aloin, and emodin, have significant implications for cosmetics. These investigations often use cell lines such as HaCaT, HEKa, and various fibroblast lines to assess the protective effects of *Aloe vera* on the skin. The observed enhancement of TGFβ1, bFGF, and Vegf-A expression in fibroblasts, as well as the promotion of keratinocyte proliferation and differentiation through lysosomal membrane stabilization, highlights the potential cosmetic benefits of *Aloe vera* (Dey & Dubey, 2023). These molecular processes contribute to skin protection, emphasizing the antioxidant and anti-inflammatory mechanisms involved. However, conflicting findings, particularly regarding acemannan's (beta-1,4-polymannose present in the gel of the *Aloe vera* plant) wound-healing properties, suggest a need for a more nuanced understanding (Ma *et al.*, 2020). Further exploration, especially through the application of proteomics, can unravel the protein-level changes and interactions that underlie the complexities of the impact of *Aloe vera* on skin repair. This deeper insight into the molecular mechanisms involved could inform the development of cosmetic products designed to optimize skin protection and wound healing. Clinical and experimental evidence has indicated the potential benefits of *Aloe vera* in wound care and healing, particularly for skin moisturization, promoting wound healing, managing thermal skin injuries, and addressing inflammatory skin conditions. As a result, there is significant interest in conducting in vitro and in vivo trials using *Aloe vera* gel as a scaffold or support matrix for tissue engineering (Epstein, 2005). These studies explore the application of *Aloe vera* gel either independently or in conjunction with conventional medications or natural extracts to enhance wound care and tissue regeneration. Proteomic techniques are employed for quality control in Aloe-based cosmetics. By identifying and quantifying specific protein markers, manufacturers can ensure the authenticity and effectiveness of *Aloe vera* extracts used in cosmetic formulations.

It follows that proteomics in Aloe has far-reaching applications across agriculture, medicine, and cosmetics. From enhancing crop resilience to identifying therapeutic proteins and guiding cosmetic product formulations, proteomic insights into the protein composition of *Aloe vera* contribute to the sustainable and diverse utilization of this versatile plant in various industries.

5.5 CHALLENGES AND FUTURE DIRECTIONS

Studying proteins presents unique challenges compared to studying nucleic acids such as DNA and RNA. Proteins are complex molecules with secondary and tertiary structures that must be maintained during analysis, unlike the linear structure of nucleic acids. Factors such as enzymes, heat, light, or vigorous mixing can denature proteins, altering their structure and function. Additionally, some proteins are difficult to work with due to their poor solubility. The presence of intricate carbohydrates in *Aloe vera* makes extracting proteins challenging (Deutsch *et al.*, 2017). It is crucial to optimize methods for extracting proteins to ensure accurate quantitative data. The various challenges faced in Aloe research include, but are not limited to:

a) Complexity of the Aloe Proteome: *Aloe vera* has a complex and dynamic proteome with a wide range of proteins involved in various physiological processes. The challenge lies in comprehensively characterizing and identifying these proteins, particularly those present in low abundance or involved in specific pathways.
b) Limited Genomic Information: The genome of *Aloe vera* is not fully sequenced, which hinders the ability to link proteomic data to specific genes. The absence of a complete genomic dataset poses challenges in understanding the genetic basis of protein expression and regulation in Aloe.
c) Sample Variability: Aloe proteomics may face challenges due to inherent variability in Aloe samples based on factors such as age, growing conditions, and geographical location. Standardizing sample collection and preparation methods is crucial for ensuring the reproducibility and reliability of proteomic analyses.
d) Identification of Post-Translational Modifications: Proteomic studies in Aloe need to address the identification and characterization of post-translational modifications (PTMs) in proteins. PTMs play a crucial role in protein function, and their accurate detection is essential for a comprehensive understanding of the biology of Aloe.

Analyzing the extensive datasets from quantitative proteomics experiments needs specialized bioinformatics tools and expertise to understand the intricate information about protein abundance. Unlike DNA, proteins cannot be easily amplified, making it challenging to detect less abundant species. For example, while albumin constitutes about half of the total protein content in plasma, less abundant proteins such as cytokines are typically found at much lower concentrations, making their detection more challenging (Dąbrowska *et al.*, 2023). Furthermore, each proteomics technology has limitations in analyzing proteins within a specific concentration range, usually focusing on higher concentration proteins. This poses challenges in detecting and analyzing proteins present at lower concentrations using these methods. Despite these challenges, the progress in proteomics technology is invaluable for identifying disease-related biomarkers and drug targets. Proteomics aims to discover and confirm these crucial elements, although hurdles exist due to the complexity of body fluids and the low abundance of certain protein biomarkers. Overcoming these challenges in biomarker discovery is crucial. To address these challenges, additional "omics" methods, such as peptidomics and metabolomics, along with bioinformatics, offer a systematic approach to studying diseases and introducing ideal biomarkers. The future of Aloe proteomics involves integrating data from genomics, transcriptomics, and metabolomics. Combining multi-omics approaches will provide a holistic understanding of the relationship between the Aloe genome, gene expression, proteins, and metabolites. Integrating various "omics" approaches provides new insights across different aspects of proteomics research. In recent decades, mass spectrometry-based proteomics has emerged as a powerful tool in studying proteins. This technology allows for the analysis of various protein aspects, including quantity and modifications, even from small tissue samples or a few cells. Despite these advancements, proteomic studies face challenges, particularly when dealing with complex tissues and high dynamic ranges of proteins (Kaur *et al.*, 2021). Continued advancements in mass spectrometry technologies, such as improved sensitivity and resolution, will enhance the depth of proteome coverage in Aloe. High-resolution mass spectrometers will allow for more accurate identification and quantification of proteins.

The ongoing efforts toward complete genomic sequencing of *Aloe vera* will be a game-changer for Aloe proteomics. Access to a comprehensive genomic dataset will enable researchers to correlate proteomic data with specific genes, facilitating a deeper understanding of the biology of Aloe. Standardizing protocols for sample collection, preparation, and analysis is critical for ensuring the reproducibility and comparability of Aloe proteomic studies. Additionally, promoting data sharing and collaboration within the scientific community will accelerate progress and overcome challenges.

Future research in Aloe proteomics would focus on functional proteomics to understand the roles of specific proteins in therapeutic pathways. This could lead to the discovery of new bioactive compounds with potential applications in medicine and skincare.

The future of Aloe proteomics holds promise with advancements in technology, genomics, and interdisciplinary collaborations. Efforts are underway to improve techniques and instrumentation, leading to rapid technological developments that significantly benefit research. However, challenges remain in sharing and comparing data among collaborations and making data publicly accessible. In the field of proteomics, it is imperative to establish common data formats and guidelines (Newton *et al.*, 2019). By adopting common formats and guidelines, researchers can collaborate more effectively, ensuring seamless data sharing and utilization across various software platforms within the proteomics field. These efforts aim to accelerate progress and enhance collaboration in proteomics research. Overcoming current challenges and embracing innovative approaches will unlock the full potential of the proteome of *Aloe vera*, leading to applications in agriculture, medicine, and cosmetics.

5.6 CONCLUSION

Modern medicine, a more recent development, can benefit from integrating traditional and contemporary approaches. *Aloe vera*, renowned for its abundant bioactive compounds, carries substantial importance in traditional medicine. Its therapeutic properties have been utilized for centuries, spanning cultures, as a natural remedy for various health conditions. The diverse array of bioactive compounds of the plant, such as polysaccharides and antioxidants, contributes to its healing and soothing attributes, making it a valuable resource in traditional medicinal practices. Proteomics, a sophisticated study of proteins, unveils the therapeutic potential of the plant. The diverse uses of Aloe vera, from gastrointestinal issues to wound healing, showcase its pharmacological activities. Proteomic techniques delve into the protein composition of the plant, aiding in understanding its intricate healing mechanisms. This holistic approach, which combines ancient knowledge with modern science, promises to enhance the cultivation and processing of Aloe vera for optimized therapeutic benefits. Proteomics is a dynamic and rapidly evolving field. Utilizing advanced techniques such as Liquid Chromatography-Mass Spectrometry (LC-MS), Matrix-Assisted Laser Desorption/Ionization Mass Spectrometry (MALDI-MS), Tandem Mass Spectrometry (MS/MS), and Quantitative Proteomics, researchers have gained unprecedented insights into the complex world of cellular processes and environmental responses. Label-free quantification, coupled with protein abundance metrics, has further refined our understanding. These methods not only unravel the mysteries of protein dynamics but also hold promise for biomarker discovery, disease research, and optimizing therapeutic applications in *Aloe vera* and beyond. The continuous refinement of proteomic methodologies ensures ongoing advancements in biological research. Overall, proteomics, a powerful tool in scientific exploration, spans various fields, including agriculture, medicine, and cosmetics. By deciphering protein dynamics, researchers gain insights into stress responses in crops, which aids in breeding and fungicide development. In medicine, proteomics accelerates drug discovery and enhances cancer research. Furthermore, proteomics plays a role in the expanding cosmetics industry by exploring the skincare and haircare benefits of peptides in *Aloe vera*. While facing obstacles in extracting proteins and sharing data, continuous progress holds the promise of transforming research approaches and encouraging cooperation to deepen our understanding of intricate biological processes. The future of proteomics holds immense potential for transformative breakthroughs in diverse scientific domains.

REFERENCES

Al-Amrani, S., Al-Jabri, Z., Al-Zaabi, A., Alshekaili, J., & Al-Khabori, M. (2021). Proteomics: Concepts and applications in human medicine. *World Journal of Biological Chemistry*, *12*(5), 57.

Ankney, J. A., Muneer, A., & Chen, X. (2018). Relative and absolute quantitation in mass spectrometry–based proteomics. *Annual Review of Analytical Chemistry*, *11*, 49–77.

Aslam, B., Basit, M., Nisar, M. A., Khurshid, M., & Rasool, M. H. (2016). Proteomics: Technologies and their applications. *Journal of Chromatographic Science, 55*(2), 1–15.

Babu, S. N., Govindarajan, S., & Noor, A. (2021). *Aloe vera* and its two bioactive constituents in alleviation of diabetes–proteomic & mechanistic insights. *Journal of Ethnopharmacology*, *280*, 114445.

Bąchor, R., Waliczek, M., Stefanowicz, P., & Szewczuk, Z. (2019). Trends in the design of new isobaric labeling reagents for quantitative proteomics. *Molecules*, *24*(4), 701.

Bantscheff, M., Schirle, M., Sweetman, G., Rick, J., & Kuster, B. (2007). Quantitative mass spectrometry in proteomics: A critical review. *Analytical and Bioanalytical Chemistry, 389*, 1017–1031.

Barkla, B. J. (2016). Identification of abiotic stress protein biomarkers by proteomic screening of crop cultivar diversity. *Proteomes*, *4*(3), 26.

Baruah, A., Bordoloi, M., & Baruah, H. P. D. (2016). *Aloe vera*: A multipurpose industrial crop. *Industrial Crops and Products, 94*, 951–963.

Bennike, T., Birkelund, S., Stensballe, A., & Andersen, V. (2014). Biomarkers in inflammatory bowel diseases: Current status and proteomics identification strategies. *World Journal of Gastroenterology: WJG*, *20*(12), 3231.

Bian, Y., Zheng, R., Bayer, F. P., Wong, C., Chang, Y.-C., Meng, C., Zolg, D. P., Reinecke, M., Zecha, J., & Wiechmann, S. (2020). Robust, reproducible and quantitative analysis of thousands of proteomes by micro-flow LC–MS/MS. *Nature Communications*, *11*(1), 157.

Bodnar, W. M., Blackburn, R. K., Krise, J. M., & Moseley, M. A. (2003). Exploiting the complementary nature of LC/MALDI/MS/MS and LC/ESI/MS/MS for increased proteome coverage. *Journal of the American Society for Mass Spectrometry, 14*(9), 971–979.

Burger, N., & Chouchani, E. T. (2024). A new era of cysteine proteomics–Technological advances in thiol biology. *Current Opinion in Chemical Biology, 79*, 102435.

Cabello-Ruiz, E. D., Torres-de la Cruz, V. M., Rivas-Morales, C., Molina-Salinas, G. M., Núñez-González, M. A., Verde-Star, M. J., & Leos-Rivas, C. (2019). Proteomic analysis of a bioactive *Aloe vera* extract. *Current Proteomics, 16*(3), 181–187.

Cabello Ruiz, E. D., Molina Salinas, G. M., Torres de la Cruz, V. M., Núñez González, M. A., Oranday Cárdenas, A., Verde Star, M. J., Martínez de Villarreal, L. E., & Rivas Morales, C. (2015). Antimicrobial activity of protein extract of *Aloe vera* leaves. *Revista Mexicana de Ciencias Farmacéuticas, 46*(1), 41–46.

Cai, M., Wang, Z., Ni, X., Hou, Y., Peng, Q., Gao, X., & Liu, X. (2019). Insights from the proteome profile of Phytophthora capsici in response to the novel fungicide SYP-14288. *PeerJ, 7*, e7626.

Chen, Y., Wang, Y., Yang, J., Zhou, W., & Dai, S. (2021). Exploring the diversity of plant proteome. *Journal of Integrative Plant Biology*, *63*(7), 1197–1210.

Choi, S., Son, B., Son, Y., Park, Y., Lee, S., & Chung, M. (2001). The wound-healing effect of a glycoprotein fraction isolated from *Aloe vera. British Journal of Dermatology*, *145*(4), 535–545.

Cong, Y., Liang, Y., Motamedchaboki, K., Huguet, R., Truong, T., Zhao, R., Shen, Y., Lopez-Ferrer, D., Zhu, Y., & Kelly, R. T. (2020). Improved single-cell proteome coverage using narrow-bore packed NanoLC columns and ultrasensitive mass spectrometry. *Analytical Chemistry*, *92*(3), 2665–2671.

Dąbrowska, J., Sroka, J., & Cencek, T. (2023). Investigating Cryptosporidium spp. using genomic, proteomic and transcriptomic techniques: Current progress and future directions. *International Journal of Molecular Sciences*, *24*(16), 12867.

Das, A., Paudel, B., & Rohila, J. S. (2015). Potentials of proteomics in crop breeding. In Jameel M. Al-Khayri, Shri Mohan Jain, and Dennis V. Johnson (eds.) *Advances in Plant Breeding Strategies: Breeding, Biotechnology and Molecular Tools* (pp. 513–537). Cham: Springer.

Das, S., Mishra, B., Gill, K., Ashraf, M. S., Singh, A. K., Sinha, M., Sharma, S., Xess, I., Dalal, K., & Singh, T. P. (2011). Isolation and characterization of novel protein with anti-fungal and anti-inflammatory properties from *Aloe vera* leaf gel. *International Journal of Biological Macromolecules*, *48*(1), 38–43.

de Ávila, R. I., Santos, S. C., Siino, V., Levander, F., Lindstedt, M., & Zeller, K. S. (2023). A proteomics dataset capturing myeloid cell responses upon cellular exposure to fungicides, adjuvants and fungicide formulations. *Data in Brief*, *46*, 108878.

Deutsch, E. W., Orchard, S., Binz, P.-A., Bittremieux, W., Eisenacher, M., Hermjakob, H., Kawano, S., Lam, H., Mayer, G., & Menschaert, G. (2017). Proteomics standards initiative: Fifteen years of progress and future work. *Journal of Proteome Research*, *16*(12), 4288–4298.

Deutzmann, R. (2004). Structural characterization of proteins and peptides. In Jochen Decler and Udo Reischl (eds.) *Molecular Diagnosis of Infectious Diseases* (pp. 269–297). US: Humana Press.

Dey, A., & Dubey, S. K. (2023). Cosmetics science and skin care: History and concepts. In P. Kesharwani and S. Dubey (eds.) *Nanocosmetics* (pp. 1–15). Florida: CRC Press.

Doll, S., Gnad, F., & Mann, M. (2019). The case for proteomics and phospho-proteomics in personalized cancer medicine. *Proteomics–Clinical Applications, 13*(2), 1800113.

Dos Santos, B. M., & Balbuena, T. S. (2017). Carbon assimilation in Eucalyptus urophylla grown under high atmospheric CO2 concentrations: A proteomics perspective. *Journal of Proteomics, 150*, 252–257.

dos Santos, L., Arita, L. S., Perez, J. D., Carvalho, V. M., Tashima, A. K., Cunha, T. S., Casarini, D. E., & Arita, D. Y. (2021). Exploring the beneficial effects of *Aloe vera* on the kidneys of diabetic rats at the protein level. *Medicine in Omics, 3*, 100013.

El-Shemy, H. A., Aboul-Soud, M. A. M., Nassr-Allah, A. A., Aboul-Enein, K. M., Kabash, A., & Yagi, A. (2010). Antitumor properties and modulation of antioxidant enzymes' activity by *Aloe vera* leaf active principles isolated via supercritical carbon dioxide extraction. *Current Medicinal Chemistry, 17*(2), 129–138.

Epstein, H. A. (2005). The influence of proteomics on cosmetic science. *Skinmed, 4*(1), 44–46.

Fabrik, I., Link, M., Härtlova, A., Dankova, V., Rehulka, P., & Stulik, J. (2014). Application of SILAC labeling to primary bone marrow-derived dendritic cells reveals extensive GM-CSF-dependent arginine metabolism. *Journal of Proteome Research, 13*(2), 752–762.

Filiou, M. D., Soukupova, M., Rewerts, C., Webhofer, C., Turck, C. W., & Maccarrone, G. (2015). Variability assessment of 15 N metabolic labeling-based proteomics workflow in mouse plasma and brain. *Molecular BioSystems, 11*(6), 1536–1542.

Filiou, M. D., Webhofer, C., Gormanns, P., Zhang, Y., Reckow, S., Bisle, B., Teplytska, L., Frank, E., Kessler, M. S., & Maccarrone, G. (2012). The 15 N isotope effect as a means for correlating phenotypic alterations and affected pathways in a trait anxiety mouse model. *Proteomics, 12*(15–16), 2421–2427.

Ghatak, A., Chaturvedi, P., & Weckwerth, W. (2017). Cereal crop proteomics: Systemic analysis of crop drought stress responses towards marker-assisted selection breeding. *Frontiers in Plant Science, 8*, 239189.

Gong, F., Hu, X., & Wang, W. (2015). Proteomic analysis of crop plants under abiotic stress conditions: Where to focus our research? *Frontiers in Plant Science, 6*, 138163.

Gopal, J., Muthu, M., Chun, S., & Wu, H. (2015). State-of-the-art nanoplatform-integrated MALDI-MS impacting resolutions in urinary proteomics. *Proteomics–Clinical Applications, 9*(5–6), 469–481.

Haq, S. A. U., Bashir, T., Roberts, T. H., & Husaini, A. M. (2024). Ameliorating the effects of multiple stresses on agronomic traits in crops: Modern biotechnological and omics approaches. *Molecular Biology Reports, 51*(1), 41.

He, J., & Jia, Y. (2022). Application of omics technologies in dermatological research and skin management. *Journal of Cosmetic Dermatology, 21*(2), 451–460.

Hossain, Z., Yasmeen, F., & Komatsu, S. (2020). Nanoparticles: Synthesis, morphophysiological effects, and proteomic responses of crop plants. *International Journal of Molecular Sciences, 21*(9), 3056.

Im, S.-A., Lee, Y.-R., Lee, Y.-H., Lee, M.-K., Park, Y. I., Lee, S., Kim, K., & Lee, C.-K. (2010). In vivo evidence of the immunomodulatory activity of orally administered *Aloe vera* gel. *Archives of Pharmacal Research, 33*, 451–456.

Jadhav, A. S., Patil, O. A., Kadam, S. V., & Bhutkar, M. A. (2020). Review on *Aloe vera* is used in medicinal plant. *Asian Journal of Research in Pharmaceutical Science, 10*(1), 26–30.

Jain, K. K. (2016). Role of proteomics in the development of personalized medicine. *Advances in Protein Chemistry and Structural Biology, 102*, 41–52.

Jan, N., Rather, A. M.-U.-D., John, R., Chaturvedi, P., Ghatak, A., Weckwerth, W., Zargar, S. M., Mir, R. A., Khan, M. A., & Mir, R. R. (2023). Proteomics for abiotic stresses in legumes: Present status and future directions. *Critical Reviews in Biotechnology, 43*(2), 171–190.

Jangra, A., Sharma, G., Sihag, S., & Chhokar, V. (2022). The dark side of miracle plant-*Aloe vera*: A review. *Molecular Biology Reports, 49*(6), 5029–5040.

Jiang, P., Peng, W., Zhao, J., Goli, M., Huang, Y., Li, Y., & Mechref, Y. (2023). Glycan/Protein-stable isotope labeling in cell culture for enabling concurrent quantitative glycomics/proteomics/glycoproteomics. *Analytical Chemistry, 95*(44), 16059–16069.

Kaur, P., Singh, A., & Chana, I. (2021). Computational techniques and tools for omics data analysis: State-of-the-art, challenges, and future directions. *Archives of Computational Methods in Engineering, 28*(7), 4595–4631.

Kaur, S., & Bains, K. (2023). Aloe Barbadensis Miller (*Aloe vera*). *International Journal for Vitamin and Nutrition Research, 94*(3–4), 308–321.

Koike, T., Beppu, H., Kuzuya, H., Maruta, K., Shimpo, K., Suzuki, M., Titani, K., & Fujita, K. (1995). A 35 kDa mannose-binding lectin with hemagglutinating and mitogenic activities from "Kidachi Aloe"(Aloe arborescens Miller var. natalensis Berger). *Journal of Biochemistry*, *118*(6), 1205–1210.

Komatsu, S., & Hossain, Z. (2013). Organ-specific proteome analysis for identification of abiotic stress response mechanism in crop. *Frontiers in Plant Science*, *4*, 44308.

Kumar, C., & Mann, M. (2009). Bioinformatics analysis of mass spectrometry-based proteomics data sets. *FEBS Letters*, *583*(11), 1703–1712.

Kumar, S., Kalita, S., Basumatary, I. B., Kumar, S., Ray, S., & Mukherjee, A. (2024). Recent advances in therapeutic and biological activities of *Aloe vera*. *Biocatalysis and Agricultural Biotechnology*, *57*, 103084.

Li, Z., Adams, R. M., Chourey, K., Hurst, G. B., Hettich, R. L., & Pan, C. (2012). Systematic comparison of label-free, metabolic labeling, and isobaric chemical labeling for quantitative proteomics on LTQ Orbitrap Velos. *Journal of Proteome Research*, *11*(3), 1582–1590.

Liu, Y., Lu, S., Liu, K., Wang, S., Huang, L., & Guo, L. (2019). Proteomics: A powerful tool to study plant responses to biotic stress. *Plant Methods*, *15*(1), 135.

Ma, J., Liu, M., Wang, Y., Xin, C., Zhang, H., Chen, S., Zheng, X., Zhang, X., Xiao, F., & Yang, S. (2020). Quantitative proteomics analysis of young and elderly skin with DIA mass spectrometry reveals new skin aging-related proteins. *Aging (Albany NY)*, *12*(13), 13529.

Majeran, W., Friso, G., Asakura, Y., Qu, X., Huang, M., Ponnala, L., Watkins, K. P., Barkan, A., & van Wijk, K. J. (2012). Nucleoid-enriched proteomes in developing plastids and chloroplasts from maize leaves: A new conceptual framework for nucleoid functions. *Plant Physiology*, *158*(1), 156–189.

Marghany, F., Ayobahan, S. U., Salinas, G., Schäfers, C., Hollert, H., & Eilebrecht, S. (2024). Transcriptomic and proteomic fingerprints induced by the fungicides difenoconazole and metalaxyl in zebrafish embryos. *Environmental Toxicology and Pharmacology*, *105*, 104348.

Martínez-Burgos, W. J., Serra, J. L., MarsigliaF, R. M., Montoya, P., Sarmiento-Vásquez, Z., Marin, O., Gallego-Cartagena, E., & Paternina-Arboleda, C. D. (2022). *Aloe vera*: From ancient knowledge to the patent and innovation landscape–A review. *South African Journal of Botany*, *147*, 993–1006.

Maus, A., Mignon, R., & Basile, F. (2018). Enhanced protein identification using graphite-modified MALDI plates for offline LC-MALDI-MS/MS bottom-up proteomics. *Analytical Biochemistry*, *545*, 31–37.

Moreno-Hernández, J. M., Benítez-García, I., Mazorra-Manzano, M. A., Ramírez-Suárez, J. C., & Sánchez, E. (2020). Strategies for production, characterization and application of protein-based biostimulants in agriculture: A review. *Chilean Journal of Agricultural Research*, *80*(2), 274–289.

Mueller, L. N., Brusniak, M.-Y., Mani, D. R., & Aebersold, R. (2008). An assessment of software solutions for the analysis of mass spectrometry based quantitative proteomics data. *Journal of Proteome Research*, *7*(01), 51–61.

Mustafa, G., & Komatsu, S. (2021). Plant proteomic research for improvement of food crops under stresses: A review. *Molecular Omics*, *17*(6), 860–880.

Nahnsen, S., Bielow, C., Reinert, K., & Kohlbacher, O. (2013). Tools for label-free peptide quantification. *Molecular & Cellular Proteomics*, *12*(3), 549–556.

Neagu, A.-N., Whitham, D., Seymour, L., Haaker, N., Pelkey, I., & Darie, C. C. (2023). Proteomics-Based identification of dysregulated proteins and biomarker discovery in invasive ductal carcinoma, the most common breast cancer subtype. *Proteomes*, *11*(2), 13.

Neilson, K. A., Ali, N. A., Muralidharan, S., Mirzaei, M., Mariani, M., Assadourian, G., Lee, A., Van Sluyter, S. C., & Haynes, P. A. (2011). Less label, more free: Approaches in label-free quantitative mass spectrometry. *Proteomics*, *11*(4), 535–553.

Newton, V. L., Riba-Garcia, I., Griffiths, C. E. M., Rawlings, A. V, Voegeli, R., Unwin, R. D., Sherratt, M. J., & Watson, R. E. B. (2019). Mass spectrometry-based proteomics reveals the distinct nature of the skin proteomes of photoaged compared to intrinsically aged skin. *International Journal of Cosmetic Science*, *41*(2), 118–131.

Nyalwidhe, J. O., Grzesik, W. J., Burch, T. C., Semeraro, M. L., Waseem, T., Gerling, I. C., Mirmira, R. G., Morris, M. A., & Nadler, J. L. (2017). Comparative quantitative proteomic analysis of disease stratified laser captured microdissected human islets identifies proteins and pathways potentially related to type 1 diabetes. *PloS One*, *12*(9), e0183908.

Rajeswari, R., Umadevi, M., Rahale, C. S., Pushpa, R., Selvavenkadesh, S., Kumar, K. P. S., & Bhowmik, D. (2012). *Aloe vera*: The miracle plant; its medicinal and traditional uses in India. *Journal of Pharmacognosy and Phytochemistry*, *1*(4), 118–124.

Ramírez, O., Pomareda, F., Olivares, B., Huang, Y.-L., Zavala, G., Carrasco-Rojas, J., Álvarez, S., Leiva-Sabadini, C., Hidalgo, V., & Romo, P. (2024). *Aloe vera* peel-derived nanovesicles display anti-inflammatory properties and prevent myofibroblast differentiation. *Phytomedicine, 122*, 155108.

Shekhova, E., Ivanova, L., Krüger, T., Stroe, M. C., Macheleidt, J., Kniemeyer, O., & Brakhage, A. A. (2019). Redox proteomic analysis reveals oxidative modifications of proteins by increased levels of intracellular reactive oxygen species during hypoxia adaptation of Aspergillus fumigatus. *Proteomics, 19*(5), 1800339.

Simões, J., Nunes, F. M., Domingues, P., Coimbra, M. A., & Domingues, M. R. (2012). Mass spectrometry characterization of an *Aloe vera* mannan presenting immunostimulatory activity. *Carbohydrate Polymers, 90*(1), 229–236.

Simova-Stoilova, L. P., Romero-Rodríguez, M. C., Sánchez-Lucas, R., Navarro-Cerrillo, R. M., Medina-Auñón, A., & Jorrin Novo, J. V. (2015). 2-DE proteomics analysis of drought treated seedlings of Quercus ilex supports a root active strategy for metabolic adaptation in response to water shortage. *Frontiers in Plant Science, 6*, 144338.

Su, M., Zhang, Z., Zhou, L., Han, C., Huang, C., & Nice, E. C. (2021). Proteomics, personalized medicine and cancer. *Cancers, 13*(11), 2512.

Thakur, S. S., Geiger, T., Chatterjee, B., Bandilla, P., Fröhlich, F., Cox, J., & Mann, M. (2011). Deep and highly sensitive proteome coverage by LC-MS/MS without prefractionation. *Molecular & Cellular Proteomics, 10*(8), M110-003699.

Ushasree, M. V., Jia, Q., Do, S. G., & Lee, E. Y. (2024). New opportunities and perspectives on biosynthesis and bioactivities of secondary metabolites from *Aloe vera*. *Biotechnology Advances, 72*, 108325.

Vanderschuren, H., Lentz, E., Zainuddin, I., & Gruissem, W. (2013). Proteomics of model and crop plant species: Status, current limitations and strategic advances for crop improvement. *Journal of Proteomics, 93*, 5–19.

Walport, L. J., Low, J. K. K., Matthews, J. M., & Mackay, J. P. (2021). The characterization of protein interactions–what, how and how much? *Chemical Society Reviews, 50*(22), 12292–12307.

Wang, W.-Q., Jensen, O. N., Møller, I. M., Hebelstrup, K. H., & Rogowska-Wrzesinska, A. (2018). Evaluation of sample preparation methods for mass spectrometry-based proteomic analysis of barley leaves. *Plant Methods, 14*, 1–13.

Woo, J., Sudhir, P., & Zhang, Q. (2020). Pancreatic tissue proteomics unveils key proteins, pathways, and networks associated with type 1 diabetes. *PROTEOMICS–Clinical Applications, 14*(6), 2000053.

Yan, S., Bhawal, R., Yin, Z., Thannhauser, T. W., & Zhang, S. (2022). Recent advances in proteomics and metabolomics in plants. *Molecular Horticulture, 2*(1), 17.

Yang, Y., Saand, M. A., Huang, L., Abdelaal, W. B., & Sirohi, M. H. (2021). Applications of multi-omics technologies for crop improvement. *Frontiers in Plant Science, 12*, 563953.

Yates, J. R., Ruse, C. I., & Nakorchevsky, A. (2009). Proteomics by mass spectrometry: Approaches, advances, and applications. *Annual Review of Biomedical Engineering, 11*, 49–79.

Yin, X., He, D., Gupta, R., & Yang, P. (2015). Physiological and proteomic analyses on artificially aged Brassica napus seed. *Frontiers in Plant Science, 6*, 118504.

6 Nanotechnology Advances in Aloe Research

Shweta Mehrotra, Vinod Goyal, Sarita, Bhupnesh, Vaibhav Baliyan, and Christian O. Dimkpa

6.1 INTRODUCTION

Aloe vera (*Aloe barbadensis Miller*) is a perennial plant belonging to the Liliaceae family, with turgid green leaves arranged in a rosette formation at the stem. The name *Aloe vera* originates from the Arabic word "Alloeh," signifying a "shining bitter substance," and "vera," meaning "true" (Joseph and Raj, 2010; Surjushe *et al.*, 2008). *Aloe vera* has a rich history of medicinal use spanning millennia across various cultures, including Greece, Egypt, India, Mexico, Japan, and China (Marshall, 1990). The Egyptians had quoted *Aloe vera* as the "Plant of Immortality." In the early 18th century, the US Pharmacopeia officially endorsed *Aloe vera* as a potent purgative and skin protectant. Furthermore, the inaugural clinical application of *Aloe vera* in treating radiotherapy burns and mucous membranes was documented in 1930.

Aloe vera leaves are divided into three sections: the first layer is Aloe-gel, which is made up of parenchyma cells and generates clear, tasteless mucilage (Nourozi and Sayyari, 2020). Long-chain hydrocarbons, proteins, minerals, ß-100 sitosterols, polysaccharides, and esters are among its constituents (Liu *et al.*, 2021). Exudate is the second component, made up of colorless mucus and yellow sap. Many substances, including glycosides, aloin, barbaloin, aloe-emodin, and anthraquinone and its derivatives, are found in yellow mucus (Chauhan and Kumar, 2020). Anthraquinone is a phenolic chemical possessing antibacterial, antioxidant, and anti-inflammatory properties. Moreover, several kinds of phenolic components, which function as antioxidants to prevent free radicals, fat peroxidation, and enzymes, are present in colorless mucus. The third component is the outer layer of skin, which is made up of 15–20 cells that have a protective function and synthesize carbohydrates and proteins. The bioactive compounds in *Aloe vera*, especially those in *Aloe vera* gel, have been exploited for various applications that fall into the following clusters: cosmetic, healthcare, and food industries. *Aloe vera* gel primarily contains 98.5%–99.5% water, with the remaining solids containing over 200 distinct components, with polysaccharides being the predominant compounds (Femenia *et al.*, 1999).

Aloe vera is mostly made up of a huge variety of organic and inorganic components (Eshun and He, 2004). Because of its richness, this plant is frequently utilized as a medicinal herb. (It exhibits various properties such as antifungal, antioxidant, and anticancer effects, accelerating skin regeneration after injury (Danish *et al.*, 2020; Iqbal and Ahmed, 2021; Hęś *et al.*, 2019; Shalabi *et al.*, 2015). In the food industry, *Aloe vera* extends shelf life, prevents food deterioration, and reduces lipid oxidation due to its antioxidant properties (Hęś *et al.*, 2019). Additionally, it serves as a foundation for cosmetics like creams, lotions, and face cleansers. Recently, it has been used as a natural dye in solar energy applications (Kaotar *et al.*, 2022). Nanoparticle synthesis using *Aloe vera* involves essential compounds like terpenoids, flavonols, and phenols, acting as reducers and stabilizers for nanoparticle formation (Ahmed *et al.*, 2023; Arsène *et al.*, 2023; Munirathnam *et al.*,

 DOI: 10.1201/9781003603108-6

2023; Vélez *et al.*, 2018). This process enhances nanoparticle biological activity and allows for precise size and shape regulation, potentially boosting their therapeutic efficacy (Jeevanandam *et al.*, 2020; Logaranjan *et al.*, 2016).

Nanotechnology, a rapidly advancing field in the 21st century, profoundly impacts various sectors, including everyday life, societal ecology, and industries (Thangadurai *et al.*, 2020). Nanomaterials, a significant outcome of nanotechnology, manipulate materials at a nanometer scale, leveraging their unique properties stemming from altered physical, chemical, and biological characteristics (Jeevanandam *et al.*, 2018). With sizes ranging from one to 100 nanometers, nanomaterials possess extensive surface areas, offering innovative solutions across diverse domains (Nile *et al.*, 2020). Nanotechnology has a wide application in medicine (Preetha and Balakrishnan, 2017), environment (Shi *et al.*, 2007), energy (Das *et al.*, 2007), information and communication (Preetha and Balakrishnan, 2017), heavy industry (Lo *et al.*, 2010) and consumer goods (Schneider, 2010). Agriculture, benefiting from the interdisciplinary nature of nanotechnology, which integrates engineering, chemistry, biology, physics, and materials science, experiences numerous applications (Bayda *et al.*, 2019). These applications enhance crop productivity, facilitate food processing, and bolster food security levels (Khan *et al.*, 2019). India, relying heavily on agriculture, particularly harnesses the critical applications of nanotechnology, reflecting its pivotal role in agricultural advancement (Shang *et al.*, 2019).

Nanotechnology research in Aloe explores the application of nanoscale properties in unlocking the therapeutic nature of plant. This innovative approach enhances the bioactive compounds of Aloe for improved nutrient absorption in plants, improved drug delivery, biomedical applications, wound healing, and cosmetic formulations for humans. The significance lies in creating novel, advanced, sustainable solutions for the pharmaceutical and cosmetic industries, harnessing the inherent properties of Aloe at the molecular level, paving the way for environmentally friendly and efficient applications in medicine, skincare, and agriculture. Nanotechnology research indicates the potential of *Aloe vera* for diverse applications, leveraging its renowned medicinal properties. The antibacterial and antimicrobial attributes of *Aloe vera* make it a compelling candidate for nanotechnology endeavors (Radha and Laxmipriya, 2015).

6.2 NANOPARTICLES

Nanoparticles are particles with dimensions ranging from one to 100 nanometers. They have gained significant attention due to their unique properties and diverse applications. Their size is comparable to that of a diameter of hair, which is approximately 80 nm (Srivastava and Bhargava, 2022). They have better qualities than bulk materials because of the enhanced specific surface area and the clearly defined shape and size (Elahi *et al.*, 2018; Pandit *et al.*, 2022; Ying *et al.*, 2022). Nanoparticles play a crucial role in various fields due to their unique properties and applications.

6.2.1 Types of Nanoparticles

Various types of nanoparticles play a pivotal role in diverse fields, ranging from medicine to materials science. Nanoparticles can be broadly categorized into metallic, semiconductor, carbon-based, polymeric, lipid-based, ceramic nanoparticles, and quantum dots, each with unique properties and applications (Park *et al.*, 2021; Perni *et al.*, 2011; Mieszawska *et al.*, 2013). Various studies have focused on synthesizing different types of nanoparticles utilizing *Aloe vera* extract, such as gold nanotriangles, silver nanoparticles, zinc oxide nanoparticles, copper oxide nanoparticles, and quantum dots (Adhikari *et al.*, 2023; Dar *et al.*, 2021; Chandran *et al.*, 2006; Primo *et al.*, 2020; Sarkar *et al.*, 2023). Their unique properties at the nanoscale make them valuable in drug delivery, imaging, and catalysis.

Carbon-based NPs: Fullerenes and carbon nanotubes (CNTs) are the two essential sub-categories of carbon-based NPs. NPs of globular hollow cages, like allotropic forms of carbon, are found in fullerenes. Due to their electrical conductivity, high strength, structure, electron affinity, and adaptability, they have sparked significant economic interest (Astefanei *et al.*, 2015; Elliott *et al.*, 2013).

Ceramics NPs: Ceramic NPs are tiny particles made up of inorganic, non-metallic materials that are heat-treated and cooled in a specific way to give particular properties. Ceramic NPs are used in various applications, including coating, catalysts, and batteries (Sigmund *et al.*, 2006).

Lipid-based NPs: These NPs are helpful in several biological applications because they include lipid moieties. Lipid NPs typically have a diameter of 10–1,000 nm and are spherical. Lipid NPs, i.e., polymeric NPs, have a solid lipid core and a matrix consisting of soluble lipophilic molecules (Khan *et al.*, 2019).

Semiconductor NPs: Semiconductor NPs have qualities similar to metals and non-metals. That is why Semiconductor NPs have unique physical and chemical properties that make them useful for various applications. For example, semiconductor NPs can absorb and emit light and can be used to make more efficient solar cells or brighter light-emitting diodes (LEDs). They can make smaller and faster electronic devices, such as transistors, and can be used in bioimaging and cancer therapy (Biju *et al.*, 2008).

Polymeric NPs: Polymeric NPs with a size between one and 1,000 nm can have active substances surface-adsorbed onto the polymeric core or entrapped inside the polymeric body. These NPs are often organic, and the term polymer nanoparticle (PNP) is commonly used in the literature to refer to them. They resemble nanospheres or nanocapsules for the most part (Khan *et al.*, 2019; Zielinska *et al.*, 2020).

Metal NPs: Metal NPs are purely made of metals. These NPs have distinctive electrical properties due to well-known localized surface Plasmon resonance (LSPR) features. Cu, Ag, and Au nanoparticles exhibit a broad absorption band in the visible region of the solar electromagnetic spectrum. Metal NPs are used in several scientific fields because of their enhanced features, such as facet, size, and shape-controlled synthesis of metal NPs (Khan *et al.*, 2019).

The synthesis of silver nanoparticles using *Aloe vera* extract has been particularly explored for its potential antibacterial and catalytic applications (Ahmad *et al.*, 2022; Sarkar *et al.*, 2023; Xavier *et al.*, 2022). The silver nanoparticles synthesized exhibited specific particle size distributions and surface plasmon resonance peaks, indicating their unique characteristics (Ahmad *et al.*, 2022; Riaz *et al.*, 2022). Additionally, the synthesis of zinc oxide nanoparticles using *Aloe vera* extract has been investigated for applications in antibacterial activity and wastewater treatment (Kavyashree *et al.*, 2015; Primo *et al.*, 2020). These nanoparticles have shown promising results in terms of their physical and chemical properties, making them suitable for various environmental and biomedical applications.

Furthermore, the use of *Aloe vera* extract in synthesizing copper oxide nanoparticles has been studied for its potential in dye removal and cytotoxicity against cancer cells (Batool *et al.*, 2019; Froushani *et al.*, 2021). The nanoparticles synthesized exhibited specific shapes and sizes, contributing to their adsorption and biological activities (Batool *et al.*, 2019; Froushani *et al.*, 2021). Additionally, the incorporation of *Aloe vera* extract in the synthesis of ZnS: Mn quantum dots has been explored to enhance biocompatibility for therapeutic applications (Anilkumar *et al.*, 2016).

6.2.2 Synthesis of Nanoparticles

Nanoparticles can be synthesized through chemical, physical, or biological methods (Figure 6.1). Common techniques involve chemical reduction, precipitation, and physical methods such as laser ablation or grinding. Each method influences the size, shape, and surface characteristics of

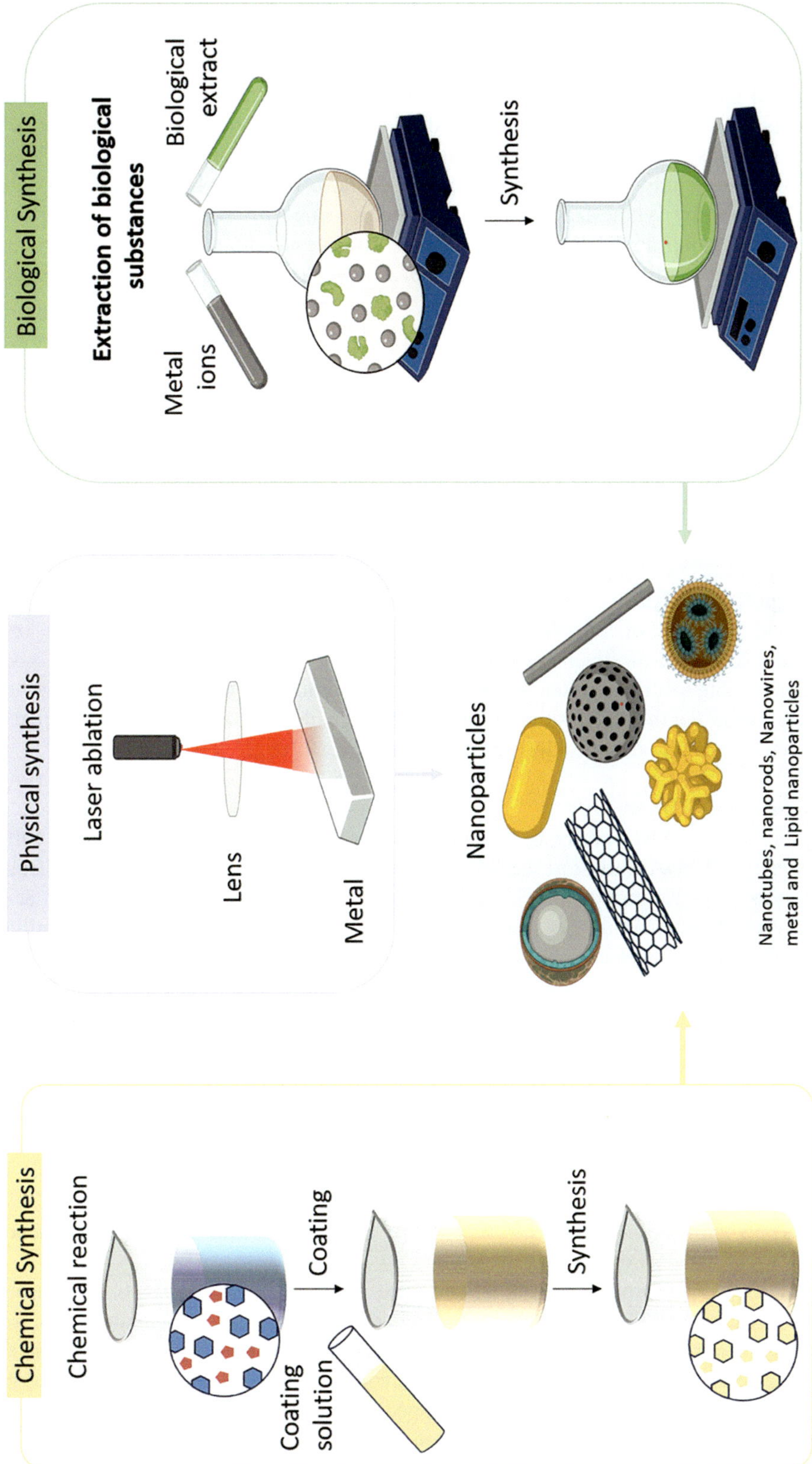

FIGURE 6.1 Basic methods of nanoparticle synthesis.

the nanoparticles, affecting their performance in specific applications. Biological approaches are gaining attention for their eco-friendliness and reduced adverse effects compared to chemical synthesis (Konishi *et al.*, 2007), enzymes (Willner *et al.*, 2006), fungi (Vigneshwaran *et al.*, 2007), and plant-based methods (Ahmad *et al.*, 2022). The green synthesis of NPs has been elaborated by De Matteis *et al.* (2023) using hot plate and microwave methods. The development of these eco-friendly methods for the synthesis of nanoparticles is evolving into an important branch of nanotechnology, especially silver nanoparticles, which have many applications (Kyriacou *et al.*, 2004). The successfully synthesized silver nanoparticles using *Aloe vera* leaf extract for catalytic organic dye degradation applications (Sarkar *et al.*, 2023) used the sol-gel synthesis method to create zinc oxide nanoparticles with the assistance of *Aloe vera* as a dispersing medium (Seifunnisha and Shanthi, 2021).

6.2.3 Characterization of Nanoparticles

Characterization is crucial to understand and optimize nanoparticle properties. Techniques like UV-Visible spectra, TEM (Transmission Electron Microscopy), and SEM (Scanning Electron Microscopy) with DLS (Dynamic Light Scattering), XRD (X-ray Diffraction), HPLC, and FTIR (Fourier Transform Infrared Spectroscopy) provide insights into size distribution, crystallinity, and chemical composition.

UV-Visible spectra analysis

Initial characterization of the AgNPs involved UV-visible spectroscopy. The process involved monitoring the reduction of silver ions to nanoparticles by examining the UV-visible spectra of diluted samples using Millipore water. A UV-Vis spectrophotometer (Varian Inc., USA) was utilized to analyze the spectra of the AgNP solution across the range of 300 to 600 nm, with Millipore water serving as a reference for baseline adjustment (Ashraf *et al.*, 2016).

X-ray diffraction (XRD) analysis

The X-ray diffraction (XRD) analysis of silver nanoparticles (AgNPs) was conducted following the method outlined by Ansari *et al.* (2015). The XRD pattern was obtained using a Bruker D8 diffractometer with CuKα radiation ($\lambda = 1.54056$Å), covering the range of $20° \leq 2\theta \leq 80°$ at 40 kV. Subsequently, the lattice parameters were determined using PowderX software. The particle size (D) of the AgNPs sample was calculated utilizing Scherrer's equation:

$$D = \frac{0.9\lambda}{B \cos \theta}$$

where λ is the wavelength of the x-ray, B is the broadening of the diffraction line measured as half of its maximum intensity in radians, and θ is Bragg's diffraction angle. The particle size of the sample was estimated from the line width of the (111) XRD peak (Ashraf *et al.*, 2016).

High-resolution transmission electron microscopy (HR-TEM) and dynamic light scattering (DLS) analysis of AgNPs

The size and shape of the produced nanoparticles were analyzed using high-resolution transmission electron microscopy (HR-TEM). Samples were prepared by depositing a small amount of a highly diluted nanoparticle solution onto a carbon-coated copper grid, followed by air-drying and storage in a desiccator. Subsequently, the specimens were mounted onto a holder and examined using a Tecnai G2 transmission electron microscope equipped with a tungsten (W) source and an ultra-high-resolution pole piece, operating at an acceleration voltage of 200 kV. Additionally, dynamic light scattering was employed to assess the size distribution or average dimensions of the synthesized silver nanoparticles (Ashraf *et al.*, 2016).

High-performance liquid chromatography (HPLC)
HPLC analysis was conducted using a Hitachi analytical HPLC system from Japan, comprising L-7100 low-pressure gradient pumps, an L-7200 sequential auto-sampler, and a high-sensitivity diode array detector (190–800 nm) controlled by D-7000 HPLC System Manager software. The Phenomenex Luna 5 μm, 100 Å, HPLC column (150 mm × 4.6 mm × 5 μm) was utilized to identify the products formed in the incubated mixtures of HSA and MG. Mobile phases A and B consisted of 0.1 M aqueous ammonium acetate and 100% HPLC analytical grade acetonitrile, respectively. Aqueous ammonium acetate solution was prepared using deionized water, and both mobile phases were degassed and sonicated for 15 minutes before use. Before HPLC analysis, all samples and solvents underwent filtration using Millipore 0.22 μm syringe filters. The accuracy and reproducibility of the HPLC method were validated through repeated testing (Ashraf *et al.*, 2016).

Fourier transformation infrared (FTIR) spectroscopy
FTIR spectroscopy was used to examine alterations in the structure of MG-modified protein across different concentrations of AgNPs. Buffer-subtracted transmission spectra were captured within the wavenumber range of 600–4,000 cm^{-1} using a Perkin Elmer Spectrum 100 FTIR spectrometer manufactured by PerkinElmer, Inc., based in the USA (Ashraf *et al.*, 2016).

6.2.4 Bioavailability

Bioavailability is a critical aspect, especially in medical applications. Bioavailability pertains to the proportion of a substance from a dosage that reaches the target site of the body. Nanoparticles can enhance bioavailability by improving drug solubility and stability. They enable targeted drug delivery, minimizing side effects and maximizing therapeutic efficacy. Factors influencing bioavailability include nanoparticle size, surface charge, and interaction with biological environments. Understanding the biological mechanisms governing uptake and bioavailability is crucial for devising efficient nanoparticle delivery systems for nutrients, nutraceuticals, and related active compounds (Routray and Patra, 2021).

To summarize, the diverse types of nanoparticles, their synthesis methods, characterization techniques, and the impact on bioavailability collectively contribute to the versatility and applicability of nanotechnology in addressing challenges across various scientific and industrial domains. The intersection of *Aloe vera* and nanotechnology holds great promise for sustainable and effective applications in various fields.

6.3 NANOPARTICLE-BASED FORMULATIONS

Nanoparticle-based drug delivery systems, including nanogels, nanosuspensions, nanoliposomes, nanocapsules, nanoemulsions, and nanofibers, represent diverse platforms with distinct properties and functions and innovative approaches that have demonstrated promising outcomes in improving drug availability and therapeutic effectiveness. Recent studies have highlighted their benefits, such as enhancing drug solubility, stability, controlled release, and absorption rates. These advancements contribute to their potential for revolutionizing drug delivery and optimizing treatment outcomes (Chinthaginjala *et al.*, 2020; Jacob *et al.*, 2020).

Nanogels
Nanogels have emerged as suitable vehicles for delivering and releasing medications to patients in recent years as one of the many dimensions of nanomedicine: the junction of nanotechnology,

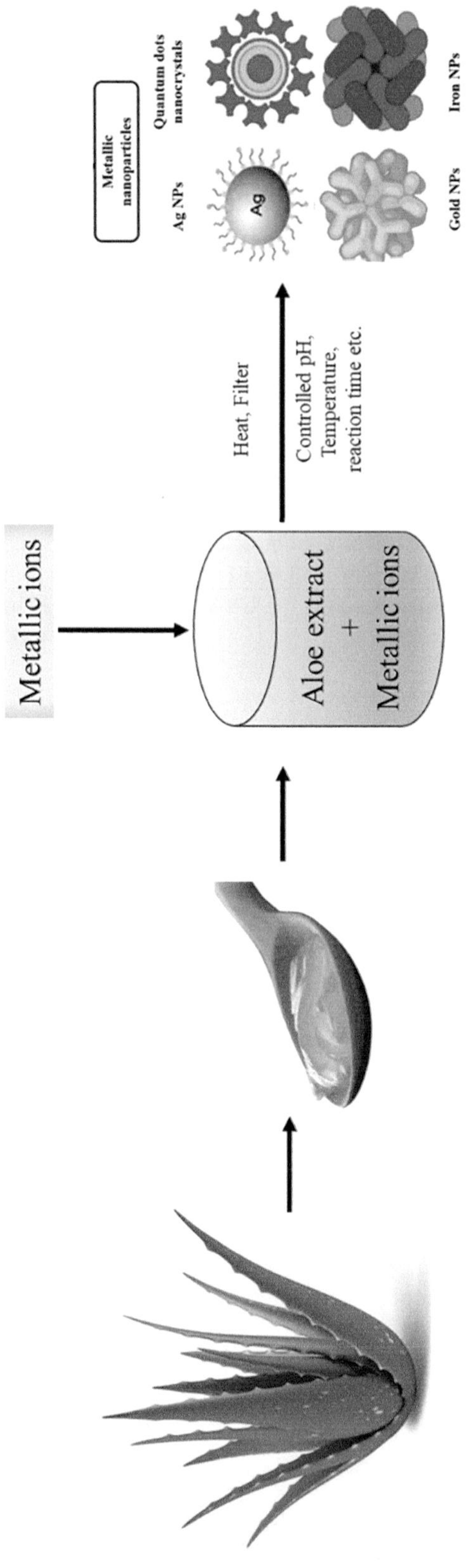

FIGURE 6.2 Generalized procedure of metallic nanoparticle synthesis using *Aloe vera* extract.

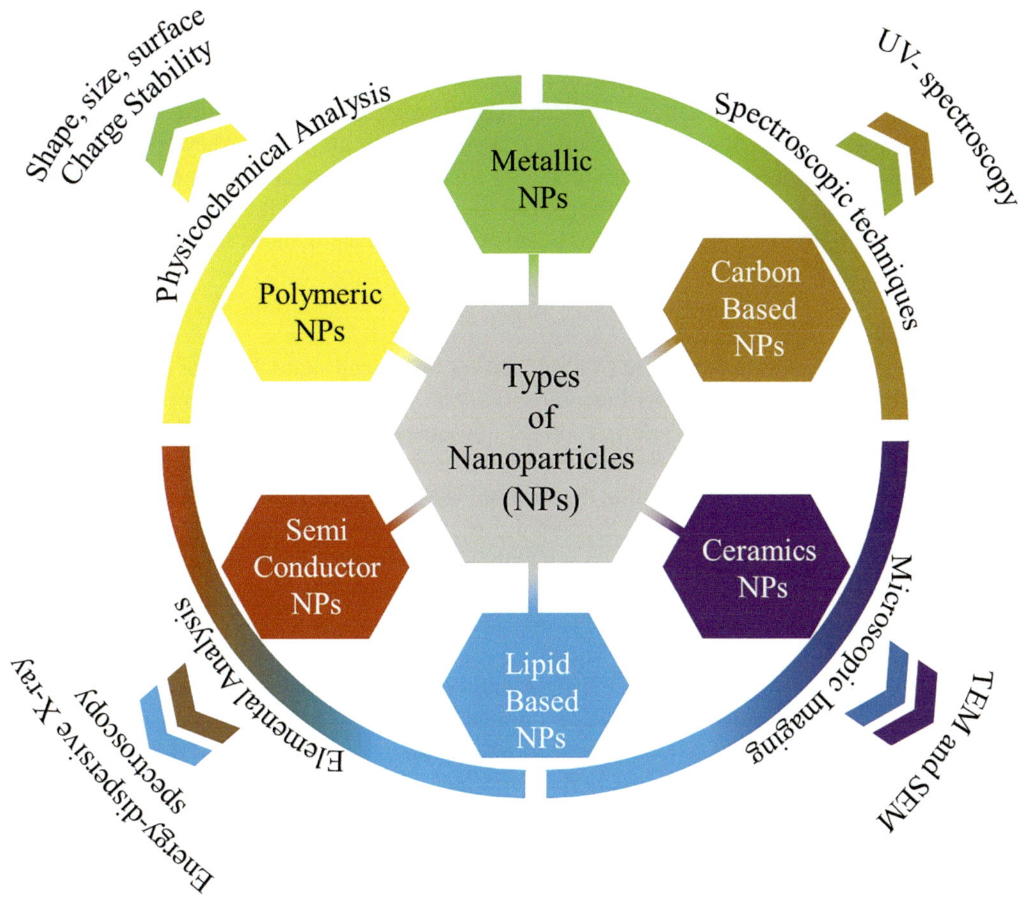

FIGURE 6.3 Types of nanoparticles and their characterization techniques.

medicine, and pharmaceuticals. Nanogels are three-dimensional cross-linked networks that can encapsulate drugs or other bioactive compounds. Nanogels are capable of absorbing enormous amounts of water (Molina *et al.*, 2015) to form hydrogels with dimensions in the nanometer range. The hydrogel structure provides a stable environment for controlled drug release. Hydrogels, being polymer-based gels, are created by linking polymer chains to establish a large molecular network. The fabrication of hydrogels involves several methods, all of which necessitate the synthesis of polymeric monomers followed by polymerization with cross-linker molecules to develop a mesh-like polymer framework (Annabi *et al.*, 2013). Nanogels can be synthesized using materials like chitosan, ethylcellulose, methylcellulose, and various polysaccharide-based polymers, including dextran, pullulan, and dextrin (Mamatha *et al.*, 2017). Nanogels exhibit swelling behavior in aqueous solutions due to their small, soft nature. This property significantly impacts their effectiveness in drug delivery mechanisms, influenced by factors such as nanogel structure, drug loading capacity, particle size, permeability, and colloidal stability (Iwasaki *et al.*, 2016). Nanogels find applications in drug delivery, wound healing, and tissue engineering.

Nanosuspensions

Nanosusp ensions, characterized by colloidal dispersions of drug particles at the nanoscale in a liquid medium, are stabilized by surfactants (Patravale *et al.*, 2004). They represent a biphasic system wherein pure drug particles, smaller than 1 μm in diameter, are dispersed in an aqueous medium. Reducing drug particles to the nanometer range enhances dissolution rates due to increased surface area and saturation solubility. The rise in saturation solubility and solution velocity of nanoparticles is attributed to the increase in the vapor pressure of the particles. Nanosuspensions address challenges associated with poorly water-soluble and lipid-soluble drugs, offering simplicity and advantages over other delivery methods (Kesisoglou *et al.*, 2007). They effectively tackle poor bioavailability by addressing issues related to solubility and permeability across membranes. Nanosuspensions are employed in pharmaceuticals to improve drug dissolution and therapeutic efficacy. For instance, the nanosuspension formulation of diclofenac demonstrated increased dissolution rates compared to coarse suspensions (Lai *et al.*, 2009).

Nanoliposomes

Nanoliposomes are vesicles composed of lipid bilayers that encapsulate drugs or other substances. These lipid-based nanoparticles offer biocompatibility and can be engineered to release their payload at specific sites. Alec Bangham introduced liposomes in 1960 as polar lipid nanoparticles, typically spherical and measuring between 50 and 450 nanometers. These structures have the capability to contain an aqueous core within one or more layers of lipid bilayers, whether derived from natural or synthetic sources, enabling the free diffusion of substances within them (Ganesan and Narayanasamy, 2017). Liposomes possess a membrane structure resembling that of cells and consist of compounds with both lipophilic and hydrophilic properties. This unique composition enables liposomes to encapsulate both hydrophobic and hydrophilic pharmacological compounds within their structure (Robson *et al.*, 2018). Liposomes, owing to their phospholipid bilayers, possess unique properties that enhance various aspects of drug delivery. These include increased drug solubility, improved drug delivery efficiency, enhanced bioavailability of the encapsulated drug, facilitated drug absorption within cells, and improved drug distribution both in vivo and in vitro (Hong *et al.*, 2015). The ADME profiles of drugs, including herbal remedies, enzymes, and proteins, can be tailored as required, particularly for the development of vaccines, nutraceuticals, and cosmetics. Moreover, liposomes offer exclusive advantages such as environmental protection for the encapsulated drug, efficient delivery ensuring minimal degradation, cost-effectiveness, and rapid treatment with reduced systemic side effects, making them highly desirable in biomedicine applications (Aqil *et al.*, 2013). Antibodies or ligands can be incorporated into liposomes to enhance target specificity. This strategy allows for precise targeting of specific cell types or receptors, increasing the effectiveness of drug delivery and minimizing off-target effects. By conjugating antibodies or ligands to the surface of liposomes, they can selectively bind to corresponding antigens or receptors on the target cells. Thangapazham *et al.* (2008) conducted a study where they developed curcumin-loaded liposomes coated with PSMA antibodies for the treatment of human prostate cancer cell lines LNCaP and C4-2B. This approach led to enhanced targeted administration, resulting in a remarkable 70–80% suppression of cell growth. Additionally, the use of PSMA antibodies conferred a ten-fold dosage advantage, highlighting the efficacy and precision of this treatment strategy.

Nanocapsules

Polymeric nanoparticles have dual roles in pharmacokinetics and polymer characteristics. They influence drug properties due to their subcellular size, impacting pharmacokinetics. Additionally, they affect the biocompatibility and biodegradability of the employed polymers. These nanoparticles are classified based on internal structure as nanospheres or nanocapsules, each with distinct compositions and functions in drug delivery systems (Lukasiewicz *et al.*, 2016). Polymeric nanospheres typically have a uniform spherical shape comprising a solid polymeric matrix. Conversely, polymeric nanocapsules feature a core, either liquid or solid, enclosed by a polymeric shell, a characteristic absent in nanospheres (Frank *et al.*, 2019). Nanocapsules are nano-sized capsules with a core-shell structure. The core can encapsulate drugs, while the shell provides protection and controlled release. Polymeric nanocapsules have garnered increased attention in drug delivery due to their core-shell microstructure. Unlike polymeric nanospheres, nanocapsules have a solid or liquid core, enhancing drug-loading efficiency. This core reduces the polymeric matrix content, optimizing drug encapsulation and delivery (Pohlmann *et al.*, 2002). The polymeric shell surrounding the encapsulated payload in nanocapsules serves multiple functions. Firstly, it isolates the payload from the tissue environment, preventing degradation or burst release triggered by various factors such as pH, temperature, and enzymes. Secondly, the polymeric shell offers opportunities for functionalization with smart molecules. These molecules can interact with targeted biomolecules, enabling precise targeting in drug delivery applications (Trindade *et al.*, 2018). Polymeric nanocapsules offer promising benefits in drug delivery, leading to heightened attention within the pharmaceutical industry. Extensive specialized reviews have delved into the intricate details of nanotechnologies and the selection of polymeric materials crucial for formulating effective nano-capsules (Kothamasu *et al.*, 2012). Nanocapsules are utilized in drug delivery for enhanced stability and targeted release, contributing to improved therapeutic outcomes.

Nano-emulsions

Nano-emulsions (NE) are unique colloidal dispersion systems or stable dispersions of oil droplets in water or vice versa, with droplet sizes in the nanometer range, rendering them transparent and thermodynamically stable. These emulsions consist of water and oil phases, with the addition of co-surfactants and surfactants to enhance stability (Vickers, 2017). Lipophilic drugs, entrapped within oil droplets in both oil-in-water (o/w) and water-in-oil (w/o) suspensions, are readily engulfed by macrophages. Consequently, these drugs accumulate in higher concentrations in organs like the liver, spleen, and kidneys. In contrast, hydrophilic drugs are present in the aqueous phase of water-in-oil (w/o) or water-in-oil-in-water (w/o/w) nano-emulsions. Due to their enhanced internal membrane permeability, hydrophilic drugs preferentially transit to the lymphatic system, particularly when administered via intramuscular or subcutaneous routes (Jaiswal *et al.*, 2015). The intestinal lymphatic transport system plays a significant role in the absorption of nano-emulsions (NE), facilitating their passage into the systemic circulation. This process can potentially enhance the oral bioavailability of drugs encapsulated within NE formulations (Lovelyn and Attama, 2011). Nano-emulsions (NE) exhibit key features such as stabilizing encapsulated substances, enabling sustained release, improving membrane permeability, solubilizing diverse lipophilic components, enhancing drug absorption, alleviating pain and allergies, reducing viscosity, facilitating easy production, and minimizing contamination risks. These advantageous properties have made NE suitable for delivering essential oils, nucleic acids, drugs, antimicrobial agents, repellents, and serving as imaging agents (Mahato, 2017). Nano-emulsions enhance the stability and delivery of hydrophobic compounds. They are used in food, pharmaceuticals, and cosmetics for improved solubility, bioavailability, and sensory attributes. Over recent times, nano-emulsions have been modified for therapeutic use through the skin, incorporating substances such as phospholipids, Transcutol®P, fatty alcohols, alkyl poly-glycosides, and PEGylated fatty acid esters (Kotta *et al.*, 2012). Various nano-emulsion formulations containing herbal drugs such as camptothecin, genistein, rutin, *Brucea javanica* oils, resveratrol, coixenolide, and others have been documented in scientific literature, offering diverse health advantages (Khani *et al.*, 2016).

Nanofibers

Nanofibers are ultrafine fibers with diameters in the nanometer range. They can be produced through techniques like electrospinning. Their high surface area and porosity contribute to efficient drug loading and release. Nanofibers act as drug delivery systems for topical and transdermal applications. This strategy enhances drug concentration in formulations, thereby increasing drug flux. Nanofibers extend beyond drug delivery, finding applications in wound dressings, tissue scaffolding, and controlled drug release matrices. Electrospun nanofibers, functionalized to integrate various structures and materials, broaden their potential applications (Arecchi *et al.*, 2010). Nanofibers enable the delivery of both hydrophobic and hydrophilic drugs, ensuring controlled release over an extended duration (Gupta *et al.*, 2013). Nanofibers find applications in tissue engineering, wound healing, and drug delivery.

To summarize, the nanoparticle-based formulations offer innovative solutions to challenges associated with drug solubility, stability, and bioavailability. The unique properties of nanogels, nanosuspensions, nanoliposomes, nanocapsules, nano-emulsions, and nanofibers contribute to advancements in medicine, biotechnology, and materials science, promising innovative solutions for various industries.

6.4 NANOTECHNOLOGY IN *ALOE VERA*-BASED COSMETIC PRODUCTS

Nanotechnology has revolutionized the cosmetic industry, particularly in skincare and haircare, introducing nanoparticles with unique properties and enhanced performance. The integration of nanotechnology across industries, including cosmetics, aims to boost product effectiveness and delivery methods. For *Aloe vera*-based cosmetics, nanotechnology presents numerous advantages.

Nanoparticles in skincare

Nanotechnology holds great potential in revolutionizing the diagnosis and treatment of dermatological conditions due to its ability to interact at the sub-atomic level with skin tissue. The skin serves as an excellent platform for exploring the application of nanomaterials in drug delivery, facilitating enhanced delivery of active ingredients and improving efficacy. Given its exposure to external factors like radiation and ultraviolet rays, the susceptibility of skin to damage underscores the importance of innovative approaches enabled by nanotechnology (Hadgraft *et al.*, 2003). The metallic nanoparticles, like zinc oxide and titanium dioxide, offer superior UV protection without the chalky residue associated with traditional sunscreens. Lipid-based nanoparticles, such as liposomes, facilitate targeted delivery of active ingredients, promoting better absorption and moisturization. Polymeric nanoparticles contribute to sustained release formulations, ensuring prolonged efficacy of skincare products. Nanotechnology plays a crucial role in enhancing topical drug delivery for dermatological conditions. Due to its impermeability, the skin poses a challenge for drug penetration. Nanotechnology offers solutions by modifying drug permeation, controlling active substance release, and prolonging skin contact time. This innovation optimizes treatment efficacy while minimizing systemic side effects associated with traditional systemic treatments (Hadgraft *et al.*, 2003), besides ensuring direct contact with the stratum corneum and skin appendages (Guterres *et al.*, 2007) and protecting the drug against chemical or physical instability. To maintain the natural barrier function of skin, it is preferable to deliver therapeutic agents without relying on chemical enhancers. Chemical enhancers like surfactants and organic solvents not only diminish the barrier function of skin but also lead to skin irritation and damage (Lademann *et al.*, 2007).

Nanoparticles in haircare

The structure of hair follicles allows for manipulation through various hair preparations, including cosmetics and medical formulations. Nanotechnology integration in these formulations offers benefits such as improved efficacy, reduced toxicity, and enhanced stability. Nanocarriers facilitate targeted compound delivery to the hair shaft and follicle, controlling release and maximizing interactions with hair fibers. Metallic nanoparticles like silver and gold exhibit antimicrobial properties, promoting scalp health. Nanoliposomes and nanocapsules enable the encapsulation of nourishing agents, delivering them directly to hair follicles for improved strength and vitality. Nanoparticles contribute to the development of advanced formulations that address specific haircare needs, including enhanced shine, volume, and texture. Additionally, nanostructures like silver, gold, and titanium dioxide nanoparticles may enhance optical properties and provide biomedical benefits (Mihranyan *et al.*, 2012; Santos *et al.*, 2019).

Aloe Nanocomposites

Nanotechnology has revolutionized the cosmetic industry by introducing novel formulations that improve the effectiveness and attributes of cosmetic products. *Aloe vera*, renowned for its skin-soothing and healing qualities, has been integrated into nanocomposites to create aloe nanocomposites advanced skincare and haircare solutions (Dar *et al.*, 2021). Nanocomposites like *Aloe vera*-based zinc oxide nanocomposites and chitosan-*Aloe vera* nanoparticulate systems provide distinct advantages such as heightened antibacterial effectiveness and regulated release of active components (Mohamad *et al.*, 2023). Incorporating nanoparticles into *Aloe vera*-based cosmetics has demonstrated the potential to enhance skin wellness and aesthetics. For instance, SiO_2 nanoparticles have been employed in beauty products due to their ability to prevent makeup from settling into fine lines (Yoshioka *et al.*, 2017). Furthermore, employing *Aloe vera* gel in creating nanocomposites, such as TiO_2-Fe_2O_3 nanocomposites and biodegradable films incorporating aluminum oxide nanoparticles, has shown improved characteristics suitable for skincare purposes (Kumar *et al.*, 2020; Kumar *et al.*, 2021). Additionally, the integration of *Aloe vera* with various materials in nanocomposites has resulted in the creation of innovative formulations with wide-ranging uses. An instance is the combination of *Aloe vera* gel with nano-hydroxyapatite nanorods for bone tissue engineering, highlighting the adaptability of *Aloe vera* in biomedical fields (Wang *et al.*, 2017).

To summarize, the integration of nanoparticles in skincare and haircare, along with the development of Aloe nanocomposites, represents a cutting-edge approach in the cosmetic industry. These advancements promise enhanced product performance, improved bioavailability of active ingredients, and a more personalized and effective beauty care experience for consumers.

6.5 APPLICATIONS IN AGRICULTURE

Nanotechnology, primarily utilized in health and medicine, is now emerging in agriculture and food systems with significant potential for growth. It promises to transform the agricultural sector by introducing advanced tools for disease treatment at the molecular level, rapid disease detection, and improving nutrient absorption in plants. The implementation of smart sensors and delivery systems will aid in combating crop pathogens effectively. Additionally, the forthcoming availability of nanostructured catalysts is expected to enhance the efficacy of pesticides and herbicides, enabling reduced usage. Furthermore, nanotechnology will contribute to environmental protection through the utilization of renewable energy sources and the development of filters and catalysts to mitigate pollution and address existing pollutants.

Nanofertilizers incorporating *Aloe vera* represent a breakthrough in sustainable agriculture. *Aloe vera*, known for its nutrient-rich composition and growth-promoting compounds, is integrated into nanofertilizer formulations to enhance nutrient uptake and overall plant health. The nanoscale size facilitates efficient nutrient absorption by plant roots, promoting robust growth and improved resistance to stressors. The nanoscale delivery system ensures precise and sustained release of nutrients, optimizing their availability to plants, reducing environmental impact, and enhancing the longevity of fertilizer in the soil.

6.5.1 Nanofertilizers for Soil and Plant Health

Nanofertilizer technology represents a promising innovation, although scientific literature on this topic remains limited. Traditional fertilizers have relatively low nutrient use efficiencies, with rates of 30–35% for nitrogen (N), 18–20% for phosphorus (P), and 35–40% for potassium (K). Nano-fertilizers, developed using substrates with nano-dimensions of 1–100 nm, offer a solution. Nanoparticles possess extensive surface areas, enabling them to hold abundant nutrients and release them slowly and steadily. This controlled release matches crop nutrient requirements, minimizing the negative effects associated with conventional fertilizers. The current awareness of nanofertilizer technology, coupled with the availability of inexpensive natural zeolites globally, has sparked significant commercial interest in the development of zeolite-based nanofertilizers. The study by Chuprova *et al.* (2004) highlighted the positive impact of zeolite fertilizers on the mobile humus substances of Chernozem soil and the biological productivity of maize. Additionally, another research effort revealed the efficacy of a patented nano-composite containing nitrogen (N), phosphorus (P), potassium (K), micronutrients, mannose, and amino acids in enhancing the uptake and utilization of nutrients by grain crops. The results highlight the capacity of fertilizers containing zeolite and nano-composite blends to enhance both soil quality and agricultural productivity (Jinghua, 2004). According to Bhattacharya *et al.* (2004), an optimal combination of NPK fertilizers supplemented with sulfur, zinc, boron, and molybdenum proves beneficial for enhancing pulse yields in red and lateritic soils. Additionally, adequate application of NPK fertilizers alone led to a notable increase of 13% and 38% in green and black gram yields, respectively, compared to untreated controls. According to Liu *et al.* (2006), when polystyrene, an organic material, is inserted between the layers of kaolinite clay, it creates nano and subnano-composites that act as a binding agent. These composites have the ability to control the release of nutrients from fertilizer capsules. This suggests that nanoparticles hold potential for regulating the membrane-controlled release of nutrients. Subramanian *et al.* (2008) found that nano-fertilizers and nanocomposites have the potential to regulate the nutrient release from fertilizer granules. This regulation enhances nutrient utilization efficiency and reduces the risk of nutrient fixation or loss into the environment. Rahale (2011) conducted research on nanofertilizer formulations containing nitrogen, observing their nutrient release patterns. The study found that nano-clay-based fertilizers, incorporating zeolite and montmorillonite with dimensions of 30–40 nm, released nitrogen over a longer duration (> 1000 hrs) compared to conventional fertilizers (< 500 hrs). Utilizing nanotechnology can enhance the efficacy of both natural and added nutrient sources, thereby improving crop fertilizer utilization efficiency. The growing interest in employing nanoporous zeolites in agriculture stems from concerns about the adverse impacts of chemical fertilizers on the agricultural ecosystem (Ramesh *et al.*, 2010). The available reports and patented innovations strongly indicate significant potential for the development of nano-fertilizers, suggesting extensive opportunities for their formulation and application in agriculture (Liu *et al.*, 2006; Subramanian *et al.*, 2008).

Nanofertilizers incorporating *Aloe vera* represent a breakthrough in sustainable agriculture. *Aloe vera*, known for its nutrient-rich composition and growth-promoting compounds, is integrated into nanofertilizer formulations to enhance nutrient uptake and overall plant health. The nanoscale size facilitates efficient nutrient absorption by plant roots, promoting robust growth and improved resistance to stressors. The nanoscale delivery system ensures precise and sustained release of nutrients, optimizing their availability to plants, reducing environmental impact, and enhancing the longevity of fertilizer in the soil. Nanofertilizers could serve as a viable approach to enhance crop productivity and optimize the efficiency of nutrient utilization in crops.

6.5.2 Nanosensors for Soil and Plant Health

Nanotechnology advancements have significantly impacted sensor technology with the introduction of nanosensors, leading to profound changes in science and technology. Nanosensors are now widely integrated into everyday devices like phones and electronics due to their compact size and large surface area, enabling them to efficiently detect bulk materials (Guisbiers *et al.*, 2012). Nanosensors composed of materials like graphene, metal oxides, and carbon nanotubes exhibit high efficiency, sensitivity, specificity, and rapid response time. They find diverse applications across agriculture, medicine, and industry due to their versatility and effectiveness (Joseph and Morrison, 2006). Advancements in agricultural technology led to the adoption of greenhouse management practices. Wireless sensors, a product of technological progress, can now detect and display water pH levels.

Additionally, wireless sensor networks find utility in the food industry, particularly in maintaining food freshness by monitoring cold storage facilities to prevent expiration (Mishra *et al.*, 2014).

Nanosensors incorporating *Aloe vera* components serve as advanced tools for real-time monitoring of soil and plant health. These nanosensors are designed to detect changes in soil composition, nutrient levels, and the presence of pathogens, providing valuable insights for precision agriculture.

The natural bioactive compounds of *Aloe vera*, when integrated into nanosensors, enhance their sensitivity and responsiveness to environmental factors. Nanosensors equipped with *Aloe vera* components can detect variations in soil moisture, nutrient levels, and the presence of harmful substances, allowing for timely and targeted interventions.

The nanoscale dimensions of these sensors enable them to be easily integrated into the soil, plant tissues, or irrigation systems. The real-time data generated by Aloe-based nanosensors empowers farmers to make informed decisions, optimizing resource use and crop yield while minimizing environmental impact.

To summarize, Aloe-based nanofertilizers and nanosensors represent innovative solutions for sustainable agriculture. The integration of *Aloe vera* into nanotechnology applications enhances nutrient delivery, promotes plant health, and facilitates precise monitoring of soil conditions. These advancements contribute to the development of smart and environmentally conscious farming practices, aligning with the evolving needs of modern agriculture.

6.6 THERAPEUTIC APPLICATIONS

6.6.1 Nanomedicine and Applications in the Pharmaceutical Industry

Nanotechnology has revolutionized drug delivery and therapeutic interventions in the pharmaceutical industry, offering precision and efficiency in the treatment of various diseases. Nanotechnology offers fresh avenues for leveraging *Aloe vera* in therapeutic contexts, notably within pharmaceuticals. *Aloe vera*, with its diverse bioactive compounds, has found a niche in nanomedicine, contributing to the development of advanced drug delivery systems. *Aloe vera*, long valued in traditional medicine, has undergone extensive examination due to its diverse therapeutic potential, spanning antibacterial, antiviral, anticancer, immunoregulatory, and hepatoprotective properties (Gao *et al.*, 2018). *Aloe vera* has a long history of therapeutic use, and its incorporation into pharmaceutical formulations enhances the efficacy of various medications. Using *Aloe vera* in nanomedicine holds promise for improving its therapeutic effectiveness. In the pharmaceutical industry, *Aloe vera* is utilized for its anti-inflammatory, antioxidant, and immunomodulatory properties. Specifically, *Aloe vera* has been integrated into the creation of nanomaterials for tissue engineering purposes, particularly in the context of wound healing (Kumar and Abrahamse, 2020). *Aloe vera* has been tailored for various therapeutic uses within nanocomposites, showcasing its versatility in biomedical contexts (Barud *et al.*, 2020). *Aloe vera* has been explored for its wound-healing properties, resulting in the development of *Aloe vera*-based hydrogels tailored for wound-healing applications (Chelu *et al.*, 2023; Puscaselu *et al.*, 2020). Hydrogels used for therapeutic purposes demonstrate various mechanisms for releasing therapeutic agents, including diffusion, swelling, and degradation. These mechanisms play crucial roles in enhancing wound healing outcomes (Chelu *et al.*, 2023). *Aloe vera* extracts are utilized in nanotechnology to synthesize silver nanoparticles, showcasing their antibacterial traits and promising prospects in healthcare (Ahmad *et al.*, 2022; Medina-Cruz *et al.*, 2021). Additionally, *Aloe vera* has found application in pharmaceutical formulations, serving as an ingredient in antiseptic herbal products. Furthermore, it holds promise as a therapeutic remedy for diverse ailments, such as irritable bowel syndrome and oral lichen planus (Stefi *et al.*, 2022). The wide range of therapeutic properties of *Aloe vera*, including wound healing and anti-inflammatory effects, positions it as a valuable natural asset for pharmaceutical and biomedical purposes and an effective platform for drug delivery (Ahluwalia *et al.*, 2020; Massoud *et al.*, 2022). Aloe-based nanocarriers, such as liposomes and nanogels, are designed to encapsulate pharmaceutical agents for targeted delivery. These nanocarriers enhance drug solubility, stability, and bioavailability while minimizing side effects. Therapeutic applications of Aloe in pharmaceuticals can be summarized as follows:

Wound healing: The regenerative properties of *Aloe vera* make it valuable in wound healing formulations. Nanoparticles, such as nanogels, carrying *Aloe vera* extracts can accelerate the healing process by promoting cell proliferation and reducing inflammation.

Anti-inflammatory medications: The anti-inflammatory compounds of *Aloe vera*, such as acemannan, are incorporated into pharmaceuticals for conditions like arthritis. Nanocarriers ensure the targeted delivery of these compounds to the affected areas, enhancing their therapeutic impact.

Dermatological treatments: The benefits of *Aloe vera* for skin health are harnessed in pharmaceutical products for dermatological conditions. Nanoparticles in creams or ointments provide controlled release, improving the absorption of the bioactive components of *Aloe vera.*

Gastrointestinal health: The potential of *Aloe vera* in gastrointestinal health is explored in pharmaceuticals. Nanotechnology aids in creating formulations for controlled release of *Aloe vera* compounds, contributing to the treatment of conditions like irritable bowel syndrome.

The integration of *Aloe vera* into nanomedicine and pharmaceutical formulations offers a promising avenue for a shift toward more targeted, efficient, and patient-friendly therapeutic interventions to maximize its therapeutic benefits. Nanotechnology enables the enhancement of *Aloe vera*-based products, optimizing their efficacy for wound healing, antibacterial properties, and addressing diverse medical conditions. By harnessing the natural healing properties of *Aloe vera* within nanocarriers, pharmaceuticals can offer enhanced efficacy and reduced side effects, promising a new era of advanced and personalized medicine. This innovative approach holds significant potential for advancing therapeutic solutions within the pharmaceutical industry.

6.7 BIOSAFETY CONSIDERATIONS

6.7.1 Toxicology and Environmental Considerations

Nanotechnology presents extensive opportunities in aloe research, emphasizing the importance of thoroughly assessing biosafety for the safe utilization of nanomaterials. In the realm of Aloe nanotechnology, ensuring biosafety is paramount to the responsible development and application of these innovative technologies. Biosafety considerations encompass the potential impact of Aloe-based nanoparticles on biological systems, both at the cellular and organismal levels. This evaluation requires a multidisciplinary approach, encompassing toxicology, environmental factors, and detailed material knowledge (Cheng, 2023). Nanotechnology holds significant promise in agriculture by effectively managing phytopathogens, improving nutrient uptake, and facilitating controlled pesticide and fertilizer release (Kim *et al*., 2017). Understanding the potential toxicity and adverse effects of nanomaterials, including those derived from *Aloe vera*, is vital for ensuring their safe application across different sectors, including aloe research. Therefore, conducting thorough biosafety assessments of nanomaterials is essential to guarantee their safe utilization in various fields. This includes evaluating the potential toxicological implications associated with the use of nanomaterials to mitigate any adverse effects (Guo and Mei, 2016). Furthermore, the environmental and human health impacts of emerging engineered nanomaterials, especially in the realm of nanomedicine, prompt the need for comprehensive scrutiny and risk appraisal (Burgum *et al*., 2018). The significance of conducting thorough toxicity analyses and biosafety evaluations cannot be overstated in light of the potential risk nanotechnologies pose to public health and the environment (Huang *et al*., 2022). The incorporation of nanotechnology into aloe research necessitates a thorough evaluation of toxicological and environmental factors to guarantee the safe and efficient utilization of nanomaterials. Furthermore, the advancement of nanotheranostics necessitates meticulous attention to biosafety factors to mitigate potential adverse impacts on human health (Burgum *et al.*, 2018). The potential risks associated with carbon nanotubes and other nanomaterials in drug delivery underscore the continuous need for rigorous assessment methods to ensure their safe application (Srinivasan and Palanisamy, 2022). Ensuring biosafety on biological systems requires various considerations.

a) Biocompatibility of Aloe-based nanoparticles involves the assessment of the compatibility of nanomaterials with biological systems to minimize adverse effects on living organisms. Comprehensive studies are essential to understand how Aloe nanoparticles interact with cells, tissues, and organs.
b) Cytotoxicity and immunogenicity studies to evaluate the impact of Aloe nanoparticles on cell viability and immune responses to determine if these nanoparticles induce cytotoxic effects or elicit immune reactions that could pose risks to human health.
c) Metabolism and biodistribution of Aloe nanoparticles are crucial for predicting their systemic behavior within the body, to know about the potential accumulation in specific organs, and guide dosage considerations.
d) Assessment of the long-term effects of Aloe nanoparticles is critical for predicting any chronic health implications. Studies should explore prolonged exposure scenarios to uncover potential cumulative effects that may emerge over time.
e) The responsible integration of Aloe nanotechnology also involves a thorough evaluation of potential toxicological and environmental impacts. Toxicology and environmental considerations include:
f) Environmental fate: Aloe nanoparticles released into the environment, whether through agricultural applications or cosmetic products, should be studied for their fate in soil and water systems. Understanding the potential for bioaccumulation and environmental persistence is crucial.
g) Ecotoxicity: Assessing the impact of Aloe nanoparticles on ecosystems and non-target organisms is vital. This includes studying the effects on aquatic organisms, soil microorganisms, and other components of the ecosystem.
h) Biodegradability: Evaluating the biodegradability of Aloe nanoparticles is essential for determining their environmental impact. Nanoparticles designed for sustainability should demonstrate the ability to break down into non-toxic components over time.
i) Risk mitigation: Strategies for risk mitigation, such as encapsulation techniques that enhance stability and minimize environmental release, should be explored. Understanding the potential risks allows for the development of preventive measures to safeguard both human health and the environment.

To summarize, with the advancement of Aloe nanotechnology, thorough biosafety assessments and considerations for toxicology and environmental impact are imperative. A responsible and holistic approach to research and development ensures the safe integration of Aloe-based nanoparticles in various applications, fostering sustainable and ethical nanotechnological practices.

6.8 CHALLENGES AND FUTURE DIRECTIONS

The progress in nanotechnology within aloe research presents both prospects and hurdles, shaping the trajectory of this multidisciplinary domain. Incorporating nanotechnology into aloe studies requires a comprehensive grasp of the potential uses and effects of nanomaterials. Functionalized nanoparticles play a pivotal role in propelling substantial breakthroughs in nanomaterial-driven innovations, notably in pharmaceuticals and biomedicine (Subbiah *et al.*, 2010). Nanotechnology applications in aloe research are gaining significance due to the growing priority placed on the development of nanobiosystems globally. As nanobiosystems become increasingly relevant, there is a heightened anticipation for their future impact, emphasizing the need to delve into the potential of nanotechnology within aloe research (Roco, 2003). The challenges to be addressed in aloe nanotechnology are:

a) Standardization and quality control: One significant challenge in aloe nanotechnology is the lack of standardized protocols for the synthesis, characterization, and application of aloe-based nanoparticles. Achieving consistency in quality across different formulations is essential for their widespread adoption.
b) Biosafety concerns: While aloe is generally considered safe, the introduction of nanoscale materials raises biosafety concerns. Understanding the potential cytotoxicity and immunogenicity of aloe nanoparticles is crucial for ensuring their safe use in various applications.
c) Scale-up issues: Moving from laboratory-scale synthesis to large-scale production poses challenges. Maintaining the stability and quality of aloe nanoparticles during scale-up processes is essential for their commercial viability.
d) Regulatory compliance: Aloe nanotechnology faces regulatory challenges due to the evolving nature of nanomaterial regulations. Establishing clear guidelines and standards for the production and use of aloe nanoparticles is necessary to navigate regulatory frameworks effectively.

Ensuring the safety of nano-based medications presents a significant challenge in the integration of nanotechnology into aloe research, despite considerable progress in nano-based drug delivery systems (Juliana *et al.*, 2019). The synergy of biological barcoding and nanotechnology poses both challenges and prospects for pioneering research endeavors, highlighting the imperative to investigate the applications of nanotechnology alongside DNA barcoding and biobarcoding (Munir *et al.*, 2020). The successful integration of nanotechnology in aloe research hinges on mitigating toxicity risks and conducting comprehensive biosafety evaluations of nanomaterials.
The future directions in Aloe nanotechnology include:

a) Advanced formulations for drug delivery: The future of aloe nanotechnology lies in developing advanced formulations for drug delivery. Tailoring nanoparticles for specific therapeutic applications, such as targeted cancer therapy or controlled release of bioactive compounds, holds immense potential.
b) Precision agriculture: Aloe-based nanofertilizers can revolutionize precision agriculture. Future directions involve refining formulations to address specific nutrient deficiencies, enhancing crop resilience, and promoting sustainable farming practices.
c) Smart cosmetics: Aloe nanocomposites in cosmetics are expected to evolve toward "smart" formulations. These could dynamically respond to environmental factors, providing personalized skincare solutions and adapting to individual skin needs.
d) Therapeutic nanomedicine: Aloe nanotechnology in medicine is poised for growth, with the development of nanocarriers that can navigate biological barriers for targeted drug delivery. This could lead to more effective treatments for various diseases.
e) Biodegradable nanomaterials: The future of aloe nanotechnology involves the exploration of biodegradable nanomaterials. Creating nanoparticles that break down naturally after fulfilling their intended purpose aligns with sustainability goals and minimizes environmental impact.
f) Interdisciplinary collaborations: Bridging gaps between nanotechnology, biology, and medicine will be crucial. Interdisciplinary collaborations can drive innovations in aloe nanotechnology by integrating expertise from various fields to address complex challenges.

Moving forward, realizing the full potential of nanotechnology in aloe research necessitates tackling challenges and leveraging opportunities for further advancements. As the field of aloe nanotechnology matures, addressing current challenges and embracing future directions will unlock the full potential of this innovative intersection between natural remedies and cutting-edge technology. The ongoing commitment to research, standardization, and ethical practices will shape a sustainable and impactful future for aloe nanotechnology across diverse industries.

6.9 CONCLUSION

Aloe vera research stands at the forefront of nanotechnology advancements, particularly in medicine and biology. Investigations have unveiled the potential of *Aloe vera* in nanotech drug delivery systems, yielding *Aloe vera* nanopowder and integrating it into biomaterial-based nanostructures. *Aloe vera* exhibits diverse biological functions like anti-tumor, anti-acid, tyrosinase inhibition, and antioxidant properties. Incorporating *Aloe vera* extract into bacterial nanocellulose membranes hints at its promise for skin repair. Studies underscore the role of *Aloe vera* in nanotech drug delivery, biomaterial nanostructures, food preservation, and healthcare. These interdisciplinary inquiries advocate for further exploration of the nano-applications of *Aloe vera* in diverse sectors and health domains.

REFERENCES

Adhikari, S. P., Paudel, A., Sharma, A., Thapa, B., Khanal, N., Shastri, N., & Adhikari, R. (2023). Development of decellularized fish skin scaffold decorated with biosynthesized silver nanoparticles for accelerated burn wound healing. *International Journal of Biomaterials, 2023*, 1–18.

Ahluwalia, B., Magnusson, M., Böhn, L., Störsrud, S., Larsson, F., Savolainen, O., & Öhman, L. (2020). Randomized clinical trial: Effects of aloe barbadensis mill. extract on symptoms, fecal microbiota and fecal metabolite profiles in patients with irritable bowel syndrome. *Neurogastroenterology & Motility, 32*(8), e13860.

Ahmad, B., Li, C., Satti, U. Q., Rehman, S. U., Arshad, H., Mustafa, G., & Tong, C. (2022). Phyto-synthesis, characterization, and in vitro antibacterial activity of silver nanoparticles using various plant extracts. *Bioengineering, 9*(12), 779.

Ahmed, N. K., Abbady, A., Elhassan, Y. A., & Said, A. H. (2023). Green synthesized titanium dioxide nanoparticle from *Aloe vera* extract as a promising candidate for radiosensitization applications. *BioNanoScience, 13*(2), 730–743.

Anilkumar, M., Bindu, K., Saj, A., & Anila, E. (2016). Enhanced biocompatibility of zns:mn quantum dots encapsulated with *Aloe vera* extract for therapeutic applications. *Chinese Physics B, 25*(8), 088103.

Annabi, N., Mithieux, S. M., Zorlutuna, P., Camci-Unal, G., Weiss, A. S., & Khademhosseini, A. (2013). Engineered cell-laden human protein-based elastomer. *Biomaterials, 34*(22), 5496–5505.

Ansari, M. A., Khan, H. M., Khan, A. A., Cameotra, S. S., & Alzohairy, M. A. (2015). Anti-biofilm efficacy of silver nanoparticles against MRSA and MRSE isolated from wounds in a tertiary care hospital. *Indian Journal of Medical Microbiology, 33*(1), 101–109.

Aqil, F., Munagala, R., Jeyabalan, J., & Vadhanam, M. V. (2013). Bioavailability of phytochemicals and its enhancement by drug delivery systems. *Cancer Letters, 334*(1), 133–141.

Arecchi, A., Scampicchio, M., Drusch, S., & Mannino, S. (2010). Nanofibrous membrane based tyrosinase-biosensor for the detection of phenolic compounds. *Analytica Chimica Acta, 659*(1–2), 133–136.

Arsène, M. M. J., Viktorovna, P. I., Alla, M., Mariya, M., Nikolaevitch, S. A., Davares, A. K. L., & Milana, D. (2023). Antifungal activity of silver nanoparticles prepared using *Aloe vera* extract against Candida albicans. *Veterinary World, 16*(1), 18.

Ashraf, J. M., Ansari, M. A., Khan, H. M., Alzohairy, M. A., & Choi, I. (2016). Green synthesis of silver nanoparticles and characterization of their inhibitory effects on AGEs formation using biophysical techniques. *Scientific Reports, 6*(1), 1–10.

Astefanei, A., Núñez, O., & Galceran, M. T. (2015). Characterisation and determination of fullerenes: A critical review. *Analytica Chimica Acta, 882*, 1–21.

Barud, H., Silva, R., Borges, M., Castro, G., Ribeiro, S., & Barud, H. (2020). Bacterial nanocellulose in dentistry: Perspectives and challenges. *Molecules, 26*(1), 49.

Batool, M., Qureshi, M. Z., Hashmi, F., Mehboob, N., & Shah, A. (2019). Congo red azo dye removal and study of its kinetics by *Aloe vera* mediated copper oxide nanoparticles. *Indonesian Journal of Chemistry, 19*(3), 626.

Bayda, S., Adeel, M., Tuccinardi, T., Cordani, M., & Rizzolio, F. (2019). The history of nanoscience and nanotechnology: from chemical–physical applications to nanomedicine. *Molecules, 25*(1), 112.

Bhattacharya, S. S., Mandal, D., Chattopadhyay, G. N., & Majumdar, K. (2004). Effect of balanced fertilization on pulse crop production in red and lateritic soils. *Better Crops, 88*(4), 25–27.

Biju, V., Itoh, T., Anas, A., Sujith, A., & Ishikawa, M. (2008). Semiconductor quantum dots and metal nanoparticles: Syntheses, optical properties, and biological applications. *Analytical and Bioanalytical Chemistry, 391*, 2469–2495.

Burgum, M. J., Evans, S. J., Jenkins, G. J., Doak, S. H., & Clift, M. J. (2018). Considerations for the human health implications of nanotheranostics. In Joao Conde (ed.) *Handbook of nanomaterials for cancer theranostics* (pp. 279–303). Elsevier.

Chandran, S. P., Chaudhary, M., Pasricha, R., Ahmad, A., & Sastry, M. (2006). Synthesis of gold nanotriangles and silver nanoparticles using *Aloe vera* plant extract. *Biotechnology Progress, 22*(2), 577–583.

Chauhan, P., & Kumar, A. (2020). Development of a microbial coating for cellulosic surface using *Aloe vera* and silane. *Carbohydrate Polymer Technologies and Applications, 1*, 100015.

Chelu, M., Musuc, A., Popa, M., & Moreno, J. (2023). *Aloe vera*-based hydrogels for wound healing: Properties and therapeutic effects. *Gels*, 9(7), 539.

Cheng, J. (2023). Nanomedicine for diagnosis and treatment of atherosclerosis. *Advanced Science, 10*(36), e2304294.

Chinthaginjala, H., Abdul, H., Reddy, A. P. G., Kodi, K., Manchikanti, S. P., & Pasam, D. (2020). Nanosuspension as promising and potential drug delivery: A review. 2020. *International Journal of Life Science and Pharm Research, 11*(1), P59–P66.

Chuprova, V. V., Ul'yanova, O. A., & Kulebakin, V. G. (2004). *The effect of bark-zeolites fertilizers on mobile humus sub-stances of chernozem and on biological productivity of maize. VV Chuprova*, OA Ulyanova, VG Kulebakin.

Danish, P., Ali, Q., Hafeez, M. M., & Malik, A. (2020). Antifungal and antibacterial activity of *Aloe vera* plant extract. *Biological and Clinical Sciences Research Journal, 2020*(1), e004.

Dar, A., Rehman, R., Zaheer, W., Shafique, U., & Anwar, J. (2021). Synthesis and characterization of zno-nanocomposites by utilizing *Aloe vera* leaf gel and extract of terminalia arjuna nuts and exploring their antibacterial potency. *Journal of Chemistry, 2021*, 1–7.

Das, S., Gates, A. J., Abdu, H. A., Rose, G. S., Picconatto, C. A., & Ellenbogen, J. C. (2007). Designs for ultra-tiny, special-purpose nanoelectronic circuits. *IEEE Transactions on Circuits and Systems I: Regular Papers, 54*(11), 2528–2540.

De Matteis, V., Cascione, M., Costa, D., Martano, S., Manno, D., Cannavale, A., & Rinaldi, R. (2023). *Aloe vera* silver nanoparticles addition in chitosan films: Improvement of physicochemical properties for eco-friendly food packaging material. *Journal of Materials Research and Technology, 24*, 1015–1033.

Elahi, N., Kamali, M., & Baghersad, M. H. (2018). Recent biomedical applications of gold nanoparticles: A review. *Talanta, 184*, 537–556.

Elliott, J. A., Shibuta, Y., Amara, H., Bichara, C., & Neyts, E. C. (2013). Atomistic modelling of CVD synthesis of carbon nanotubes and graphene. *Nanoscale, 5*(15), 6662–6676.

Eshun, K., & He, Q. (2004). *Aloe vera*: A valuable ingredient for the food, pharmaceutical and cosmetic industries–A review. *Critical Reviews in Food Science and Nutrition, 44*(2), 91–96.

Femenia, A., Sánchez, E. S., Simal, S., & Rosselló, C. (1999). Compositional features of polysaccharides from *Aloe vera* (Aloe barbadensis Miller) plant tissues. *Carbohydrate Polymers, 39*(2), 109–117.

Frank, L. A., Gazzi, R. P., de Andrade Mello, P., Buffon, A., Pohlmann, A. R., & Guterres, S. S. (2019). Imiquimod-loaded nanocapsules improve cytotoxicity in cervical cancer cell line. *European Journal of Pharmaceutics and Biopharmaceutics, 136*, 9–17.

Froushani, S. M. A., Zamani, A., & Abbasi, A. (2021). Cytotoxic effects of combined walnut shell-extracted copper oxide: Nanoparticles and hydroalcoholic extracts of *Aloe vera* against human leukemic cell line. *Zahedan Journal of Research in Medical Sciences, 23*(1), e94572.

Ganesan, P., & Narayanasamy, D. (2017). Lipid nanoparticles: Different preparation techniques, characterization, hurdles, and strategies for the production of solid lipid nanoparticles and nanostructured lipid carriers for oral drug delivery. *Sustainable Chemistry and Pharmacy, 6*, 37–56.

Gao, Y., Kuok, K., Jia, Y., & Wang, R. (2018). Biomedical applications of *Aloe vera*. *Critical Reviews in Food Science and Nutrition, 59*(sup1), S244–S256.

Guisbiers, G., Mejía-Rosales, S., & Deepak, F. L. (2012). Nanomaterial properties: Size and shape dependencies. *Journal of Nanomaterials, 2012*, 20–20.

Guo, X., & Mei, N. (2016). *Aloe vera*: A review of toxicity and adverse clinical effects. *Journal of Environmental Science and Health Part C, 34*(2), 77–96.

Gupta, S., Bansal, R., Gupta, S., Jindal, N., & Jindal, A. (2013). Nanocarriers and nanoparticles for skin care and dermatological treatments. *Indian Dermatology Online Journal, 4*(4), 267–272.

Guterres, S. S., Alves, M. P., & Pohlmann, A. R. (2007). Polymeric nanoparticles, nanospheres and nanocapsules, for cutaneous applications. *Drug Target Insights, 2*, 117739280700200002.

Hadgraft, J., Goosen, C., Du Plessis, J., & Flynn, G. (2003). Predicting the dermal absorption of thalidomide and its derivatives. *Skin Pharmacology and Physiology, 16*(2), 123–129.

Hęś, M., Dziedzic, K., Górecka, D., Jędrusek-Golińska, A., & Gujska, E. (2019). *Aloe vera* (L.) Webb.: Natural sources of antioxidants–A review. *Plant Foods for Human Nutrition, 74*, 255–265.

Hong, S. S., Kim, S. H., & Lim, S. J. (2015). Effects of triglycerides on the hydrophobic drug loading capacity of saturated phosphatidylcholine-based liposomes. *International Journal of Pharmaceutics, 483*(1–2), 142–150.

Huang, Z., Xie, L., Zhang, J., Li, Q., Liu, Y., Fu, X., & Li, Q. (2022). RNA-seq based toxicity analysis of mesoporous polydopamine nanoparticles in mice following different exposure routes. *Frontiers in Bioengineering and Biotechnology, 10*, 893608.

Iqbal, F., & Ahmed, A. (2021). Antibacterial activity of Aloe barbadensis Mill. *Polish Journal of Environmental Studies, 30*(4), 3637–3643.

Iwasaki, Y., Kondo, J. I., Kuzuya, A., & Moriyama, R. (2016). Crosslinked duplex DNA nanogels that target specified proteins. *Science and Technology of Advanced Materials, 17*(1), 285–292.

Jacob, S., Nair, A. B., & Shah, J. (2020). Emerging role of nanosuspensions in drug delivery systems. *Biomaterials Research, 24*(1), 3.

Jaiswal, M., Dudhe, R., & Sharma, P. K. (2015). Nanoemulsion: An advanced mode of drug delivery system. *3 Biotech, 5*, 123–127.

Jeevanandam, J., Barhoum, A., Chan, Y. S., Dufresne, A., & Danquah, M. K. (2018). Review on nanoparticles and nanostructured materials: History, sources, toxicity and regulations. *Beilstein Journal of Nanotechnology, 9*(1), 1050–1074.

Jeevanandam, J., San Chan, Y., Wong, Y. J., & Hii, Y. S. (2020). Biogenic synthesis of magnesium oxide nanoparticles using Aloe barbadensis leaf latex extract. In *IOP Conference Series: Materials Science and Engineering* (Vol. 943, No. 1, p. 012030). IOP Publishing.

Jinghua, G. (2004). Synchrotron radiation, soft X-ray spectroscopy and nano-materials. *Journal of Nanotechnology, 1*, 193–225.

Joseph, B., & Raj, S. J. (2010). Pharmacognostic and phytochemical properties of *Aloe vera* linn an overview. *International Journal of Pharmaceutical Sciences Review and Research, 4*(2), 106–110.

Joseph, T., & Morrison, M. (2006). *Nanotechnology in agriculture and food: A nanoforum report*. www.Nanoforum.org.

Juliana, F., Kesse, S., Boakye-Yiadom, K., Veroniaina, H., & Sun, M. (2019). Promising approach in the treatment of glaucoma using nanotechnology and nanomedicine-based systems. *Molecules, 24*(20), 3805.

Kaotar, S., Mansouri, E., & Ali, A. (2022). Investigation of Aloe vera latex used as natural dye in TiO2-based heterojunction and liquid-electrolyte dye-sensitized solar cells. *Journal of Solar Energy Engineering, 144*(5), 051002.

Kavyashree, D., Shilpa, C., Nagabhushana, H., Prasad, B. D., Sreelatha, G. L., Sharma, S., & Premkumar, H. (2015). Zno superstructures as an antifungal for effective control of malassezia furfur, dermatologically prevalent yeast: prepared by *Aloe vera* assisted combustion method. *ACS Sustainable Chemistry &Amp; Engineering, 3*(6), 1066–1080.

Kesisoglou, F., Panmai, S., & Wu, Y. (2007). Nanosizing-oral formulation development and biopharmaceutical evaluation. *Advanced Drug Delivery Reviews, 59*(7), 631–644.

Khan, I., Saeed, K., & Khan, I. (2019). Nanoparticles: Properties, applications and toxicities. *Arabian Journal of Chemistry, 12*(7), 908–931.

Khani, S., Keyhanfar, F., & Amani, A. (2016). Design and evaluation of oral nanoemulsion drug delivery system of mebudipine. *Drug Delivery, 23*(6), 2035–2043.

Kim, D., Kadam, A., Shinde, S., Saratale, R., Patra, J., & Ghodake, G. (2017). Recent developments in nanotechnology transforming the agricultural sector: A transition replete with opportunities. *Journal of the Science of Food and Agriculture*, *98*(3), 849–864.

Konishi, Y., Ohno, K., Saitoh, N., Nomura, T., Nagamine, S., Hishida, H., & Uruga, T. (2007). Bioreductive deposition of platinum nanoparticles on the bacterium Shewanella algae. *Journal of Biotechnology*, *128*(3), 648–653.

Kothamasu, P., Kanumur, H., Ravur, N., Maddu, C., Parasuramrajam, R., & Thangavel, S. (2012). Nanocapsules: The weapons for novel drug delivery systems. *BioImpacts: BI*, *2*(2), 71.

Kotta, S., Khan, A. W., Pramod, K., Ansari, S. H., Sharma, R. K., & Ali, J. (2012). Exploring oral nanoemulsions for bioavailability enhancement of poorly water-soluble drugs. *Expert Opinion on Drug Delivery*, *9*(5), 585–598.

Kumar, H., Gehlaut, A., Gupta, H., Gaur, A., Kamsonlian, S., & Kumar, D. (2021). Facile synthesis and application of aluminum oxide nanoparticle based biodegradable film. *Polymer Composites*, *42*(8), 3899–3910.

Kumar, M., Abebe, B., Nagaswarupa, H., Murthy, H., Ravikumar, C., & Sabir, F. (2020). Enhanced photocatalytic and electrochemical performance of tio2-fe2o3 nanocomposite: Its applications in dye decolorization and as supercapacitors. *Scientific Reports*, *10*(1), 1249.

Kumar, S. & Abrahamse, H. (2020). Advancement of nanobiomaterials to deliver natural compounds for tissue engineering applications. *International Journal of Molecular Sciences*, *21*(18), 6752.

Kyriacou, S. V., Brownlow, W. J., & Xu, X. H. N. (2004). Using nanoparticle optics assay for direct observation of the function of antimicrobial agents in single live bacterial cells. *Biochemistry*, *43*(1), 140–147.

Lademann, J., Richter, H., Teichmann, A., Otberg, N., Blume-Peytavi, U., Luengo, J., & Sterry, W. (2007). Nanoparticles–an efficient carrier for drug delivery into the hair follicles. *European Journal of Pharmaceutics and Biopharmaceutics*, *66*(2), 159–164.

Lai, F., Sinico, C., Ennas, G., Marongiu, F., Marongiu, G., & Fadda, A. M. (2009). Diclofenac nanosuspensions: Influence of preparation procedure and crystal form on drug dissolution behaviour. *International Journal of Pharmaceutics*, *373*(1–2), 124–132.

Liu, H., Liu, S., Du, B., Dong, K., Wang, Y., & Zhang, Y. (2021). *Aloe vera* gel coating aggravates superficial scald incidence in 'Starking' apples during low-temperature storage. *Food Chemistry*, *339*, 128151.

Liu, M., Liang, R., Zhan, F., Liu, Z., & Niu, A. (2006). Synthesis of a slow-release and superabsorbent nitrogen fertilizer and its properties. *Polymers for Advanced Technologies*, *17*(6), 430–438.

Lo, P. K., Karam, P., Aldaye, F. A., McLaughlin, C. K., Hamblin, G. D., Cosa, G., & Sleiman, H. F. (2010). Loading and selective release of cargo in DNA nanotubes with longitudinal variation. *Nature Chemistry*, *2*(4), 319–328.

Logaranjan, K., Raiza, A. J., Gopinath, S. C., Chen, Y., & Pandian, K. (2016). Shape-and size-controlled synthesis of silver nanoparticles using *Aloe vera* plant extract and their antimicrobial activity. *Nanoscale Research Letters*, *11*, 1–9.

Lovelyn, C., & Attama, A. A. (2011). Current state of nanoemulsions in drug delivery. *Journal of Biomaterials and Nanobiotechnology*, *2*(05), 626.

Łukasiewicz, S., Szczepanowicz, K., Podgórna, K., Błasiak, E., Majeed, N., Ogren, S. O. Ö., & Dziedzicka-Wasylewska, M. (2016). Encapsulation of clozapine in polymeric nanocapsules and its biological effects. *Colloids and Surfaces B: Biointerfaces*, *140*, 342–352.

Mahato, R. (2017). Nanoemulsion as targeted drug delivery system for cancer therapeutics. *Journal of Pharmaceutical Sciences and Pharmacology*, *3*(2), 83–97.

Mamatha, B., Srilatha, D., Sivanarayani, C. H., Kumar Desu, P., & Venkateswara Rao, P. (2017). Co-processed excipients: An overview. *World Journal of Pharmaceutical Research*, *6*, 224–237.

Marshall, J. M. (1990). *Aloe vera* gel: What is the evidence. *Pharmaceutical Journal*, *24*, 360–362.

Massoud, D., Alrashdi, B., Fouda, M., El-kott, A., Soliman, S., & Abd-Elhafeez, H. (2022). *Aloe vera* and wound healing: A brief review. *Brazilian Journal of Pharmaceutical Sciences*, *58*, e20837.

Medina-Cruz, D., Vernet-Crua, A., Mostafavi, E., González, M., Martínez, L., Jones, A., & Webster, T. (2021). *Aloe vera*-mediated te nanostructures: Highly potent antibacterial agents and moderated anticancer effects. *Nanomaterials*, *11*(2), 514.

Mieszawska, A. J., Mulder, W. J. M., Fayad, Z. A., & Cormode, D. P. (2013). Multifunctional gold nanoparticles for diagnosis and therapy of disease. *Molecular Pharmaceutics*, *10*(3), 831–847.

Mihranyan, A., Ferraz, N., & Strømme, M. (2012). Current status and future prospects of nanotechnology in cosmetics. *Progress in Materials Science*, *57*(5), 875–910.

Mishra, V., Mishra, R. K., Dikshit, A., & Pandey, A. C. (2014). Interactions of nanoparticles with plants: An emerging prospective in the agriculture industry. In Parvaiz Ahmad and Saiema Rasool (eds.) *Emerging technologies and management of crop stress tolerance* (pp. 159–180). Academic press.

Mohamad, E., Gad, A., El-Rhman, R., & Darwish, M. (2023). Chitosan and *Aloe vera* decorated nanoparticulate system loaded with minoxidil as a suggested topical formulation for alopecia therapy. *Advances in Natural Sciences Nanoscience and Nanotechnology*, *14*(2), 025002.

Molina, M., Asadian-Birjand, M., Balach, J., Bergueiro, J., Miceli, E., & Calderón, M. (2015). Stimuli-responsive nanogel composites and their application in nanomedicine. *Chemical Society Reviews*, *44*(17), 6161–6186.

Munir, S., Ahmed, S., Ibrahim, M., Khalid, M., & Ojha, S. (2020). A spellbinding interplay between biological barcoding and nanotechnology. *Frontiers in Bioengineering and Biotechnology*, *8*, 556291.

Munirathnam, R., Vidya, Y. S., Manjunatha, H. C., Seenappa, L., Sridhar, K. N., Manjunatha, S., & Kumar, S. (2023). *Aloe vera* mediated synthesis of silver doped cerium oxide nanoparticles for display and supercapacitor applications. *Surfaces and Interfaces*, *41*, 103148.

Nile, S. H., Baskar, V., Selvaraj, D., Nile, A., Xiao, J., & Kai, G. (2020). Nanotechnologies in food science: Applications, recent trends, and future perspectives. *Nano-Micro Letters*, *12*, 1–34.

Nourozi, F., & Sayyari, M. (2020). Enrichment of *Aloe vera* gel with basil seed mucilage preserve bioactive compounds and postharvest quality of apricot fruits. *Scientia Horticulturae*, *262*, 109041.

Pandit, C., Roy, A., Ghotekar, S., Khusro, A., Islam, M. N., Emran, T. B., ... & Bradley, D. A. (2022). Biological agents for synthesis of nanoparticles and their applications. *Journal of King Saud University-Science*, *34*(3), 101869.

Park, H. J., Shin, D. J., & Yu, J. (2021). Categorization of quantum dots, clusters, nanoclusters, and nanodots. *Journal of Chemical Education*, *98*(3), 703–709.

Patravale, V. B., Date, A. A., & Kulkarni, R. M. (2004). Nanosuspensions: A promising drug delivery strategy. *Journal of Pharmacy and Pharmacology*, *56*(7), 827–840.

Perni, S., Prokopovich, P., Pratten, J., Parkin, I. P., & Wilson, M. (2011). Nanoparticles: their potential use in antibacterial photodynamic therapy. *Photochemical Photobiological Sciences*, *10*(5), 712–720.

Pohlmann, A. R., Weiss, V., Mertins, O., da Silveira, N. P., & Guterres, S. S. (2002). Spray-dried indomethacin-loaded polyester nanocapsules and nanospheres: Development, stability evaluation and nanostructure models. *European Journal of Pharmaceutical Sciences, 16*(4–5), 305–312.

Preetha, P. S., & Balakrishnan, N. (2017). A review of nano fertilizers and their use and functions in soil. *International Journal of Current Microbiology and Applied Sciences*, *6*(12), 3117–3133.

Primo, J. d. O., Bittencourt, C., Acosta, S., Sierra-Castillo, A., Colomer, J., Jaerger, S., & Anaissi, F. J. (2020). Synthesis of zinc oxide nanoparticles by ecofriendly routes: Adsorbent for copper removal from wastewater. *Frontiers in Chemistry, 8. Frontiers in Chemistry*, 8, 571790.

Puscaselu, R., Lobiuc, A., Dimian, M., & Covașă, M. (2020). Alginate: from food industry to biomedical applications and management of metabolic disorders. *Polymers*, *12*(10), 2417.

Radha, M. H., & Laxmipriya, N. P. (2015). Evaluation of biological properties and clinical effectiveness of Aloeávera: A ásystematic review. *Journal of Traditional and Complementary Medicine*, *5*(1), 21–26.

Rahale, S. (2011). Nutrient release pattern of nanofertilizer formulation. PhD (Agri.) Thesis. Tamilnadu Agricultural University.

Ramesh, K., Biswas, A. K., Somasundaram, J., & Rao, A. S. (2010). Nanoporous zeolites in farming: Current status and issues ahead. *Current Science*, *99*(6), 760–764.

Riaz, M., Sharafat, U., Zahid, N., Ismail, M., Park, J., Ahmad, B., & Tabassum, A. (2022). Synthesis of biogenic silver nanocatalyst and their antibacterial and organic pollutants reduction ability. *ACS Omega*, *7*(17), 14723–14734.

Robson, A. L., Dastoor, P. C., & Hua, S. (2018). Advantages and limitations of current imaging techniques for characterizing liposome morphology. *Frontiers in Pharmacology*, *9*, 328115.

Roco, M. (2003). Nanotechnology: convergence with modern biology and medicine. *Current Opinion in Biotechnology*, *14*(3), 337–346.

Routray, B. S., & Patra, N. (2021). Preparation and characterization of solid lipid nanoparticles of cinnacalcet HCl. *Journal of Pharmaceutical Research International*, 173–184. 10.9734/jpri/2021/v33i49B33354.

Santos, A. C., Panchal, A., Rahman, N., Pereira-Silva, M., Pereira, I., Veiga, F., & Lvov, Y. (2019). Evolution of hair treatment and care: Prospects of nanotube-based formulations. *Nanomaterials*, *9*(6), 903.

Sarkar, P. K., Paul, S., Sahu, K., Mandal, S., Islam, Q. A., & Chakrabarty, R. (2023). A greener synthesis of silver nanoparticles using tea leaf and *Aloe vera* leaf extract for catalytic organic dye degradation application. *ChemistrySelect*, *8*(28), e202301906.

Schneider, A. (2010). *Amid nanotech's dazzling promise, health risks grow*. AOL News.

Seifunnisha, O., & Shanthi, J. (2021). Influence of Aloe vera and pluronic f127 on the enhancement of photocatalytic degradation and antibacterial activity of ZnO by sol-gel synthesis method. *Optik*, 248, 168064.

Shalabi, M., Khilo, K., Zakaria, M. M., Elsebaei, M. G., Abdo, W., & Awadin, W. (2015). Anticancer activity of *Aloe vera* and Calligonum comosum extracts separetely on hepatocellular carcinoma cells. *Asian Pacific Journal of Tropical Biomedicine*, *5*(5), 375–381.

Shang, Y., Hasan, M. K., Ahammed, G. J., Li, M., Yin, H., & Zhou, J. (2019). Applications of nanotechnology in plant growth and crop protection: A review. *Molecules*, *24*(14), 2558.

Shi, X., Wang, S., Meshinchi, S., Van Antwerp, M. E., Bi, X., Lee, I., & Baker Jr, J. R. (2007). Dendrimer-entrapped gold nanoparticles as a platform for cancer-cell targeting and imaging. *Small*, *3*(7), 1245–1252.

Sigmund, W., Yuh, J., Park, H., Maneeratana, V., Pyrgiotakis, G., Daga, A., ... & Nino, J. C. (2006). Processing and structure relationships in electrospinning of ceramic fiber systems. *Journal of the American Ceramic Society*, *89*(2), 395–407.

Srinivasan, V., & Palanisamy, P. (2022). Carbon nanotubes: an optimistic nanomaterial with superfluity characteristics in drug delivery for the treatment of arthritis. *Indian Journal of Pharmaceutical Education and Research*, *56*(4s), s627–s650.

Srivastava, S., & Bhargava, A. (2022). *Green nanoparticles: The future of nanobiotechnology* (pp. 1–352). Springer.

Stefi, S., Joseph, J., A, M., Kumar, V., Suresh, A., & Thirunavukarasu, R. (2022). Evaluation and formulation of antiseptic herbal hand wipes. *Asia-Pacific Journal of Pharmacotherapy & Toxicology,* 11–17.

Subbiah, R., Veerapandian, M., & Yun, K. (2010). Nanoparticles: functionalization and multifunctional applications in biomedical sciences. *Current Medicinal Chemistry*, *17*(36), 4559–4577.

Subramanian, K. S., Paulraj, C., & Natarajan, S. (2008). Nanotechnological approaches in nutrient management. *Nanotechnology Applications in Agriculture*, *61*, 37–42.

Surjushe, A., Vasani, R., & Saple, D. (2008). *Aloe vera*: A short review. *Indian Journal of Dermatology*, *53*(4), 163–166.

Thangadurai, D., Sangeetha, J., & Prasad, R. (Eds.). (2020). *Nanotechnology for food, agriculture, and environment*. Springer.

Thangapazham, R. L., Puri, A., Tele, S., Blumenthal, R., & Maheshwari, R. K. (2008). Evaluation of a nanotechnology-based carrier for delivery of curcumin in prostate cancer cells. *International Journal of Oncology*, *32*(5), 1119–1123.

Trindade, I. C., Pound-Lana, G., Pereira, D. G. S., de Oliveira, L. A. M., Andrade, M. S., Vilela, J. M. C., & Mosqueira, V. C. F. (2018). Mechanisms of interaction of biodegradable polyester nanocapsules with non-phagocytic cells. *European Journal of Pharmaceutical Sciences, 124*, 89–104.

Vélez, E., Campillo, G., Morales, G., Hincapié, C., Osorio, J., & Arnache, O. (2018). Silver nanoparticles obtained by aqueous or ethanolic *Aloe vera* extracts: An assessment of the antibacterial activity and mercury removal capability. *Journal of Nanomaterials, 2018*, 7215210.

Vickers, N. J. (2017). Animal communication: When I'm calling you, will you answer too?. *Current Biology*, *27*(14), R713–R715.

Vigneshwaran, N., Ashtaputre, N. M., Varadarajan, P. V., Nachane, R. P., Paralikar, K. M., & Balasubramanya, R. H. (2007). Biological synthesis of silver nanoparticles using the fungus Aspergillus flavus. *Materials Letters*, *61*(6), 1413–1418.

Wang, H., Zhang, X., Mani, M., Jaganathan, S., Huang, Y., & Wang, C. (2017). Microwave-assisted dip coating of *Aloe vera* on metallocene polyethylene incorporated with nano-rods of hydroxyapaptite for bone tissue engineering. *Coatings*, *7*(11), 182.

Willner, I., Baron, R., & Willner, B. (2006). Growing metal nanoparticles by enzymes. *Advanced Materials*, *18*(9), 1109–1120.

Xavier, S. A., Wahab, A., & Rajeshkumar, S. (2022). Cytotoxic effects of fresh *Aloe vera* gel mediated silver nano composite. *International Journal of Health Sciences*, *6*(S4), 3594–3600.

Ying, S., Guan, Z., Ofoegbu, P. C., Clubb, P., Rico, C., He, F., & Hong, J. (2022). Green synthesis of nanoparticles: Current developments and limitations. *Environmental Technology & Innovation*, *26*, 102336.

Yoshioka, Y., Kuroda, E., Hirai, T., Tsutsumi, Y., & Ishii, K. (2017). Allergic responses induced by the immunomodulatory effects of nanomaterials upon skin exposure. *Frontiers in Immunology*, *8*, 169.

Zielińska, A., Carreiró, F., Oliveira, A. M., Neves, A., Pires, B., Venkatesh, D. N., & Souto, E. B. (2020). Polymeric nanoparticles: Production, characterization, toxicology and ecotoxicology. *Molecules*, *25*(16), 3731.

7 *Aloe vera* Germplasm Characterization and Gene Discovery

Sonia Sihag, Alka Jangra, Garima Sharma, Khushbu Dawra, and Vinod Chhokar

7.1 INTRODUCTION

The utilization of *Aloe vera* as a medicinal plant dates back to 6000 B.C. Archaeological evidence from plates dating to the Sumerian period around 2200 years B.C. suggests the plant's use for medicinal purposes (Ijinu, George, and Pushpangadan 2022). *Aloe vera*, also known by its synonym Aloe barbadensis Miller, belongs to the Liliaceae family, which comprises approximately 360 species. In contrast, Aloe capensis, commonly referred to as Cape aloes, belongs to a distinct species (Vogler and Ernst 1999). *Aloe vera*, a semi-tropical plant known for its versatile herbal properties, has regained prominence in the global market. Extracts derived from its leaves, containing therapeutic and nutritive substances, are extensively utilized in commercial formulations for pharmaceuticals, cosmetics, and as a fresh food source (Cristiano et al. 2016). The plant has a rich history of medicinal use spanning millennia across various cultures, including Greece, Egypt, India, Mexico, Japan, and China. It is essential to note that the majority of these claims are rooted in historical practices rather than being supported by substantial scientific evidence. *Aloe vera* is well known for its marvelous medicinal properties (Jadhav et al. 2020). *Aloe vera* is renowned globally as a premier natural agent, exhibiting a spectrum of properties such as antiseptic, antibiotic, antimicrobial, antiviral, detoxifying, antibacterial, anti-allergic, anti-inflammatory, anti-ulcer, anti-diabetic, anti-tumor, antifungal, immunostimulant, and a potent intracellular antioxidant (Mangaiyarkarasi et al. 2015). Analgesic, tonic, pain reliever, antiasthmatic, antileukopenic, antipyretic, anthelmintic, anticoagulant, demulcent, diuretic, emollient, hair stimulant, hypoglycemic, insecticide, liver stimulant, local anaesthetic, rejuvenating, appetite stimulant, stomachache remedy, uterine stimulant, and parasite killer are some of the many uses for this amazing plant (Miko Ajczak 2018). These plants have been widely grown as ornamental plants. Due to its versatile applications, Aloe has been cultivated as a houschold plant. It has served as a traditional remedy for generations.

Although *Aloe vera* is now extensively found in arid and semi-arid places all over the world, it is said to have originated in the Arabian Peninsula. It is often found in Africa, the Middle East, and some regions of Asia, and it grows well in a range of climates, from tropical to subtropical (Martínez-Burgos et al. 2022). For its therapeutic and cosmetic qualities, *Aloe vera* is grown both locally and commercially. It can be spread by leaf cuttings, offshoots (pups), and seeds (Alamgir and Alamgir 2017). *Aloe vera* germplasm represents the genetic diversity of this versatile plant, which has been valued for its medicinal and cosmetic uses throughout history. Conservation efforts and ongoing research contribute to the sustainable cultivation and utilization of *Aloe vera* for various purposes (Gantait, Sinniah, and Das 2014). Due to the vegetative mode of reproduction, plants can retain spontaneous cytological and genetic changes in their genome. Given the extensive cultivation of Aloe for over a century in India, the potential for genotypic variability among different clones

DOI: 10.1201/9781003603108-7

cannot be dismissed. It has become essential to thoroughly investigate Aloe germplasm, ranging from household potted plants to those in commercial cultivation, to identify any existing natural variability (Jadhav et al. 2020). This study aims to analyze *Aloe vera* germplasm for variations in morphology, cytology, and biochemistry.

7.2 GENETIC RESOURCES AND GERMPLASM COLLECTION

7.2.1 Sources for Germplasm Collection

Building an extensive germplasm collection for *Aloe vera* entails assembling a variety of genetic resources from different parts of the world. These collections are essential for breeding programs, research, conservation, and sustainable use (Cowell et al. 2022). According to various sources, Grace et al. (2009) suggest that there are 548 accepted species within the genus Aloe. Earlier, Steenkamp and Stewart (2007) mentioned that there are approximately 500 species within the genus Aloe, with 160 of them being indigenous to South Africa. Recent studies by Grace et al. (2009) have described over 500 species within the Aloe genus, all belonging to the family Xanthorrhoeaceae. However, the International Aloe Science Council (IASC) has listed 449 species in its Aloe Scientific Primer. Despite the large number of Aloe species, only about ten are commonly used in herbal medicine, and fewer than ten are considered safe for consumption. It is noteworthy that approximately four percent of Aloe species are classified as "most endangered" according to the criteria of the International Union for Conservation of Nature (IUCN) Red List (X. Liu et al. 2011).

Aloe species, including *Aloe vera*, are kept in collections by numerous botanical gardens and arboreta across the world. These organizations frequently house living collections, specimens from herbaria, and related databases that provide information about the genetic diversity and geographical origins of the plants (Demissew 2005). Universities and research institutions with botanical or agricultural departments may have established collections of *Aloe vera* germplasm. These collections may include living plants, preserved specimens, and genetic material such as seeds, tissue cultures, or DNA samples (Bedigian 2005). Various countries maintain germplasm repositories or gene banks dedicated to preserving plant genetic resources. Institutions like the International Plant Genetic Resources Institute (IPGRI) or regional gene banks often have collections of Aloe species, including *Aloe vera*, sourced from different regions around the world (Priyanka et al. 2021). The Germplasm Resources Information Network (GRIN) and the Global Biodiversity Information Facility (GBIF) are two examples of online databases and germplasm repositories that offer digital records and distribution data on *Aloe vera* germplasm collections held by different institutions throughout the world (Bedigian 2005).

7.2.2 Genetic Diversity Assessment

Understanding germplasm diversity and genetic relationships within a germplasm collection is crucial for conservation and crop improvement strategies. Genetic diversity constitutes a crucial aspect of biodiversity examination, encompassing the full range of genetically driven distinctions observed among individuals within a species regarding the manifestation of specific traits. This variability in genes is shaped by various environmental factors such as geographical conditions, seasonal variations, and soil characteristics (Lal et al. 2012). The extent of genetic diversity within a population directly impacts its ability to adapt to diverse environmental conditions (Briggs and Walters 2016). Understanding Aloe species holds significant potential for various aspects of biodiversity conservation, sustainable commercial use, health applications, ecosystem restoration, genetic diversity analysis, and ecotourism (Briggs and Walters 2016). Molecular markers provide a means to assess diversity independent of environmental influences and can be applied to DNA from any growth stage (Mondini, Noorani, and Pagnotta 2009).

Lattoo et al. (2008) suggest that the best estimation of genetic diversity is achieved when utilizing a combination of agro-morphological, biochemical, and marker studies together. In contemporary times, molecular markers have emerged as vital tools for efficiently evaluating genetic diversity in plants. Among the array of molecular markers available, Random Amplified Polymorphic DNA (RAPD) stands out as the most widely utilized. RAPD is advantageous as it does not necessitate prior genomic information and is simpler and more cost-effective compared to other DNA marker techniques (Lattoo et al. 2008; Khan et al. 2009).

SSR (Simple Sequence Repeat), also known as microsatellite markers, is deemed to be highly informative and powerful for assessing the genetic variability of *Aloe vera*. Genetic diversity offers estimates regarding the level of genetic variation among various materials. These estimates can be invaluable in evaluating the stability and purity of genotypes used in breeding or multiplication initiatives. The importance of the SSR marker system is shown by its effectiveness, which supposedly encompasses the entire genome. For medicinal plants like *Aloe vera*, SSR marker analysis might therefore be extremely important for developing collection plans and germplasm conservation initiatives (Kumar et al. 2015).

Among the various marker systems available, AFLP (Amplified Fragment Length Polymorphism) stands out as a highly effective technique. AFLP is a multilocus DNA fingerprinting method that is considered close to ideal for evaluating the genetic variability of species, groups, and individuals. Its main advantages include high levels of polymorphism and discriminative capacity, particularly beneficial for closely related accessions (Tripathi, Saini, and Tiwari 2011). AFLPs have been widely utilized for estimating genetic relationships and authenticating many medicinal plant species, such as Swertia (Misra et al. 2010), Tribulus terrestris (Sarwat, Das, and Srivastava 2008). Comparative studies involving various marker systems like Restriction Fragment Length Polymorphism (RFLP), Random Amplification of Polymorphic DNA (RAPD), AFLP, and microsatellites have consistently demonstrated AFLP's efficiency in estimating genetic diversity. This is attributed to its high reproducibility and multiplex ratio, making AFLP a preferred choice for genetic diversity assessment in many plant species (Tripathi, Saini, and Tiwari 2011).

7.3 MORPHOLOGICAL CHARACTERIZATIONS

Morphological characterization of *Aloe vera* involves a detailed examination and description of its physical traits, which is essential for taxonomy, identification, and understanding of its genetic diversity. These include characteristics such as plant height and width, leaf size and shape, fruit size, shape, and color, seed size, pigmentation, and other visually distinguishable features, collectively referred to as morphological markers. Due to the vast array of morphological variations present in plants, traditional morphological characterization remains a valuable technique for assessing plant diversity. It serves as a conventional method for species identification and is instrumental in breeding programs and plant improvement efforts (Rana, Sharma, and Bakshi 2018). Morphological characterization provides valuable information for plant breeders, horticulturists, and researchers studying *Aloe vera*. By documenting and analyzing physical traits, they can classify different varieties, assess genetic diversity, and develop cultivation techniques tailored to specific characteristics or desired traits (A. Das et al. 2016).

Alagukannan and Ganesh (2006) examined twenty-one *Aloe vera* ecotypes sourced from various locations in South India, including wild habitats and research institutions. Their investigation revealed significant diversity in growth, yield, and quality attributes among the tested ecotypes. One standout ecotype, named TCY, exhibited remarkable characteristics such as tall plant height (79.31 cm), expansive North-South spread (85.38 cm), rapid growth rates in terms of height rise (8.64 cm per month) and leaf production (3.94 leaves per month), as well as a high leaf count per plant (29.50). Additionally, TCY displayed impressive leaf dimensions (63.10 cm in length, 10.92 cm in width), substantial leaf weight (434.30 g), and notable aloin yield per leaf (0.32 g).

On the other hand, another ecotype known as TVM showcased superior traits in yield and quality parameters. Noteworthy attributes of TVM included thick leaves (2.74 cm), substantial leaf volume (429.27 cm3), high leaf yield per plant (1597.37 g), generous gel yield per leaf (282.57 g), as well as elevated levels of total solids (0.81%) and soluble solids in the gel (0.71%). Although TCY demonstrated exceptional growth characteristics, TVM outshone in terms of yield and quality parameters.

Kaur (2011) conducted a screening of 60 *Aloe vera* accessions from North India, assessing morphological, Cytological, biochemical, and molecular diversity. Morphologic analysis categorized the 60 accessions into distinct groups based on height, color and orientation. Furthermore, accessions were divided into two groups depending on the color of their spines (green and red). In order to analyze the diversity within and between species, Nayanakantha et al. (2010) evaluated the morphometric variables of 11 Aloe germplasm accessions. For comparison, measurements of height of the plant, leaf length, leaf breadth, and thickness of leaf were taken for each Aloe accession. The New Ornamentals Society (NOS) color scheme was used to select the color of the leaves. The morphological assessment revealed qualitative variations among the studied accessions. *Aloe zeylanicum* emerged as the tallest species (53 cm), characterized by a different caulescent stem with elongated spaces between nodes and linear leaf shape. Conversely, Aloe lotus exhibited the shortest height (17 cm), resembling common cactus plants with ovate-lanceolate-shaped and broad leaves (4.5 cm). *Aloe perryi* was distinguished as a blue-green variety due to its leaf coloration, featuring a caulescent growth habit with shorter internodal lengths compared to A. zeylanicum. Notable variations were observed within *Aloe vera* strains, particularly in terms of plant height, leaf size, thickness, and width, with minimal fluctuations in leaf color. Furthermore, DARL-3 displayed the highest leaf thickness (1.1 cm) among the accessions studied. Subsequently, these accessions underwent RAPD analysis for further characterization.

Nayanakantha et al. (2010) examined the morphometric characteristics of eleven Aloe germplasm accessions to evaluate diversity within and among species. They documented measurements such as height of the plant, length, width, and thickness of the leaves, in addition to recording leaf color using the New Ornamentals Society (NOS) color chart. Morphological evaluation revealed a variation in quality among the different accessions. Aloe zeylanicum measured the tallest at 53 cm due to its caulescent stem with long internodes and linear leaf shape, unlike others, except A. lotus. It had the minimum leaf thickness (0.4 cm) and width (1.0 cm), indicating lower gel content. Aloe lotus, the shortest (17 cm), resembled a common cactus with ovate-lanceolate leaves, the broadest (4.5 cm) among Aloe accessions. *A. perryi* displayed a blue-green leaf color, similar to *A. zeylanicum* but with shorter internodes. Leaf dimensions slightly differed from those of other *Aloe vera* strains. Noticeable variations in plant morphological characteristics were detected among *Aloe vera* strains, with minimal variation in leaf color. DARL-3 had the highest thickness (1.1 cm) of leaf. These accessions underwent RAPD analysis to assess genetic variation at the molecular level.

In 2018, Kiran and Tirkey's research was to explore the genetic variability, heritability (h^2), and genetic advance for eight traits in *Aloe vera*. Thirteen genotypes of *Aloe barbadensis* were evaluated to analyze Phenotypic Coefficient of Variation (PCV), Genotypic Coefficient of Variation (GCV), Genetic Advance (GA%), and diversity patterns. They found significant variability in plant morphological characteristics such as height (cm), leaf length (cm), number of spines per leaf, number of leaves per plant, number of suckers, and gel content per leaf (g). PCV values slightly exceeded GCV, suggesting a modest environmental impact on genotype expression. Characteristics include plant height (cm), leaf length (cm), number of spines per leaf, number of suckers, and gel content per leaf (g). They exhibited high h^2 coupled with substantial genetic advance, suggesting effective selection possibilities for these traits. Diversity analysis using D^2 statistics grouped the genotypes into four clusters. Cluster IV accommodated the maximum number of genotypes (5), followed by clusters II, I, and III. The highest inter-cluster distance was observed between clusters I and IV (5.350), followed by clusters IV and III (3.569), indicating wider diversity among these groups. Based on these findings, it is suggested that Accessions 1, 2, 3, 12, and 13 could serve as potential donors in hybridization programs aimed at developing varieties with higher yields.

7.4 CHEMICAL CHARACTERIZATIONS

Chemical characterization of *Aloe vera* involves analyzing its constituents to understand its medicinal properties and potential applications. *Aloe vera* is a unique plant that contains a myriad of chemical substances, contributing significantly to its importance in the international market. The plant's chemical analysis revealed over 200 diverse biologically active compounds have been identified, including vitamins, minerals, enzymes, sugars, anthraquinones or phenolic compounds, lignin, saponins, sterols, amino acids, and salicylic acid. *Aloe vera* raw pulp consists of about 98.5% water, while the mucilage or gel component exhibits even higher water content at around 99.5%. The pH of *Aloe vera* gel is typically 4.5. According to research by Boudreau and Beland (2006) and Hamman (2008), the residual solid material, which makes up 0.5% to 1% of the leaf, contains a wide range of substances, such as enzymes, polysaccharides, phenolic compounds, organic acids, water-soluble and fat-soluble vitamins, minerals, and enzymes.

7.4.1 Phytochemical Composition of *Aloe vera*

Aloe vera comprises a variety of carbohydrate polymers, particularly glucomannans, alongside various other organic and inorganic elements. Phenolic compounds identified include chromone, anthraquinone, or anthrone derivatives. Three distinct preparations of aloe plants are commonly utilized in medicinal applications, each differing in their chemical makeup and therapeutic benefits. Aloe latex is a topical treatment for multiple skin conditions and has various internal uses, including as a laxative, in the treatment of diabetic and hyperlipidaemic patients, and for healing gastric ulcers, while aloe extract appears to be useful in cancer and AIDS. *Aloe vera* contains many therapeutic properties, including wound, burn, and frostbite healing, as well as antibacterial, antifungal, anti-inflammatory, hypoglycemic, and gastroprotective activities (Mukherjee et al. 2014). The phytochemical components of *Aloe barbadensis* whole leaves were examined in a previous work using solvent, ether, and aqueous extracts. The presence of a variety of chemicals, such as carbohydrates, saponins, flavonoids, tannins, alkaloids, anthraquinones, and resins, was determined by testing these extracts. After the outer rind is removed, the gel component of the aloe leaf is retrieved and is specifically utilized to assess anti-inflammatory activity. On the other hand, the crude sap was employed to assess the antimicrobial activity. Their findings revealed that the gel extracted from the Aloe leaf exhibited substantial anti-inflammatory activity, indicating its potential therapeutic benefits in this regard. Similarly, the crude sap demonstrated antimicrobial activity, suggesting its efficacy against microbial pathogens (Bista, Ghimire, and Subedi 2020).

Kaur (2011) conducted biochemical analysis on different accessions of North Indian germplasm to assess the quality of *Aloe vera*. A number of elements, such as proteins, phenols, aloin, gel, stabilized gel, total carbohydrates, and chlorophyll, were quantitatively estimated. To evaluate how maturity affected the quantity of these constituents, two samples for each accession were taken during the 2011-2012 season at different maturity levels, one pre-flowering (C) and the other post-flowering (D). In the pre-flowering phase, the sugar content (mg/g) ranged between 0.96±0.0031 mg/g to 3.30±0.0018 mg/g in leaves of Aloe germplasm accessions investigated, whereas in the post-flowering phase, the values varied from 1.20±0.004 mg/g, 3.33. The protein levels in the leaves of various Aloe accessions varied from 0.229±0.001 mg/g to 1.49±0.0017 mg/g in the vegetative phase and from 0.462±0.0097 mg/g to 1.58±0.0022 mg/g in the post-flowering phase. Phenol levels in the leaves of all gathered accessions were assessed periodically. During the pre-flowering phase (October to November), the total phenol content ranged from 87.2±5.40 mg/100 g to 149.0±2.07 mg/100 g, while during the post-flowering phase, it ranged between 98.2±2.80 mg/100 g to 152.3±1.07 mg/100 g.

Gangwar et al. (2017) conducted a study to analyze the enzymatic activity and biochemical composition of *Aloe vera* leaves across five different genotypes. Among the genotypes, Germplasm IC-112517 showcased the highest gel percentage (97.00%) and moisture percentage (97.03%), while IC-112527 exhibited the highest percentages of carbohydrates (64.20%), total minerals (11.10%), total sugar content (30.07%), and total chlorophyll content (0.20 mg/gm). Germplasm IC-285626 demonstrated the highest crude fiber content at 11.43%. IC-112527 exhibited the highest catalase enzyme activity at 1.92 units/ml and peroxidase enzyme activity at 1.67 units/ml. This study offers a detailed analysis of the enzymatic activity and biochemical composition found in the leaves of *Aloe vera*.

7.4.2 Nutritional Composition and Secondary Metabolites

The *Aloe vera* plant is well-known for its therapeutic qualities and nutritional value. It is reported that *Aloe vera* gel is predominantly composed of water, accounting for 98.5 to 99.5%, and includes 200 active chemicals and up to 75 nutrients (Babak and Nahashon 2014; Mikoajczak 2018). The nutritional composition of *Aloe vera* includes various macronutrients (carbohydrates, proteins, and fats), micronutrients (vitamins and minerals), and dietary fibers. *Aloe vera* is also known for its rich content of secondary metabolites, which include anthraquinones, phenolic compounds, polysaccharides, glycoproteins, and saponins (Surjushe, Vasani, and Saple 2008). The phytochemicals present in *Aloe vera* include secondary metabolites such as saponins, tannins, and flavonoids (P. Das and Srivastav 2015) as well as acidic substances, lectins, polymannans, terpenoids, sterols, resins, acetylated mannans, and anthrones (Mariappan and Shanthi 2012). In addition to enzymes and salicylic acid (Joseph and Raj 2010) other active ingredients could be chromones, anthraquinones, and lignin (Mariappan and Shanthi 2012), Numerous pharmacological effects, including anticancer, antifungal, anti-inflammatory, antiviral, antioxidant, immunomodulatory, antidiabetic, anti-arthritis, and wound-healing characteristics, are demonstrated by these substances (Sharma et al. 2014 and Maan et al. 2018).

According to earlier research, the chemical composition of whole *Aloe vera* leaves (on a dry matter basis) is as follows: 2.85% crude fat, 60.34–72.17% crude fiber, 14.65–16.88% ash, and 6.86–15.4% crude protein (Femenia et al. 1999; Ahmed and Hussain 2013). In particular, 2% protein and 38% carbohydrates are present in *Aloe vera* gel (Luta, Duncan, and Sinnott 2009). Along with vitamins like thiamine, folic acid, niacin, B12, A, C, and E, it is also a rich natural source of various minerals such as potassium, chromium, magnesium, calcium, copper, zinc, iron, sodium, and manganese (Shelton 1991). The synthesis of enzymes involved in metabolic pathways depends on these minerals (Sharma et al. 2014). *Aloe vera* also contains a wealth of enzymes that aid in the absorption of essential nutrients (Mulu et al. 2015). In addition to isozymes of superoxide dismutase and proteolytic enzymes like glutathione peroxidase and carboxypeptidase, notable enzymes include amylase, oxidase, carboxypeptidase, and alkaline phosphatase (Nandal and Bhardwaj 2012). These enzymes play a significant role in metabolism.

Aloe vera's rich nutritional composition and diverse array of secondary metabolites make it a valuable plant for health and wellness. Understanding the specific compounds and their effects provides insights into the numerous benefits of *Aloe vera*, guiding its use in medicinal, cosmetic, and nutritional applications. As research continues, discoveries about *Aloe vera* bioactive components will further enhance its therapeutic potential and applications.

7.5 GENETIC CHARACTERIZATIONS

Genetic characterization involves studying the genetic structure, variation, and relationships within and between *Aloe vera* populations. To fully harness and improve the benefits of *Aloe vera*, understanding its genetic makeup is crucial.

TABLE 7.1
Constituents of *Aloe* vera

Anthraquinones	Saccharides	Vitamins	Inorganic Compounds
Aloin/Barbaloin Isobarbaloin	Cellulose	B1	Calcium
Aloe-emodin	Glucose	B2	Sodium
Emodin	Mannose	B6	Chlorine
Aloetic Acid	L-Rhamnose	Choline	Manganese
Ester of Cinnamic Acid Anthranol	Aldopentose	Folic Acid	Zinc
Chrysophanic Acid Resistannol		Ascorbic Acid	Chromium
Anthracene		α-Tocopherol	Copper
Ethereal oil		β–Carotene	Magnesium Iron
Enzymes	**Nonessential Amino Acids**	**Essential Amino Acids**	**Miscellaneous**
Cyclooxygenase	Histidine	Lysine	Cholesterol Triglycerides
Oxidase	Arginine	Threonine	Steroids
Amylase	Hydroxyproline	Valine	β-Sitosterol
Catalase	Aspartic Acid	Leucine	Lignins
Lipase	Glutamic Acid	Isoleucine Phenylalanine Methionine	Uric Acid
Alkaline	Proline		Gibberellin
phosphatase Carboxypeptidase	Glycine		Lectin like substances Salicylic Acid Arachidonic Acid
	Alanine		

Sources: Shelton (1991); Vogler and Ernst (1999).

7.5.1 Molecular Markers and Genetic Diversity Analysis

According to Pradeep Reddy et al. (2002) methods of molecular markers based on DNA have developed into effective and accurate instruments for determining genotypes and evaluating genetic variation. Nonetheless, there are not many studies on the genetic diversity of Aloe species in the literature. Assessing the genetic diversity across Aloe genotypes is crucial for furthering genetic improvement initiatives because of the commercial importance of *Aloe vera* (Arolu et al. 2015). Morphological markers are frequently unreliable because of their sensitivity to developmental phases and environmental factors, but physiological markers can be employed to quantify genetic diversity in plants. Molecular markers, on the other hand, provide a more reliable and stronger method for examining genetic diversity and genotype connections (R. Kumar et al. 2022). Because molecular markers are closely linked to desirable qualities, they significantly increase the efficacy of molecular breeding. Additionally, the transcriptome sequencing of *Aloe vera* by Choudhri et al. (2018) offers important information for investigating the genes involved in the manufacture of vital chemicals and creating sophisticated marker systems. There are various benefits to using molecular markers to measure genetic variety in *Aloe vera*, chief among them being their resistance to environmental effects, and measurements can be taken using DNA from any growth stage (Tripathi, Saini, and Tiwari 2011). Four distinct molecular markers can be employed to evaluate the genetic diversity within *Aloe vera*. These include AFLP, SSR, RAPD, and SNPs.

In the past, RAPD markers were used to evaluate the genetic diversity of A*loe vera*. The phytochemical, antibacterial, and DNA profiles of Aloe species native to Mauritius and Reunion were compared to those of *Aloe vera* by Ranghoo Sanmukhiya et al. (2010). Their genetic diversity studies employing RAPD markers revealed significant differences between Mascarene Aloe species and *Aloe vera*. This pioneering study was the first to underscore the metabolic and genomic distinctions between Aloe species from Mascarene and *Aloe vera*. Kumar et al. (2015) conducted a genetic diversity analysis of *Aloe vera* (L.) Burm. f. collected from 12 states across various agro-climatic zones in India. They utilized SSR marker assays for this investigation. The CTAB technique was used to extract genomic DNA from samples of young leaves. Twenty primers were chosen; some were created using Primer 3 software from known *Aloe vera* sequences, while others were derived from the similar species Asparagus officinalis L. To show the phenetic links between samples, NTSys software was used to create similarity matrices and dendrograms. To assess genetic diversity, the Marker Index (MI), effective multiplex ratio (EMR), and polymorphic information content (PIC) were computed. With similarity ratings ranging from 46% to 100%, the twelve Aloe vera germplasm accessions were sorted into three main clusters using a neighbor-joining tree built from all of their SSR segments. Haryana and Uttar Pradesh accessions showed the most resemblance (100%), followed by Haryana, Punjab, and Rajasthan accessions with 93% similarity. On the other hand, accessions from Kerala and Gujarat showed the least amount of resemblance. In addition to demonstrating the usefulness of SSR marker analysis in evaluating the genetic diversity of medicinal plants, this work emphasized the significant genetic variety among *Aloe vera* accessions from various agro-climatic zones in India. Furthermore, Tripathi et al. (2011) used AFLP markers to assess the genetic diversity of 12 elite *Aloe vera* accessions that were gathered from different sites in Madhya Pradesh, India. They obtained a total of 400 fragments, averaging 51.2 fragments per primer combination across the 12 genotypes. The discriminatory power of eight primer combinations was assessed using marker attributes such as Polymorphism Information Content (PIC) levels, polymorphism percent, and distinct alleles. The average PIC value was 0.364±0.0257, with a range of 0.246 to 0.458. Principal Component Analysis (PCA) and a UPGMA (Unweighted pair group method with arithmetic mean) phenogram were used to cluster the accessions based on the genotypic data collected. Twelve *Aloe vera* germplasm accessions were sorted into two primary clusters by the neighbor-joining tree, which was built from all AFLP fragments. This demonstrated that there was a considerable degree of heterogeneity among the genotypes that were gathered.

7.5.2 Population Structure Analysis

Population structure refers to the division of individuals into distinct groups based on genetic similarities and differences. By elucidating the population structure of *Aloe vera*, researchers can gain insights into its demographic history, gene flow patterns, and potential breeding strategies. Through population structure analysis, we seek to uncover hidden genetic diversity, identify unique gene pools, and inform sustainable management strategies for this valuable botanical resource. Advances in molecular biology and genetic characterization tools, such as AFLP markers, have greatly enhanced our ability to assess and monitor genetic diversity within *Aloe vera* populations. These tools provide valuable insights into the genetic structure, variability, and evolutionary relationships among different genotypes.

The findings from Adienge et al. (2019) suggest that among the three Aloe species studied in the arid and semi-arid lands of Kenya, Aloe secundiflora exhibited the widest distribution, particularly in the Coast, Lake, and Central regions, indicating its dominance. Significant genetic variation was observed both among and within populations studied. A. secundiflora showed the highest genetic diversity, whereas A. turkanensis exhibited the least. The Maralal population displayed the highest genetic diversity among all populations studied. By utilizing both ISSR and RAPD markers, researchers uncovered substantial genetic variation among populations, surpassing the insights gained from using either marker alone. The research revealed a distinct population genetic structure, delineating three main clusters corresponding to the three Aloe species. Minimal reproductive crossing between species was evident. To further understand the dominance of A. secundiflora, researchers recommend conducting biophysical investigations across the three Aloe species. Additionally, ex-situ conservation efforts are advised for A. turkanensis due to its low genetic diversity and limited geographic distribution. Conservation efforts should also target the Maralal sec population owing to its high levels of genetic variations. Future research on genetic variation in Aloe species should focus on using both RAPD and ISSR markers together, as they offer comprehensive insights into genetic diversity.

Sixty *Aloe vera* genotypes from various agro-climatic regions were examined for genetic variation using AFLP markers in the study by Kumar et al. (2022). Averaging 41.9 loci per primer, ten primer combinations produced 16,578 fragments from 419 loci. The primers demonstrated resolving power ranging from 24.86 to 87.20, with an average of 55.23, and PIC ranging from 0.92 to 0.98, averaging 0.95. Clustering and principal coordinate analysis (PCA) indicated high genetic heterogeneity among the genotypes. Ne (number of different alleles), H (expected heterozygosity), I (Shannon's information index), NPL (number of polymorphic loci), PPL (percentage of polymorphic loci), Hs (gene diversity within subpopulations), Ht (total gene diversity), and Nm (gene diversity), Nm (gene flow), and Gst (genetic differentiation) are important indicators of genetic diversity. Group differences accounted for 7.46% of genetic variation, whereas genotype variation accounted for 91.46%, according to analysis of molecular variance (AMOVA). Variance components and degrees of freedom (df) were 57 and 42.858 among genotypes and two and 3.496 among groups. Three populations that matched the geographic locations of the genotypes were found using population structure analysis. The study concluded that AFLP markers are ideal for assessing genetic diversity in *Aloe vera* due to their high reproducibility, polymorphism, throughput, and cost-effectiveness, aiding in the maintenance and utilization of *Aloe vera* genotypes.

7.6 CONSERVATION AND MANAGEMENT OF *ALOE VERA* GERMPLASM

Aloe vera, renowned for its medicinal, cosmetic, and therapeutic properties, is a valuable plant species with significant economic importance. Ensuring the conservation and effective management of *Aloe vera* germplasm is critical to maintaining its genetic diversity, which is essential for the plant's adaptability, resilience, and potential for upcoming breeding initiatives. Genetic diversity within *Aloe vera* populations provides the foundation for developing new varieties with enhanced traits, such as increased resistance to pests, infections, and diseases, improved growth in diverse environmental conditions, and higher yields of bioactive compounds (Quazi, Golani, and Capuzzo 2021).

7.6.1 Germplasm Conservation and Utilization

Aloe vera faces several threats that could undermine its genetic diversity. Habitat destruction, overexploitation, climate change, and the narrow genetic base of cultivated varieties pose significant risks to its genetic resources. In this context, it is imperative to implement comprehensive conservation tactics that encompass both in-situ and ex-situ approaches. In-situ conservation involves protecting and managing *Aloe vera* in its natural habitat, while ex-situ conservation includes preserving germplasm in gene banks, botanical gardens, and through tissue culture techniques (Demissew 2005). The primary objectives of *Aloe vera* germplasm conservation and management are to preserve genetic diversity, ensure sustainable utilization, and facilitate the development of improved varieties. This involves a multifaceted approach, including the identification and protection of diverse genotypes, the establishment of germplasm repositories, the promotion of sustainable harvesting practices, and the integration of traditional knowledge with modern scientific techniques (Quazi, Golani, and Capuzzo 2021).

7.6.2 Methods of Germplasm Conservation

In situ conservation involves preserving *Aloe vera* in its natural habitat, maintaining ecological balance, and allowing the plant to evolve naturally within protected areas, biosphere reserves, and national parks. The term "ex situ conservation" describes the conservation of *Aloe vera* outside its natural habitat, which is particularly useful for safeguarding against habitat loss and environmental changes. This method includes various strategies: Seed banks, despite *Aloe vera* seeds being less commonly used due to low germination rates, store seeds under controlled conditions to ensure long-term viability; Field gene banks involve growing and maintaining *Aloe vera* plants in fields, facilitating easy access for research and breeding purposes; Tissue culture, or micropropagation, is an effective method for conserving *Aloe vera* germplasm by growing plants from tissues or cells under sterile conditions, which not only conserves genetic material but also allows for the rapid multiplication of disease-free plants. Additionally, cryopreservation involves storing *Aloe vera* tissues, such as seeds or shoot tips, in liquid nitrogen at extremely low temperatures, ensuring long-term preservation with minimal genetic changes (Nalina et al. 2022).

The utilization of *Aloe vera* germplasm encompasses various approaches in breeding, genetic improvement, pharmaceutical and cosmetic applications, and agronomic practices. In breeding and genetic improvement, the selection of superior varieties involves using the genetic diversity in germplasm collections to cultivate *Aloe vera* varieties with desirable traits, such as higher gel content, increased disease resistance, and improved growth rates (Suresh Kumar et al. 2019). Hybridization, which entails crossbreeding different *Aloe vera* varieties, combines desirable traits to develop new cultivars that enhance medicinal properties, yield, and adaptability to environmental conditions (Trojak-Goluch et al. 2021). Advances in biotechnology, including genetic engineering and molecular biology, allow the introduction of specific traits into *Aloe vera*, with techniques like CRISPR/Cas9 gene editing used to enhance disease resistance or increase the production of beneficial compounds. In pharmaceutical and cosmetic applications, *Aloe vera* gel, rich in bioactive compounds with anti-inflammatory, antimicrobial, and antioxidant properties, is utilized to produce therapeutic products for wound healing, skin treatments, and other medicinal uses (Nayeri et al. 2023). The gel's moisturizing and soothing properties make it a staple in skincare and cosmetic products, with different germplasm collections providing various formulations that enhance the development of new products. Agronomic practices and cultivation benefit from conserved germplasm by promoting sustainable agricultural practices, such as developing drought-resistant varieties and plants requiring minimal fertilizers and pesticides (Roberts and Mattoo 2018). The genetic diversity of *Aloe vera* also aids in developing Integrated Pest Management (IPM) strategies that reduce chemical pesticide reliance and promote ecological balance. Furthermore, the use of genetically diverse *Aloe vera* germplasm in organic farming systems supports the production of high-quality, chemical-free products, ensuring sustainability and environmental health (Singh 2023).

7.6.3 Germplasm Exchange and Distribution

Germplasm exchange of *Aloe vera* plays a crucial role in enhancing genetic diversity, which enables the integration of genetic material from diverse ecological zones, essential for developing new varieties with improved traits. This increased genetic diversity supports breeding programs by ensuring the development of robust plant varieties that can withstand diseases, pests, and varying environmental conditions (Shekhawat et al. 2013). Additionally, the exchange of *Aloe vera* germplasm fosters international collaboration among researchers, cultivators, and institutions, promoting the sharing of knowledge, resources, and best practices that accelerate advancements in *Aloe vera* research and cultivation techniques. Furthermore, germplasm exchange supports the adaptation of *Aloe vera* to different climatic conditions and agricultural systems by introducing traits such as drought resistance and low fertilizer requirements, which are vital for sustainable cultivation and long-term agricultural productivity in the face of climate change (Odeleye, Elujoba, and Gbolade 2009). The mechanisms facilitating germplasm exchange include international agreements and protocols, such as the Convention on Biological Diversity (CBD) and the International Treaty on Plant Genetic Resources for Food and Agriculture (ITPGRFA), which provide frameworks ensuring fair and equitable sharing of benefits. Germplasm repositories, including seed banks, botanical gardens, and research institutions, alongside networks like the Global Crop Diversity Trust, play a pivotal role in conserving and distributing *Aloe vera* genetic material to breeders and researchers worldwide (Ebert et al. 2023). To prevent the spread of diseases and pests during germplasm exchange, rigorous quarantine and phytosanitary measures, including thorough inspections, treatments, and certifications, are essential. However, challenges such as navigating intellectual property rights (IPR) and access issues, maintaining genetic integrity, and overcoming logistical and financial constraints must be addressed. Ensuring that germplasm exchange respects the rights of both originating countries and recipients, maintaining careful documentation and proper storage conditions to avoid genetic erosion, and securing funding for transportation, storage, and handling are all critical components for the effective conservation and distribution of *Aloe vera* germplasm (Demissew 2005).

7.7 EXPLORING THE WORLD OF GENE DISCOVERY

Aloe vera, a succulent plant revered for its medicinal and cosmetic properties, has been used for centuries in traditional medicine. Recent developments in genomics and molecular biology have created new opportunities in understanding the genetic underpinnings of *Aloe vera*'s beneficial traits. Gene discovery in *Aloe vera* involves identifying and understanding the genetic components that contribute to its unique properties, such as its medicinal and therapeutic benefits (Terefe and Neges 2017). Gene discovery in *Aloe vera* is pivotal for several reasons. Firstly, understanding the genetic basis of *Aloe vera*'s medicinal properties can lead to the development of enhanced varieties with increased bioactive compound production. Secondly, gene discovery can aid in improving the plant's resistance to diseases and environmental stresses, ensuring sustainable cultivation. Lastly, elucidating the genetic makeup of *Aloe vera* can provide insights into its evolutionary history and relationship with other species, contributing to broader botanical and ecological knowledge (Gurib-Fakim 2006). The process of gene discovery begins with the sequencing of DNA, which allows scientists to decode the genetic information contained in an organism. Advanced sequencing technology, such as high-throughput sequencing and bioinformatics tools, has significantly accelerated gene discovery (Delseny, Han, and Hsing 2010). Once the DNA sequence is obtained, researchers use various techniques to annotate the genome, which involves pinpointing the locations of genes and predicting their functions. Functional studies, including gene expression analysis, knockouts, and overexpression experiments, further elucidate the roles of specific genes in physiological processes and disease mechanisms (Chauhan et al. 2021). Gene discovery is crucial for advancing our knowledge of biology, enabling the development of new medical treatments, and improving agricultural practices by identifying genes that contribute to desirable traits in plants and animals. As our understanding of genetics deepens, the potential for innovative applications in medicine, biotechnology, and other fields continues to expand (Halewood et al. 2018).

7.8 GENE DISCOVERY METHODS

Gene discovery in *Aloe vera* employs advanced methods to uncover the genetic basis of its medicinal properties. The process begins with next-generation sequencing (NGS) to generate comprehensive genomic data. Bioinformatics tools then annotate the genome, identifying gene locations and functions. Functional genomics, including transcriptome analysis and RNA sequencing (RNA-seq), examines gene expression patterns. Comparative genomics identifies conserved genes and regulatory elements by comparing *Aloe vera* with other species. Techniques like gene knockout and overexpression studies further clarify gene functions. These methods collectively enhance our understanding of *Aloe vera* genetics, aiding biotechnological applications and cultivation improvements.

7.8.1 DNA Extraction and Preparation

The leaves of medicinal plants are often abundant in secondary metabolites, polysaccharides, and polyphenols; thus, it is crucial to optimize the DNA isolation process for the genetic characterization of medicinal plants. These substances can complicate the extraction of genomic DNA. Extracting and preparing DNA from *Aloe vera* for gene discovery involves several steps: collecting plant material, extracting DNA, assessing DNA quality and quantity, and preparing the DNA for downstream applications such as PCR, sequencing, or cloning (Moyo et al. 2008). Despite the fact that effective genomic DNA isolation techniques have been created for plant species with high levels of polyphenol and polysaccharide, none of them can be applied to all plants (Varma, Padh, and Shrivastava 2007). Therefore, in order to acquire DNA of the appropriate quality, researchers frequently alter or mix several techniques (Varma et al. 2007; Jabbarzadeh et al. 2009). A successful DNA isolation procedure should be easy to follow, quick, and effective, producing enough high-quality DNA for molecular analysis. However, the extraordinarily high amounts of secondary metabolites in *Aloe vera* leaves make it difficult to isolate DNA from the leaves because they obstruct the extraction process. To overcome this issue, young leaf gel was used, yielding high-quality DNA suitable for molecular biological applications (Deore, Samantaray, and Maiti 2014).

Collected plant material from fresh *Aloe vera* leaves were surface-sterilize with 70% ethanol, cut into small pieces using a sterile scalpel, and optionally freeze in liquid nitrogen for easier grinding. For DNA extraction, prepare a CTAB buffer (2% CTAB, 100 mM Tris-HCl, pH 8.0, 20 mM EDTA, 1.4 M NaCl, 1% PVP-40) with 0.2% 2-mercaptoethanol added before use. Grind 0.5-1 g of leaf tissue with liquid nitrogen or a homogenizer, add 700 µL of pre-warmed CTAB buffer, incubate at 65°C for 30 minutes, mix with chloroform: isoamyl alcohol (24:1), centrifuge at 12,000 x g for 10 minutes, transfer the aqueous phase, precipitate with isopropanol at -20°C for 30 minutes, centrifuge, use 70% ethanol to wash the DNA pellet, then let it air dry and dissolve it in TE buffer. Optionally treat with RNase A. Assess DNA quality and quantity using a Nanodrop spectrophotometer (Moyo et al. 2008; Deore et al. 2014, and Kumar et al. 2018).

7.8.2 High-throughput Sequencing Technologies

High-throughput sequencing technologies are instrumental in gene discovery within *Aloe vera*. These advanced sequencing methods, such as next-generation sequencing (NGS) or third-generation sequencing (TGS), enable rapid and comprehensive analysis of the plant's genome (Gao et al. 2023). NGS platforms, like Illumina sequencing, provide high-throughput data generation at a relatively low cost, facilitating the identification of novel genes, regulatory elements, and genetic variations in *Aloe vera*. Additionally, TGS technologies, such as PacBio or Long-read capabilities provided by Oxford Nanopore sequencing, are very helpful for resolving complicated genomic areas, detecting structural variations, and uncovering gene isoforms (Woudstra et al. 2021). Integrating these high-throughput sequencing approaches with bioinformatics analysis enhances gene discovery efforts, advancing our understanding of the genetic basis underlying *Aloe vera* medicinal properties and adaptive traits (Younessi-Hamzekhanlu et al. 2022). Utilizing emerging technologies like NGS, particularly whole genome and transcriptome sequencing, offers a precise approach for identifying key genes involved in synthesizing vital compounds. Additionally, these methods aid in developing high-throughput molecular markers (Li et al. 2021). RNA-Seq, a NGS-based technique, has gained prominence in comprehensive molecular studies of medicinal plants due to its ability to accurately detect low-abundance transcripts. It can reveal gene expression patterns in plants, whether or not a reference genome is available (Soltani Howyzeh et al. 2018). While reference genomes exist for many model plants, they are often lacking for medicinal species. Thus, RNA-Seq plays a pivotal role in establishing a de novo transcription map by clarifying the genome and transcriptome of non-model plants (B. D. Singh and Singh 2015).

Jaiswal et al. (2021) conducted a groundbreaking study, revealing the initial draft genomic sequence of *Aloe vera*. Their novel approach combines Illumina short-read and Oxford Nanopore long-read sequencing to assemble the genomic sequence successfully. Additionally, they analyzed and sequenced the transcriptomes of two essential tissues, the roots and the leaves, yielding profound insights into gene expression and facilitating precise gene set determination. Additionally, the team constructed a comprehensive *Aloe vera* genome-wide phylogeny in comparison to other monocot genomes that are accessible, shedding light on its phylogenetic position.

7.9 GENE PREDICTION AND FUNCTIONAL ANNOTATION

Gene prediction and functional annotation in *Aloe vera* provide information on the genetic foundation of its valuable properties. By utilizing a combination of sequencing technologies, bioinformatics tools, and databases, researchers can systematically identify and characterize genes, paving the way for biotechnological applications and further biological studies.

7.9.1 Identification of Unique or Novel Genes

The identification of unique or novel genes in *Aloe vera* is crucial for several reasons. *Aloe vera* is renowned for its medicinal and cosmetic properties, including its anti-inflammatory, antimicrobial, and wound-healing effects. Discovering novel genes can provide knowledge of the biosynthesis pathways responsible for these bioactive compounds, potentially resulting in the development of novel pharmaceuticals and skincare products. Additionally, unique genes may reveal mechanisms of stress resistance, growth, and development that could be leveraged to enhance crop resilience and productivity (Choi and Chung 2003). Understanding the genetic basis of *Aloe vera* beneficial traits can also facilitate genetic improvement and sustainable cultivation practices. Furthermore, novel gene identification contributes to the broader field of plant genomics, offering comparative data that can elucidate evolutionary processes and plant diversity. Overall, uncovering these unique or novel genes not only advances scientific knowledge but also has practical implications for agriculture, medicine, and industry (O Lawal et al. 2021).

Aloe vera's 199 genes were shown to be experiencing positive selection in a research by Jaiswal et al. (2021) using a false discovery rate (FDR) q-value threshold of 0.05. KEGG pathways, eggNOG categories, and Gene Ontology (GO) term categories were used to examine the distribution of these positively chosen genes. Genes linked to flowering, which is crucial for successful reproduction; calcium-ion-binding genes and transcription factors, which are involved in signal transduction in response to external stimuli; genes linked to the catabolism of carbohydrates, which are required for energy production; and genes involved in the plant's reaction to abiotic stress are among the genes that have been found to be involved in critical functions with phenotypic consequences (Yoo et al. 2019; Chakraborty et al. 2021).

7.9.2 Gene Expression Profiling

Gene expression profiling in *Aloe vera* is essential for understanding the molecular mechanisms underlying its medicinal properties, growth, and stress responses. This process involves creating a global picture of cellular function by concurrently measuring the activity of thousands of genes. Techniques such as RNA sequencing (RNA-Seq) and microarray analysis are commonly used for gene expression profiling (Zou et al. 2021). The microarray analysis technique, using hybridization to concurrently evaluate the expression levels of several genes. Quantitative real-time PCR (qRT-PCR) is often employed for validating results from RNA-seq or microarrays due to its high sensitivity and specificity. These methods help elucidate gene function, stress responses, metabolic pathways, and the plant's adaptation mechanisms (Salinas et al. 2019). By leveraging these profiling techniques, researchers can better understand *Aloe vera's* medicinal properties, such as its anti-inflammatory, wound-healing, and immunomodulatory effects, clearing the path for the development of new pharmaceuticals and improving cultivation practices (Raghunath et al. n.d.).

Differential Gene Expression (DGE) analysis revealed that the majority of genes expressed are those involved in secondary metabolism in *Aloe vera* roots compared to leaves, with root-specific transcripts for saponin biosynthesis upregulated, similar to findings in *Asparagus racemosus*. Key genes in the saponin biosynthesis pathway, such as ACT, MVK, PMVK, MDD, and CAS, were upregulated in roots, while DOXP synthase and DOXP reductoisomerase were upregulated in leaves, indicating a dominant plastid-bound MEP pathway in leaf tissue. UDP glycosyltransferase (UGT) and octaketide synthase (OKS) were highly expressed in both tissues, suggesting aloin synthesis occurs in both roots and leaves (Choudhri et al. 2018).

7.9.3 Functional Annotation and Pathway Analysis

After gene expression profiles, functional annotation, and pathway analysis are essential techniques to elucidate the biological roles and mechanisms of *Aloe vera* bioactive compounds. This involves classifying the influenced genes into functional categories through Gene Ontology (GO) analysis and identifying protein-protein interactions to understand their functional networks. Pathway analysis using databases like KEGG (Kyoto Encyclopedia of Genes and Genomes) and Reactome helps map these genes onto specific metabolic and signaling pathways, revealing how *Aloe vera* affects various biochemical processes (Zhou, Yao, and Liu 2017). Integration of data from transcriptomics, proteomics, and metabolomics through a systems biology approach provides a comprehensive understanding of *Aloe vera's* effects at a molecular level. Bioinformatics tools such as Cytoscape and DAVID facilitate network visualization and functional enrichment analysis (Kuo, Tian, and Tseng 2013).

The eggNOG-mapper tool was used to functionally annotate and classify genes into different eggNOG categories (Huerta-Cepas et al. 2017). The KAAS genome annotation server was used to assess KEGG orthology and assign genes to various KEGG pathways (Moriya et al. 2007). The WebGestalt web server was used to perform Gene Ontology (GO) enrichment analysis (Liao et al. 2019). Only GO categories with a hypergeometric test p-value of less than 0.05 were deemed functionally enriched for over-representation analysis. Furthermore, genes' functional annotation was manually curated, and genes were categorized and phenotyped according to this manual annotation. The STRING database provided the protein-protein interaction and co-expression data, while Cytoscape was used to do network analysis (Shannon et al. 2003; Szklarczyk et al. 2016).

The efficiency of *Aloe vera* sequencing was demonstrated by Choudhri et al. (2018), who reported a significant number of high-quality reads, refined transcripts, and CDS lengths using next-generation sequencing. Significant BLAST hits were obtained for both root and leaf samples using the functionally annotated CDS, suggesting that the *Aloe vera* genome was well covered. Elaeis guineensis and Phoenix dactylifera accounted for the majority of the hits, indicating a tight genetic link with *Aloe vera*. The found CDS were assigned roles by Gene Ontology (GO) mapping, which divided them into three primary domains: cellular components, molecular functions, and biological processes. According to the functional GO distribution, metabolism accounted for a maximum of 5,890 genes in the root and 6,329 genes in the leaf within biological processes. Furthermore, the high metabolic activity of *Aloe vera* was highlighted by the fact that the bulk of the top BLAST hits (980 in the root sample and 922 in the leaf sample were linked to metabolic processes).

Pathway analysis in *Aloe vera* involves mapping the gene expression data obtained from transcriptome studies to specific biochemical and signaling pathways to understand the biological functions and mechanisms of its bioactive compounds. This process starts with identifying differentially expressed genes using high-throughput sequencing technologies. These genes are then annotated and categorized into functional groups using databases like KEGG (Kyoto Encyclopedia of Genes and Genomes) and Reactome (Zhou, Yao, and Liu 2017). By mapping these genes onto established pathways, researchers can identify which metabolic and signaling routes are influenced by *Aloe vera* compounds. This analysis has revealed that *Aloe vera* impacts various pathways related to inflammation, oxidative stress response, metabolism, and immune function (Choudhri et al. 2018). For instance, *Aloe vera* has been shown to modulate the NF-κB and MAPK pathways, which are critical for inflammatory responses, and the Nrf2 pathway, which is involved in antioxidant defense (Hahn, Shin, and Bae 2020). Pathway analysis also highlights *Aloe vera's* role in regulating lipid and glucose metabolism, as well as its effects on immune system modulation. These insights not only deepen our understanding of *Aloe vera's* therapeutic potential but also guide the development of targeted applications in medicine and wellness (Gauttam et al. 2024).

7.10 VALIDATION OF CANDIDATE GENES

Validation of candidate genes in *Aloe vera* is a critical process to confirm their roles in various biological processes and pathways identified through transcriptome and functional annotation studies. This validation typically involves several molecular, genetic, and biochemical techniques. Quantitative PCR (qPCR) is used to measure the expression levels of candidate genes across different tissues or conditions, providing a precise assessment of gene activity (Manoli et al. 2012). A Study by Choudhri et al. (2018) shows that transcript validation was conducted via qRT-PCR, focusing on 16 unigenes associated with anthraquinone, saponin, and lignin biosynthesis. RNA interference (RNAi) is employed to silence specific genes, allowing researchers to observe resultant phenotypic changes and infer gene function. Overexpression studies involve introducing candidate genes into *Aloe vera* or model organisms to assess the effects of increased gene activity (Jangra et al. 2023). CRISPR-Cas9 gene editing technology is used to introduce specific mutations, enabling the study of gene function through the analysis of mutant phenotypes (Parikh et al. 2015). Protein expression and localization studies, such as western blotting and immunolocalization, help determine the presence and spatial distribution of proteins encoded by the candidate genes. Additionally, functional assays, including enzyme activity tests and stress response evaluations, further elucidate the roles of these genes. Together, these approaches provide a comprehensive validation of candidate genes, thereby enhancing our understanding of their contributions to *Aloe vera* physiology and potential therapeutic applications (Jangra et al. 2023).

7.11 GENE DISCOVERY SIGNIFICANCE IN *ALOE VERA*

The discovery of genes in *Aloe vera* holds significant potential for both scientific research and practical applications. Gene discovery can facilitate the production of synthetic versions of *Aloe vera*'s active ingredients, making them more accessible and affordable. This research can also contribute to a deeper understanding of plant biology and the mechanisms behind *Aloe vera* resilience and adaptability, providing insights that could be applied to other crops. Overall, gene discovery in *Aloe vera* represents a promising frontier in both plant science and medical applications, with the potential to yield substantial benefits in healthcare and agriculture (Coats and Ahola 2003).

7.11.1 Implications of Discovered Genes

The discovery of specific genes in *Aloe vera* holds significant implications across multiple domains such as medicine, agriculture, cosmetics, biotechnology, and environmental science. In medicine, identifying genes responsible for producing bioactive compounds like aloin and acemannan could lead to enhanced production of these substances, which are known for their anti-inflammatory, antimicrobial, and healing properties. In agriculture, genetic insights can lead to the breeding of *Aloe vera* plants with higher yields, better growth rates, and improved resistance to pests and diseases, contributing to more sustainable cultivation practices. For the cosmetic industry, genes involved in synthesizing skin-beneficial compounds can enhance product development, leading to more effective skincare products tailored to specific needs. Biotechnologically, *Aloe vera* genes can be utilized in synthetic biology and bioreactor applications to produce valuable compounds more efficiently.

The abscisic acid (ABA) responsive element binding factor ABF (ABA responsive element binding factor) gene exhibits differential expression in response to drought and other abiotic stressors alter the expression of particular target genes By attaching to the Abscisic-Acid-Response Element (ABRE), a feature of ABA-inducible genes (Liu et al. 2019). According to a prior study, this gene's mutations make people more susceptible to drought (Yoshida et al. 2015). Additionally, ABA controls solute transport and stomatal closure, two processes essential for drought tolerance. With its overexpression enhancing drought tolerance through ABA-guided stomatal closure, the trehalase 1 (TRE1) gene has been favorably chosen (Yamaguchi-Shinozaki and Shinozaki 2006; Van Houtte et al. 2013). Additionally, by generating the nitric oxide (NO) molecule, the G6PD5 gene, which has also demonstrated *Aloe vera*'s favorable selection, helps shield plants from a variety of challenges, including salinity stress. Consequently, this NO generation triggers the expression of defense response genes (Arasimowicz-Jelonek et al. 2014).

Succulent plants, like *Aloe vera*, are distinguished by their capacity to effectively absorb ambient CO2 and reduce water loss through transpiration using the CAM (Crassulacean Acid Metabolism) pathway, a specialized type of photosynthesis. Higher succulence and a greater amplitude of CAM metabolism are significantly correlated, suggesting that CAM evolved as a response to conditions of low CO2 and water. In *Aloe vera*, a number of important genes related to CAM metabolism exhibit site-specific indicators of adaptive evolution. These include phosphoenolpyruvate carboxylase (PEPC), which changes phosphoenolpyruvate into oxaloacetate and promotes CO2 assimilation; the potassium channel gene (KAT2), which controls stomatal opening and closure; and the malic enzyme (ME), which changes malic acid into pyruvate. Potassium transporters KT2/3, pyruvate orthophosphate dikinase (PPDK), phosphoenolpyruvate carboxylase kinase 1 (PPCK1), carbonic anhydrase 1 (CA1), peroxisomal NAD-malate dehydrogenase 2 (PMDH2), tonoplast dicarboxylate transporter (TDT), aluminum-activated malate transporter (ALMT9), and other CAM-related genes also showed distinct substitutions that affect *Aloe vera*'s functionality (Jaiswal et al. 2021). Among the MSA (Multiple Signs of Adaptive Evolution) genes found in *Aloe vera*, two osmotic biosensor genes, "CPA" and "AT2G42100," were discovered. Furthermore, the MSA genes also contained a variety of membrane transporters that aid in the movement of metals, osmolytes, and signaling chemicals. Among these transporters are two peroxisomal transporters: "PEX14," which transports signaling proteins with PTS1 and PTS2 domains, and "PNC1," a nucleotide carrier protein. An iron transporter called "IRT3," a lipid transporter called "AT4G17650," an amino acid transporter called "AT2G40420," a phospholipid transporter called "ALA1," a nucleobase-ascorbate transporter called "NAT8," a high-affinity nitrate transporter called "NRT2.6," and a sodium/metabolite co-transporter called "BASS6" were also described. Under drought stress, these osmotic sensors and transporters are essential for improving plant function, assisting the plant's adjustment to water scarcity (Waris et al. 2018; Jarzyniak and Jasiski 2014).

7.11.2 Potential Applications in *Aloe vera* Cultivation and Products

Aloe vera cultivation and its derived products offer a diverse array of applications across various industries. Cultivation itself provides economic opportunities through the farming of *Aloe vera* plants, which are resilient and thrive in arid climates, making them suitable for regions with limited water resources. The market for food and beverages offers a favorable environment for *Aloe vera*, which is increasingly recognized for its role in functional foods such as yogurt and health drinks (Jadhav et al. 2020). Botanical products like *Aloe vera* are widely utilized as nutritional supplements to promote health and prevent diseases. Research by Serrano et al. (2006) indicates that by delaying the loss of important functional elements, such as ascorbic acid and total phenolics, an edible coating made of *Aloe vera* gel can be used to extend the safety and quality of fresh items, including table grapes. *Aloe vera* gel extracted from the leaf pulp has emerged as a significant global industry, particularly in the food sector. It is prominently featured in functional foods, particularly health beverages that do not cause laxative effects. Additionally, *Aloe vera* gel finds application in various other food products such as milk, ice cream, and confectionery items. Moreover, it serves as both a flavoring agent and a natural preservative in certain foods (Christaki and Florou-Paneri 2010). The FDA has authorized the internal use of *Aloe vera* gel in the US as a "dietary supplement." *Aloe vera* can be used by the feed industry as a sensory addition within the "flavoring compounds" functional group in accordance with Annex I of Regulation No. 1831/2003 of the European Commission (EC), with the goal of enhancing the palatability or fragrance of feed products (World Health Organization 1999; Franz et al. 2005).

Gibberellins, a growth hormone, and polysaccharides in *Aloe vera* may enhance collagen and elastin formation, potentially reducing wrinkles. *Aloe vera*'s high healing capacity is attributed to its content of mucopolysaccharides (MPS), with concentrations typically ranging between 10,000 and 20,000 MPS per liter. Additionally, *Aloe vera* effects in treating scar tissue and preventing scar formation after skin injury are likely due to the activity of amino acids essential for new cell formation, as well as its enzymes' ability to promote regeneration in the skin's deepest layers (Surjushe et al. 2008; Eshun and He 2004; Choi and Chung 2003). Aloe vera's salicylic acid content, known for its analgesic and anti-inflammatory properties, inhibits the production of prostaglandins from arachidonic acid. As a result, *Aloe vera* has been utilized to aid in managing arthritis and joint-related issues. Furthermore, Aloe polysaccharides have been shown to enhance immune activity (Reynolds and Dweck 1999; Tizard et al. 1994). This plant supports the healing process of first to second-degree burns, although the exact mechanisms are not fully understood. It is hypothesized that lectins may play a role in the therapeutic effects observed (Hamman 2008; Maenthaisong et al. 2007). Several researchers have noted that *Aloe vera* inhibits the growth of certain microorganisms, such as Streptococcus pyogenes, Shigella flexneri, and Klebsiella species. This antimicrobial activity is particularly effective against Gram-positive bacteria that cause food poisoning and diseases in both humans and animals (Surjushe et al. 2008; Boudreau and Beland 2006). These actions of *Aloe vera* on inhibiting microorganisms may be attributed to both indirect effects, such as immune system stimulation, and direct effects involving anthraquinones (Surjushe, Vasani, and Saple 2008). Clinical trials are currently underway to gather conclusive evidence regarding the potential use of *Aloe vera* in treating conditions like HIV-AIDS or cancer (Eshun and He 2004; Franz et al. 2005).

Aloe vera also appears to play a significant role in promoting growth and health management in chickens (Bejar 2018; Mwale et al. 2005). For instance, supplementation of broiler chickens has been demonstrated to enhance their immune system and production performance by giving male broilers a basal diet containing 600 mg•kg-1 of aloe powder, water extract, ethanol extract, or a combination of these extracts. Among these, Aloe water extracts have demonstrated particularly beneficial results (Wang Chang Kang et al. 2007). Similar findings have been reported by other researchers regarding the body weight of broilers when their drinking water was supplemented with *Aloe vera* extract at concentrations ranging from 5 to 30 cm3 per dm3 of water (Bejar 2018; Olupona et al. 2010). Broilers fed with *Aloe vera* have shown significantly higher haemagglutination inhibition (HI) titre values against Newcastle disease (Mehala and Moorthy 2008). Additionally, *Aloe vera* has been reported as effective in treating and controlling coccidiosis in chickens (Chandrakesan et al. 2009). Incorporating *Aloe vera* into laying hen diets has been shown to significantly improve egg production (eggs per hen), although no significant differences were seen in the ratio of feed conversion or feed consumption (Moorthy et al. 2009). Additionally, nutritional supplementation of *Aloe vera* extracts in laying hens has demonstrated potential in preventing or treating the effects of experimentally induced lead intoxication in birds (Wagan et al. 2022).

7.12 SUMMARY

The present study focuses on the comprehensive investigation of *Aloe vera's* genetic resources, aiming to enhance knowledge on its genetic diversity and facilitate gene discovery for practical applications. *Aloe vera* Germplasm Characterization and Gene Discovery represents a multifaceted exploration into the genetic diversity, morphological traits, and chemical composition of *Aloe vera*, coupled with cutting-edge gene discovery methodologies. Beginning with the collection and documentation of genetic resources, the study evaluates the plant's morphological attributes and delves into its phytochemical and nutritional compositions, revealing insights crucial for both scientific understanding and practical applications. Genetic characterization employs molecular markers and population structure analyses to elucidate the genetic diversity within *Aloe vera* populations. Conservation strategies are outlined to ensure the sustainable management and utilization of *Aloe vera* germplasm worldwide. The introduction to gene discovery methodologies, including DNA extraction, high-throughput sequencing, and gene prediction, paves the way for identifying novel genes and understanding their functional roles through expression profiling and pathway analysis. Ultimately, the significance of discovered genes is highlighted, suggesting potential advancements in *Aloe vera* cultivation practices and product development, underscoring the study's broad implications for agriculture and biotechnology.

REFERENCES

Adienge, Asenath, Gabriel Muturi, Stanley Nadir, John Gicheru, Johnson Kinyua, and Jane Ngaira. 2019. "Genetic Diversity and Population Structure of Three Commercial Indigenous Aloe Species in Selected ASALs of Kenya." *Journal of Plant Science & Molecular Breeding* 8 (1): 1–10.

Ahmed, Muaz, and Fatma Hussain. 2013. "Chemical Composition and Biochemical Activity of Aloe Vera (Aloe Barbadensis Miller) Leaves." *International Journal of Chemical and Biochemical Sciences* 3 (5): 29–33.

Alagukannan, G., and S. Ganesh. 2006. "Variation and Correlation Studies in Aloe Vera L. Ecotypes." *Madras Agricultural Journal* 93 (Jul-Dec): 1.

Alamgir, A. N. M. 2017. "Pharmacognostical Botany: Classification of Medicinal and Aromatic Plants (MAPs), Botanical Taxonomy, Morphology, and Anatomy of Drug Plants." *Therapeutic Use of Medicinal Plants and Their Extracts: Volume 1: Pharmacognosy*, 1: 177–293.

Arasimowicz-Jelonek, Magdalena, Jolanta Floryszak-Wieczorek, Dariusz Abramowski, and Karolina Izbia ska. 2014. "Nitric Oxide and Reactive Nitrogen Species." *Nitric Oxide in Plants: Metabolism and Role in Stress Physiology* 165–84.

Arolu, I. W., M. Y. Rafii, M. M. Hanafi, T. M. M. Mahmud, and S. Askani. 2015. "Genetic Divergence and Evaluation of Yield Potential of Jatropha Curcas Accessions Collected from Peninsular Malaysia." *Pertanika Journal of Tropical Agricultural Science* 38 (1): 127–142.

Babak, Darabighane, and Samuel N. Nahashon. 2014. "A Review on Effects of Aloe Vera as a Feed Additive in Broiler Chicken Diets." *Annals of Animal Science* 14 (3): 491–500.

Bedigian, Dorothea. 2005. "A Guide to Effective Management of Germplasm Collections. IPGRI Handbooks for Genebanks No. 6." *Economic Botany* 59 (4): 403–4.

Bejar, Feleciano R. 2018. "Growth and Sensory Evaluation of Broilers Supplemented with Aloe Vera (Aloe Barbadensis) Extract in Drinking Water." *International Research Journal of Agriculture and Rural Development* 1 (7): 1–7.

Bista, Rojina, Arjun Ghimire, and Sadikshya Subedi. 2020. "Phytochemicals and Antioxidant Activities of Aloe Vera (Aloe Barbadensis)." *Journal of Nutrition Science Health Diet* 1 (1): 25–36.

Boudreau, Mary D., and Frederick A. Beland. 2006. "An Evaluation of the Biological and Toxicological Properties of Aloe Barbadensis (Miller), Aloe Vera." *Journal of Environmental Science and Health Part C* 24 (1): 103–54.

Briggs, David, and S. Max Walters. 2016. *Plant Variation and Evolution*. Cambridge University Press.

Chakraborty, Abhisek, Shruti Mahajan, Shubham K. Jaiswal, and Vineet K. Sharma. 2021. "Genome Sequencing of Turmeric Provides Evolutionary Insights into its Medicinal Properties." *Communications Biology* 4 (1): 1193.

Chandrakesan, Periasamy, Kathirvel Muralidharan, V. Dinesh Kumar, Gurusamy Ponnudurai, Tirunelveli Jeyagopal Harikrishnan, and K. S. V. N. Rani. 2009. "Efficacy of a Herbal Complex against Caecal Coccidiosis in Broiler Chickens." *Veterinarski Arhiv* 79 (2): 199–203.

Chauhan, Jagat, Mohan Kumar, Chetan Joshi, Sandeep Sharma, and Mukesh Kumar Sharma. 2021. "Functional Annotation and Network Analysis of Protein in Respect with Diabetes Targeting Aloe Vera Plant Compounds by Bioinformarics Tools." *Journal of Advanced Scientific Research* 12 (03): 261–68.

Choi, Seongwon, and Myung-Hee Chung. 2003. "A Review on the Relationship between Aloe Vera Components and their Biological Effects." *Seminars in Integrative Medicine* 1: 53–62.

Choudhri, Pragati, Muniya Rani, Rajender S. Sangwan, Ravinder Kumar, Anil Kumar, and Vinod Chhokar. 2018. "De Novo Sequencing, Assembly and Characterisation of Aloe Vera Transcriptome and Analysis of Expression Profiles of Genes Related to Saponin and Anthraquinone Metabolism." *BMC Genomics* 19: 1–21.

Christaki, Efterpi V., and Panagiota C. Florou-Paneri. 2010. "Aloe Vera: A Plant for Many Uses." *Journal of Food, Agriculture, and Environment* 8 (2): 245–49.

Coats, Bill C., and Robert Ahola. 2003. *Aloe Vera: The New Millennium: The Future of Wellness in the 21st Century*. iUniverse.

Cowell, Carly, Alan Paton, James S. Borrell, China Williams, Paul Wilkin, Alexandre Antonelli, William J. Baker, Richard Buggs, Michael F. Fay, and Roberta Gargiulo. 2022. "Uses and Benefits of Digital Sequence Information from Plant Genetic Resources: Lessons Learnt from Botanical Collections." *Plants, People, Planet* 4 (1): 33–43.

Cristiano, Giuseppe, Bernardo Murillo-Amador, and Barbara De Lucia. 2016. "Propagation Techniques and Agronomic Requirements for the Cultivation of Barbados Aloe (Aloe Vera (L.) Burm. F.)-A Review." *Frontiers in Plant Science* 7: 1410.

Das, Anusree, S. K. Moquammel Haque, Biswajit Ghosh, Krishnadas Nandagopal, and Timir Baran Jha. 2016. "Morphological and Genetic Characterization of Micropropagated Field Grown Plants of Aloe Vera L." *Plant Tissue Culture and Biotechnology* 25 (2): 231–46.

Das, Priyanka, and Alok Kumar Srivastav. 2015. "Phytochemical Extraction and Characterization of the Leaves of Aloe Vera Barbadensis for Its Anti-Bacterial and Anti-Oxidant Activity." *International Journal of Science Research* 4 (6): 658–61.

Delseny, Michel, Bin Han, and Yue Ie Hsing. 2010. "High Throughput DNA Sequencing: The New Sequencing Revolution." *Plant Science* 179 (5): 407–22.

Demissew, Sebsebe. 2005. "The Genus Aloe. Medicinal and Aromatic Plants-Industrial Profiles. Vol. 38." *Economic Botany* 59 (4): 396.

Deore, Pushpa, S. Samantaray, and S. Maiti. 2014. "Genomic DNA Isolation Protocol for Aloe Barbadensis Miller: Using Leaf Gel for Genetic Characterization." *International Journal of Scientific and Research Publications* 4: 1–5.

Ebert, Andreas W., Johannes M. M. Engels, Roland Schafleitner, Theo van Hintum, and Godfrey Mwila. 2023. "Critical Review of the Increasing Complexity of Access and Benefit-Sharing Policies of Genetic Resources for Genebank Curators and Plant Breeders-a Public and Private Sector Perspective." *Plants* 12 (16): 2992.

Eshun, Kojo, and Qian He. 2004. "Aloe Vera: A Valuable Ingredient for the Food, Pharmaceutical and Cosmetic Industries-A Review." *Critical Reviews in Food Science and Nutrition* 44 (2): 91–6.

Femenia, Antoni, Emma S. Sánchez, Susana Simal, and Carmen Rosselló. 1999. "Compositional Features of Polysaccharides from Aloe Vera (Aloe Barbadensis Miller) Plant Tissues." *Carbohydrate Polymers* 39 (2): 109–17.

Franz, C. H., R. Bauer, R. Carle, D. Tedesco, A. Tubaro, and K. Zitterl Eglseer. 2005. "Study on the Assessment of Plant/Herb Extracts and Their Naturally or Synthetically Produced Components as 'Additives' for Use in Animal Production."

Gangwar, Chetna, Pratibha Singh, Raj Narayan Kewat, and Nitin Vikram. 2017. "Biochemical Composition and Enzymatic Activity of Aloe Vera (Aloe Barbadensis L.)." *International Journal of Current Microbiology and Applied Sciences* 6 (11): 3572–76.

Gantait, Saikat, Uma Rani Sinniah, and Prakash Kanti Das. 2014. "Aloe Vera: A Review Update on Advancement of in Vitro Culture." *Acta Agriculturae Scandinavica, Section B-Soil & Plant Science* 64 (1): 1–12.

Gao, Longlong, Wenjie Xu, Tianyi Xin, and Jingyuan Song. 2023. "Application of Third-Generation Sequencing to Herbal Genomics." *Frontiers in Plant Science* 14: 1124536.

Gauttam, Vinod Kumar, Kavita Munjal, Hitesh Chopra, Aftab Ahmad, Mahesh K. Rana, and Mohammad Amjad Kamal. 2024. "A Mechanistic Review on Therapeutic Potential of Medicinal Plants and Their Pharmacologically Active Molecules for Targeting Metabolic Syndrome." *Current Pharmaceutical Design* 30 (1): 10–30.

Grace, Olwen M., Monique S. J. Simmonds, Gideon F. Smith, and Abraham E. van Wyk. 2009. "Documented Utility and Biocultural Value of Aloe L.(Asphodelaceae): A Review." *Economic Botany* 63: 167–78.

Gurib-Fakim, Ameenah. 2006. "Medicinal Plants: Traditions of Yesterday and Drugs of Tomorrow." *Molecular Aspects of Medicine* 27 (1): 1–93.

Hahn, Dongyup, Seung Ho Shin, and Jong-Sup Bae. 2020. "Natural Antioxidant and Anti-Inflammatory Compounds in Foodstuff or Medicinal Herbs Inducing Heme Oxygenase-1 Expression." *Antioxidants* 9 (12): 1191.

Halewood, Michael, Tinashe Chiurugwi, Ruaraidh Sackville Hamilton, Brad Kurtz, Emily Marden, Eric Welch, Frank Michiels, Javad Mozafari, Muhamad Sabran, and Nicola Patron. 2018. "Plant Genetic Resources for Food and Agriculture: Opportunities and Challenges Emerging from the Science and Information Technology Revolution." *New Phytologist* 217 (4): 1407–19.

Hamman, Josias H. 2008. "Composition and Applications of Aloe Vera Leaf Gel." *Molecules* 13 (8): 1599–616.

Houtte, Hilde Van, Lies Vandesteene, Lorena López-Galvis, Liesbeth Lemmens, Ewaut Kissel, Sebastien Carpentier, Regina Feil, Nelson Avonce, Tom Beeckman, and John E. Lunn. 2013. "Overexpression of the Trehalase Gene AtTRE1 Leads to Increased Drought Stress Tolerance in Arabidopsis and is Involved in Abscisic Acid-Induced Stomatal Closure." *Plant Physiology* 161 (3): 1158–71.

Huerta-Cepas, Jaime, Kristoffer Forslund, Luis Pedro Coelho, Damian Szklarczyk, Lars Juhl Jensen, Christian Von Mering, and Peer Bork. 2017. "Fast Genome-Wide Functional Annotation through Orthology Assignment by EggNOG-Mapper." *Molecular Biology and Evolution* 34 (8): 2115–22.

Ijinu, Thadiyan Parambil, Varughese George, and Palpu Pushpangadan. 2022. "History of Research on Medicinal Plants in India." In Ákos Máthé and Irfan Ali Khan (eds.) *Medicinal and Aromatic Plants of India* 1, 35–61. Springer.

Jabbarzadeh, Z., M. Khosh-Khui, H. Salehi, and A. Saberivand. 2009. "Optimization of DNA Extraction for ISSR Studies in Seven Important Rose Species of Iran." *American-Eurasian Journal of Sustainable Agriculture* 3 (4): 639–42.

Jadhav, Ashwini S., Omkar A. Patil, Sampada V. Kadam, and Mangesh A. Bhutkar. 2020. "Review on Aloe Vera Is Used in Medicinal Plant." *Asian Journal of Research in Pharmaceutical Science* 10 (1): 26–30.

Jaiswal, Shubham K., Shruti Mahajan, Abhisek Chakraborty, Sudhir Kumar, and Vineet K. Sharma. 2021. "The Genome Sequence of Aloe Vera Reveals Adaptive Evolution of Drought Tolerance Mechanisms." *Iscience* 24 (2): 102079.

Jangra, Alka, Siddhant Chaturvedi, Sonia Sihag, Garima Sharma, Siddharth Tiwari, and Vinod Chhokar. 2023. "Identification and Functional Characterization of a Novel Aldo-keto Reductase from Aloe Vera." *Planta* 258 (6): 107.

Jarzyniak, Karolina M., and Micha Jasi ski. 2014. "Membrane Transporters and Drought Resistance-A Complex Issue." *Frontiers in Plant Science* 5: 687.

Joseph, Baby, and S. Justin Raj. 2010. "Pharmacognostic and Phytochemical Properties of Aloe Vera Linn an Overview." *International Journal of Pharmaceutical Sciences Review and Research* 4 (2): 106–10.

Kaur, Ramandeep. 2011. "Evaluation and Improvement of Germplasm of *Aloe* Vera L from North India." (Ph.D. thesis, Department of Botany, Punjabi University). v, 214 pp. Retrieved from http://hdl.handle.net/10603/2340

Khan, Muhammad A., Sabine von Witzke-Ehbrecht, Brigitte L. Maass, and Heiko C. Becker. 2009. "Relationships among Different Geographical Groups, Agro-Morphology, Fatty Acid Composition and RAPD Marker Diversity in Safflower (Carthamus Tinctorius)." *Genetic Resources and Crop Evolution* 56: 19–30.

Kiran, Shashi, and Alice Tirkey. 2018. "Assessment of Genetic Diversity in Collected Accession Aloe Vera from Chhattisgarh Using Morphological Characters." *International Journal of Current Microbiology and Applied Sciences*. Special Issue-7: 3759–3767.

Kumar, Ravinder, Raj Kumar Salar, Ashok Kumar, Anil Kumar, and Vinod Chhokar. 2018. "A Low Cost, High Throughput Gel Electrophoresis Method for Separation of SSR Markers in *Aloe* Vera." *The Pharma Innovation Journal* 7 (3): 622–27.

Kumar, Ravinder, Raj Kumar Salar, Pradeep K. Naik, Manju Yadav, Anil Kumar, Ashok Kumar, Rajesh Yogi, Mukesh Kumar, and Vinod Chhokar. 2022. "Elucidation of Genetic Diversity and Population Structure of Sixty Genotypes of Aloe Vera Using AFLP Markers." *South African Journal of Botany* 147: 1146–55.

Kumar, Sandeep, Manila Yadav, Amita Yadav, and Jaya Parkash Yadav. 2015. "Molecular Assessment of Genetic Diversity in Indian Accessions of Aloe Vera Using SSR Marker." *International Journal of Pharmacy and Pharmaceutical Science* 7 (12): 149–54.

Kumar, Suresh, M. M. Azam, K. Venkatesan, Anjly Pancholy, and R. N. Kulloli. 2019. "Morphological and Biochemical Variability in Aloe Germplasm in Hot Arid Regions of India." *Journal of Herbs, Spices & Medicinal Plants* 25 (2): 158–71.

Kuo, Tien-Chueh, Tze-Feng Tian, and Yufeng Jane Tseng. 2013. "3Omics: A Web-Based Systems Biology Tool for Analysis, Integration and Visualization of Human Transcriptomic, Proteomic and Metabolomic Data." *BMC Systems Biology* 7: 1–15.

Lal, Aparna, Simon Hales, Nigel French, and Michael G. Baker. 2012. "Seasonality in Human Zoonotic Enteric Diseases: A Systematic Review." *PLoS One* 7 (4): e31883.

Lattoo, Surrinder K., Rekha S. Dhar, Shabnam Khan, S. Bamotra, Maharaj K. Bhan, Autar K. Dhar, and Kuldip K. Gupta. 2008. "Comparative Analysis of Genetic Diversity Using Molecular and Morphometric Markers in Andrographis Paniculata (Burm. f.) Nees." *Genetic Resources and Crop Evolution* 55: 33–43.

Li, Hui, Yanshi Ma, Fangyi Pei, Haiyan Zhang, Jicheng Liu, and Ming Jiang. 2021. "Large-Scale Advances in SSR Markers with High-Throughput Sequencing in Euphorbia Fischeriana Steud." *Electronic Journal of Biotechnology* 49: 50–5.

Liao, Yuxing, Jing Wang, Eric J. Jaehnig, Zhiao Shi, and Bing Zhang. 2019. "WebGestalt 2019: Gene Set Analysis Toolkit with Revamped UIs and APIs." *Nucleic Acids Research* 47 (W1): W199–205.

Liu, Chang, Yan Cui, Fuwei Pi, Yuliang Cheng, Yahui Guo, and He Qian. 2019. "Extraction, Purification, Structural Characteristics, Biological Activities and Pharmacological Applications of Acemannan, a Polysaccharide from Aloe Vera: A Review." *Molecules* 24 (8): 1554.

Liu, Xiaohong, Jun Li, Yongsi Zhang, Linhui Li, and Daowen He. 2011. "Biological Research Advancement in Aloe." *Journal of Medicinal Plants Research* 5: 1046–52.

Luta, Gabriela, Christy Duncan, and Robert Sinnott. 2009. "Chemical Characterization of Polysaccharide-Rich Ingredients from Aloe Vera, Larix Laricina and Larix Occidentalis, and Undaria Pinatifida." In *Scripps Center for Integrative Medicine's 6th Annual Natural Supplements Conference, San Diego, California*. January 22–25. San Diego, California.

Maan, Abid Aslam, Akmal Nazir, Muhammad Kashif Iqbal Khan, Tahir Ahmad, Rabia Zia, Misbah Murid, and Muhammad Abrar. 2018. "The Therapeutic Properties and Applications of Aloe Vera: A Review." *Journal of Herbal Medicine* 12: 1–10.

Maenthaisong, Ratree, Nathorn Chaiyakunapruk, Surachet Niruntraporn, and Chuenjid Kongkaew. 2007. "The Efficacy of Aloe Vera Used for Burn Wound Healing: A Systematic Review." *Burns* 33 (6): 713–18.

Mangaiyarkarasi, S. P., T. Manigandan, M. Elumalai, Priyanka K. Cholan, and Roopam Pal Kaur. 2015. "Benefits of Aloe Vera in Dentistry." *Journal of Pharmacy & Bioallied Sciences* 7 (Suppl 1): S255.

Manoli, Alessandro, Alba Sturaro, Sara Trevisan, Silvia Quaggiotti, and Alberto Nonis. 2012. "Evaluation of Candidate Reference Genes for QPCR in Maize." *Journal of Plant Physiology* 169 (8): 807–15.

Mariappan, V., and G. Shanthi. 2012. "Antimicrobial and Phytochemical Analysis of Aloe Vera. L." *International Research Journal of Pharmacy* 3: 158–61.

Martínez-Burgos, Walter Jose, Josilene Lima Serra, Ronald M. MarsigliaF, Pedro Montoya, Zulma Sarmiento-Vásquez, Oranys Marin, Euler Gallego-Cartagena, and Carlos D. Paternina-Arboleda. 2022. "Aloe Vera: From Ancient Knowledge to the Patent and Innovation Landscape-A Review." *South African Journal of Botany* 147: 993–1006.

Mehala, C., and M. Moorthy. 2008. "Effect of Aloe Vera and Curcuma Longa (Turmeric) on Carcass Characteristics and Biochemical Parameters of Broilers." *International Journal of Poultry Science* 7 (9): 857–61.

Mikoajczak, Natalia. 2018. "Potential Health Benefits of Aloe Vera." *Journal of Education, Health and Sport* 8 (9): 1420–35.

Misra, A., A. K. Shasany, A. K. Shukla, M. P. Darokar, S. C. Singh, V. Sundaresan, J. Singh, G. D. Bagchi, S. P. Jain, and D. Saikia. 2010. "AFLP Markers for Identification of Swertia Species (Gentianaceae)." *Genetics Molecular Research* 9 (3): 1535–44.

Mondini, Linda, Arshiya Noorani, and Mario A. Pagnotta. 2009. "Assessing Plant Genetic Diversity by Molecular Tools." *Diversity* 1 (1): 19–35.

Moorthy, M., S. Saravanan, C. Mehala, S. Ravi, M. Ravikumar, K. Viswanathan, and S. C. Edwin. 2009. "Performance of Single Comb White Leghorn Layers Fed with Aloe Vera, Curcuma Longa (Turmeric) and Probiotic." *International Journal of Poultry Science* 8 (8): 775–78.

Moriya, Yuki, Masumi Itoh, Shujiro Okuda, Akiyasu C. Yoshizawa, and Minoru Kanehisa. 2007. "KAAS: An Automatic Genome Annotation and Pathway Reconstruction Server." *Nucleic Acids Research* 35 (suppl_2): W182–85.

Moyo, M., S. O. Amoo, M. W. Bairu, J. F. Finnie, and J. Van Staden. 2008. "Optimising DNA Isolation for Medicinal Plants." *South African Journal of Botany* 74 (4): 771–5.

Mukherjee, Pulok K., Neelesh K. Nema, Niladri Maity, Kakali Mukherjee, and Ranjit K. Harwansh. 2014. "Phytochemical and Therapeutic Profile of Aloe Vera." *Journal of Natural Remedies* 14 (1): 1–26.

Mulu, Tewolde, Firomsa Teshale, Samuel Gemeda, and Omprakash Sahu. 2015. "Medicated Evaluation of Aloe Vera: Overview on Characteristics and Application." *World Journal of Nutrition and Health* 3 (1): 1–7.

Mwale, Marizvikuru, Evison Bhebhe, Michael Chimonyo, and Tinyiko Edward Halimani. 2005. "Use of Herbal Plants in Poultry Health Management in the Mushagashe Small-Scale Commercial Farming Area in Zimbabwe." *International Journal of Applied Research in Veterinary Medicine* 3 (2): 163–70.

Nalina, L., K. Rajamani, K. A. Shanmugasundaram, and M. Boomiga. 2022. "Breeding and Conservation of Medicinal Plants in India." In Ákos Máthé and Irfan Ali Khan (eds.) *Medicinal and Aromatic Plants of India*, (1): 201–36. Springer.

Nandal, Urvashi, and Raju Lal Bhardwaj. 2012. "Aloe Vera for Human Nutrition, Health and Cosmetic Use-A Review." *International Research Journal of Plant Science* 3 (3): 38–46.

Nayanakantha, N. M. C., B. R. Singh, and A. K. Gupta. 2010. "Assessment of Genetic Diversity in Aloe Germplasm Accessions from India Using RAPD and Morphological Markers." *Ceylon Journal of Science (Biological Sciences)* 39 (1): 1–9.

Nayeri, Shahnoush, Zahra Dehghanian, Behnam Asgari Lajayer, Allan Thomson, Tess Astatkie, and G. W. Price. 2023. "CRISPR/Cas9-Mediated Genetically Edited Ornamental and Aromatic Plants: A Promising Technology in Phytoremediation of Heavy Metals." *Journal of Cleaner Production* 428: 139512.

Niveditha, Mannem, Megha R. Meshram, Siddhnath A. Shendekar, and Venna Santhosh. n.d. "Germplasm Collection, Exchange and Quarantine." *Seed Science & Technology* (1): 230.

O Lawal, Ibraheem, Rofiat I. Oyediran, Musbau B. Olaniyi, Toluope F. Okanlawon, Ayokunmi O. Oyeleye, Funmilola O. Akanni, Basirat O. Rafiu, Temitope O. Omogbene, Anne A. Adam, and Oyinkansade Y. Babalola. 2021. "A Review on the Diverse Uses, Conservation Measures and Agronomic Aspect of Aloe Vera (L.) Burm. F." *European Journal of Medicinal Plants* 32 (9): 39–51.

Odeleye, O. M., A. A. Elujoba, and A. A. Gbolade. 2009. "Comparative Pharmacognostic Studies on Aloe Schweinfurthii and Aloe Vera (Aloeceae) Leaves." *Planta Medica* 75 (4): P-23.

Olupona, J. A., O. R. Omotoso, A. A. Adeyeye, O. D. Kolawole, A. P. Airemionkhale, and O. O. Adejinmi. 2010. "Effect of Aloe Vera Juice Application through Drinking Water on Performance, Carcass Characteristics, Hematology and Organoleptics Properties in Broilers." *Poultry Science* 88 (1): 42–3.

Parikh, Bijal A., Diana L. Beckman, Swapneel J. Patel, J. Michael White, and Wayne M. Yokoyama. 2015. "Detailed Phenotypic and Molecular Analyses of Genetically Modified Mice Generated by CRISPR-Cas9-Mediated Editing." *PloS One* 10 (1): e0116484.

Pradeep Reddy, M., N. Sarla, and E. A. Siddiq. 2002. "Inter Simple Sequence Repeat (ISSR) Polymorphism and Its Application in Plant Breeding." *Euphytica* 128: 9–17.

Priyanka, Veerala, Rahul Kumar, Inderpreet Dhaliwal, and Prashant Kaushik. 2021. "Germplasm Conservation: Instrumental in Agricultural Biodiversity-A Review." *Sustainability* 13 (12): 6743.

Quazi, Sameer, Tanya Golani, and Arnaud Martino Capuzzo. 2021. "Germplasm Conservation." *Endangered Plants*, 180: 123.

Raghunath, Kandekar Yash, Tarade Vijay, Kasar Utkarsh Balasaheb, Sonawane Vivek Rajaram, and Kandekar Mahesh Raghunath. n.d. "A Review on *Aloe* vera from Traditional Healing to Modern." *International Research Journal of Modernization in Engineering Technology and Science* 5 (12): Article 47524. https://doi.org/10.56726/IRJMETS47524

Rana, Sabina, Dipka Sharma, and Niharika Bakshi. 2018. "A Mini Review on Morphological, Biochemical and Molecular Characterization of *Aloe* era L." *International Journal of Chemical Studies* 4: 3109–15.

Ranghoo Sanmukhiya, Mala, Joyce Govinden Soulange, Christophe Lavergne, Shannoo Khoyratty, Denis Da Silva, Michel Frederich, and Hippolyte Kodja. 2010. "Molecular Biology, Phytochemistry and Bioactivity of Three Endemic Aloe Species from Mauritius and Réunion Islands." *Phytochemical Analysis* 21 (6): 566–74.

Reynolds, T., and A. C. Dweck. 1999. "*Aloe* Vera Leaf Gel: A Review Update." *Journal of Ethnopharmacology* 68 (1–3): 3–37.

Roberts, Daniel P., and Autar K. Mattoo. 2018. "Sustainable Agriculture-Enhancing Environmental Benefits, Food Nutritional Quality and Building Crop Resilience to Abiotic and Biotic Stresses." *Agriculture* 8 (1): 8.

Salinas, Pamela, Carlos Salinas, Rodrigo A. Contreras, Gustavo E. Zuñiga, Paul Dupree, and Liliana Cardemil. 2019. "Water Deficit and Abscisic Acid Treatments Increase the Expression of a Glucomannan Mannosyltransferase Gene (GMMT) in Aloe Vera Burm. F." *Phytochemistry* 159: 90–101.

Sarwat, Maryam, S. Das, and P. S. Srivastava. 2008. "Analysis of Genetic Diversity through AFLP, SAMPL, ISSR and RAPD Markers in Tribulus Terrestris, a Medicinal Herb." *Plant Cell Reports* 27: 519–28.

Serrano, María, Juan Miguel Valverde, Fabián Guillén, Salvador Castillo, Domingo Martínez-Romero, and Daniel Valero. 2006. "Use of Aloe Vera Gel Coating Preserves the Functional Properties of Table Grapes." *Journal of Agricultural and Food Chemistry* 54 (11): 3882–86.

Shannon, Paul, Andrew Markiel, Owen Ozier, Nitin S. Baliga, Jonathan T. Wang, Daniel Ramage, Nada Amin, Benno Schwikowski, and Trey Ideker. 2003. "Cytoscape: A Software Environment for Integrated Models of Biomolecular Interaction Networks." *Genome Research* 13 (11): 2498–504.

Sharma, Priyanka, Amit C. Kharkwal, Harsha Kharkwal, M. Z. Abdin, and Ajit Varma. 2014. "A Review on Pharmacological Properties of Aloe Vera." *International Journal of Pharmaceutical Sciences Review and Research* 29 (2): 31–7.

Shekhawat, Narpat S., Mangal S. Rathore, Smita Shekhawat, Sumitra K. Choudhary, Mahendra Phulwaria, Harish, Manoj K. Rai, J. B. Vibha, Nitika S. Rathore, and A. K. Patel. 2013. "Micropropagation of Aloe Vera for Improvement and Enhanced Productivity." *Climate Change and Plant Abiotic Stress Tolerance* 857–80.

Shelton, Ronald M. 1991. "Aloe Vera: Its Chemical and Therapeutic Properties." *International Journal of Dermatology* 30 (10): 679.

Singh, Brahma Deo, and Ashok Kumar Singh. 2015. *Marker-Assisted Plant Breeding: Principles and Practices* (1st ed.). India: Springer.

Singh, Devendra Pal. 2023. *Integrated Pest Management in Diverse Cropping Systems*. CRC Press.

Soltani Howyzeh, Mehdi, Seyed Ahmad Sadat Noori, Vahid J. Shariati, and Mahboubeh Amiripour. 2018. "Comparative Transcriptome Analysis to Identify Putative Genes Involved in Thymol Biosynthesis Pathway in Medicinal Plant Trachyspermum Ammi L." *Scientific Reports* 8 (1): 13405.

Steenkamp, Vanessa, and M. J. Stewart. 2007. "Medicinal Applications and Toxicological Activities of Aloe. Products." *Pharmaceutical Biology* 45 (5): 411–20.

Surjushe, Amar, Resham Vasani, and D. G. Saple. 2008. "Aloe Vera: A Short Review." *Indian Journal of Dermatology* 53 (4): 163–66.

Szklarczyk, Damian, John H. Morris, Helen Cook, Michael Kuhn, Stefan Wyder, Milan Simonovic, Alberto Santos, Nadezhda T. Doncheva, Alexander Roth, and Peer Bork. 2016. "The STRING Database in 2017: Quality-Controlled Protein-protein Association Networks, Made Broadly Accessible." *Nucleic Acids Research*, gkw937.

Terefe, Tesfahun, and Tsehaye Neges. 2017. "Review on Therapeutic and Medicinal Use of Aloe Vera." *Cancer Biology* 7 (4): 29–38.

Tizard, I., D. Busbee, B. Maxwell, and M. C. Kemp. 1994. "Effects of Acemannan, a Complex Carbohydrate, on Wound-Healing in Young and Aged Rats." *Wounds-A Compendium of Clinical Research and Practice* 6 (6): 201–9.

Tripathi, Niraj, Navinder Saini, and Sharad Tiwari. 2011. "Assessment of Genetic Diversity among Aloe Vera Accessions Using Amplified Fragment Length Polymorphism." *International Journal of Medicinal Aromatic Plants* 1: 115–21.

Trojak-Goluch, Anna, Magdalena Kawka-Lipi ska, Katarzyna Wielgusz, and Marcin Praczyk. 2021. "Polyploidy in Industrial Crops: Applications and Perspectives in Plant Breeding." *Agronomy* 11 (12): 2574.

Varma, Astha, Harish Padh, and Neeta Shrivastava. 2007. "Plant Genomic DNA Isolation: An Art or a Science." *Biotechnology Journal: Healthcare Nutrition Technology* 2 (3): 386–92.

Vogler, B. K., and E. Ernst. 1999. "Aloe Vera: A Systematic Review of Its Clinical Effectiveness." *British Journal of General Practice* 49 (447): 823–28.

Wagan, Maqsood Ali, Tanveer Fatima Miano, Bilawal Nizamani, Hussain Ali Shar, Muhammad Khan Wagan, and Farhan Ali Wagan. 2022. "Influence of Sunlight on Growth and Yield of Aloe Vera Plant." *American Journal of Bioscience and Bioengineering* 10 (4): 84–8.

Wang ChangKang, Jia HongQiang, Tong JianMing, Gao WeiWei, Sa RenNa, and Zhang Qi. 2007. "Effect of aloe powder and extract on production performance and immune function of broiler chickens." *Journal of Fujian Agriculture and Forestry*: 614–617.

Waris, Zeenat, Yousaf Iqbal, Shafqatullah Arshad Hussain, Asad Ali Khan, Akhtar Ali, and Mohammad Wasiullah Khan. 2018. "19. Proximate Composition, Phytochemical Analysis and Antioxidant Capacity of Aloe Vera, Cannabis Sativa and Mentha Longifolia." *Pure and Applied Biology (PAB)* 7 (3): 1122–30.

World Health Organization. 1999. *WHO Monographs on Selected Medicinal Plants*, Vol. 2. World Health Organization.

Woudstra, Yannick, Juan Viruel, Martin Fritzsche, Thomas Bleazard, Ryan Mate, Caroline Howard, Nina Rønsted, and Olwen M. Grace. 2021. "A Customised Target Capture Sequencing Tool for Molecular Identification of Aloe Vera and Relatives." *Scientific Reports* 11 (1): 24347.

Yamaguchi-Shinozaki, Kazuko, and Kazuo Shinozaki. 2006. "Transcriptional Regulatory Networks in Cellular Responses and Tolerance to Dehydration and Cold Stresses." *Annual Review Plant Biology* 57 (1): 781–803.

Yoo, Yo-Han, Woo-Jong Hong, and Ki-Hong Jung. 2019. "A Systematic View Exploring the Role of Chloroplasts in Plant Abiotic Stress Responses." *BioMed Research International*: 1–17.

Yoshida, Takuya, Yasunari Fujita, Kyonoshin Maruyama, Junro Mogami, Daisuke Todaka, Kazuo Shinozaki, and Kazuko Yamaguchi Shinozaki. 2015. "Four A Rabidopsis AREB/ABF Transcription Factors Function Predominantly in Gene Expression Downstream of SnRK2 Kinases in Abscisic Acid Signalling in Response to Osmotic Stress." *Plant, Cell & Environment* 38 (1): 35–49.

Younessi-Hamzekhanlu, Mehdi, Munir Ozturk, Parinaz Jafarpour, and Nasser Mahna. 2022. "Exploitation of Next-Generation Sequencing Technologies for Unraveling Metabolic Pathways in Medicinal Plants: A Concise Review." *Industrial Crops and Products* 178: 114669.

Zhou, Tao, Jun Yao, and Zhanjiang Liu. 2017. "Gene Ontology, Enrichment Analysis, and Pathway Analysis." In Liu, Zhanjiang John (ed.) *Bioinformatics in Aquaculture: Principles and Methods,* 150–68. John Wiley & Sons

Zou, Yi, Changyou Han, Fang Wang, Yanhua Tan, Saina Yang, Chuying Huang, Shengyou Xie, and Xueqin Xiao. 2021. "Integrated Metabolome and Transcriptome Analysis Reveal Complex Molecular Mechanisms Underlying Selenium Response of *Aloe* Vera L." *Journal of Plant Biology* 64:135–43.

8 Conventional to Genome-Assisted Breeding in *Aloe vera*

Soniya Goyal, Mahiti Gupta, Rajesh Yogi, Ravinder Kumar, and Vikas Beniwal

8.1 INTRODUCTION

Aloe vera, which belongs to the Asphodelaceae family, is a succulent plant species known for its thick, fleshy leaves that contain a gel-like substance. This gel is widely used for various health and skincare purposes. It is native to North Africa but cultivated in various regions around the world. The inner part of the *Aloe vera* leaf contains a clear, gel-like substance which is rich in water, vitamins, minerals, amino acids, and antioxidants (Misir et al., 2014). The gel is known for its soothing and moisturizing properties. *Aloe vera* is commonly used in traditional medicine for its potential health benefits (Nagansurkar et al., 2024). It may have anti-inflammatory, antimicrobial, and antioxidant properties. Some people use *Aloe vera* to help soothe sunburn, wounds, or skin irritations (Kumar et al., 2024). It also helps to hydrate skin, reduce inflammation, promote healing, and is an important ingredient in skincare products such as lotions, creams, and gels (Hekmatpou et al., 2019).

8.1.1 Historical Perspective on Conventional Breeding in *Aloe vera*

Aloe vera has been cultivated and bred for thousands of years. *Aloe vera* is believed to have originated in the Arabian Peninsula and North Africa. It has a long history of use in various cultures, prized for its medicinal and skin care properties (Shah et al., 2024). While the historical details of breeding practices specifically for *Aloe vera* are not as well-documented as for some other crops, the cultivation and propagation methods have evolved, as represented in Figure 8.1.

- **Ancient cultivation:** *Aloe vera* has been used since ancient times by various civilizations, including the Egyptians, Greeks, and Romans. It is mentioned in historical texts for its healing properties. In these early periods, cultivation likely involved the natural propagation of the plant through offshoots and division (Vergauwen and De Smet, 2017).
- **Traditional uses in Asia**: *Aloe vera* spread to Asia, where it found its place in traditional medicine, especially in countries like India and China. Over the centuries, local communities would have developed their own cultivation practices, which might have included selective breeding for desirable traits, even though it might not have been as formally documented as in modern agriculture (Van et al., 2022).
- **Colonial and global spread:** With the age of exploration and colonization, *Aloe vera* was introduced to different parts of the world. As it spreads, various communities adapted cultivation practices according to their climatic conditions and needs. In this process, natural selection and unintentional selection for certain traits likely occurred (Zsogon et al., 2018).
- **Twentieth-century commercialization:** The formalization of *Aloe vera* breeding and cultivation likely began in the 20th century due to its various medicinal and cosmetic properties.

 DOI: 10.1201/9781003603108-8

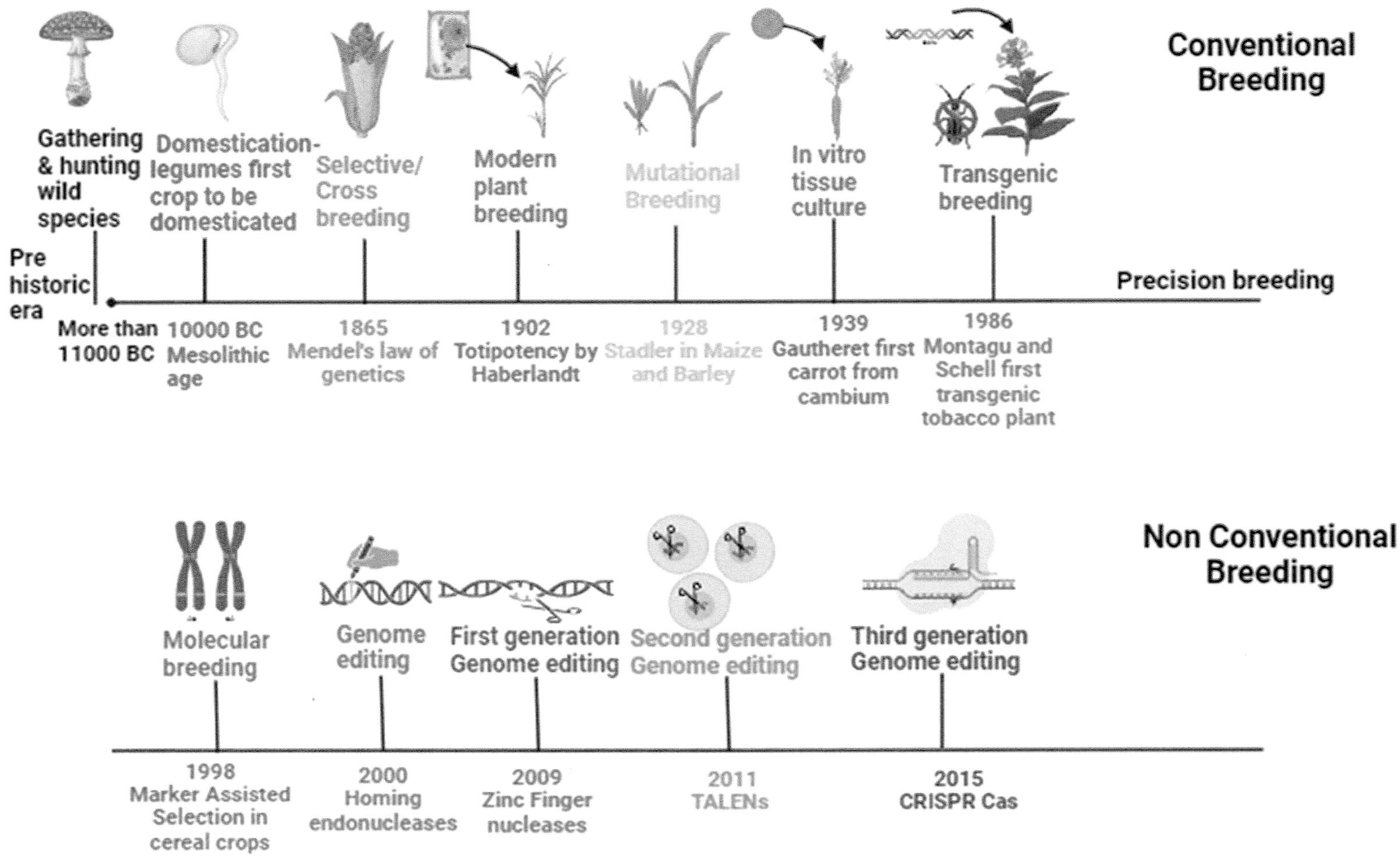

FIGURE 8.1 History of breeding in plants: conventional and non-conventional methods.

The commercial cultivation of *Aloe vera* began due to its specific characteristics, such as high gel content, better adaptability to different climates, and resistance to diseases (Swaminathan, 2009; Vergauwen and De Smet, 2017).

8.2 METHODS OF CONVENTIONAL BREEDING IN *ALOE VERA*

Plant breeding has been a significant part of an emerging human civilization since the Neolithic age, based on increasing genetic variability, enhancement of yield, and improving the nutritional value of crops by crossing wild species and landraces (Podevin et al., 2013). There are primarily three methods of conventional breeding, namely cross-breeding, mutagenesis breeding, and transgenic breeding, used in agricultural improvement for centuries, as mentioned below (Gao, 2021).

8.2.1 Cross-breeding

This method involves mating individuals from two distinct but related varieties or species to combine desirable traits found in each parental line. Cross-breeding offers the combination of favorable traits from different parental lines, such as stress resistance, yield potential, and quality attributes. An introduction of genetic diversity through the cross-breeding method can also enhance the adaptability and resilience of crop varieties to changing environmental conditions (Kyndt et al., 2015).

8.2.2 Mutagenesis Breeding

This type of breeding involves inducing random genetic mutations in plant DNA using irradiation or chemical mutagenesis to generate novel variation, which may result in desirable traits. Mutagenesis can also introduce novel genetic variation, which leads to beneficial traits and expands genetic diversity available for breeding programs, potentially unlocking new avenues for crop improvement (Holme et al., 2019). Thus, both the cross-breeding and mutation breeding techniques have some limitations, including the introduction of undesirable traits, labor intensity, and a long time period required for the selection of a rare variety that possesses the desired feature (Gao, 2021).

8.2.3 Transgenic Breeding

This involves the introduction of specific genes from one organism into the genome of another organism to confer desirable traits. A target gene is isolated and cloned from a donor organism and inserted into the genome of target organisms using different biotechnological methods, including Agrobacterium-mediated transformation or gene gun bombardment (Pinto-Carbo et al., 2016). Then, the identification, screening, and characterization of transgenic plants having introduced genes are done for confirmation of expression of desired traits. This breeding approach allows precise modification of specific genes, offering target solutions to address agronomic challenges or consumer preferences (Hickey et al., 2019; Chen et al., 2019).

However, this approach involves the random integration of foreign DNA into the plant genome and was strictly subjected to regulations imposed by the government for introducing foreign genes and products, which were known as GMOs (Genetically modified organisms) (Enfssi et al., 2021).

All of these conventional breeding approaches have their advantages as well as limitations, and their selection depends on various factors such as objectives of breeding, availability of resources, and regulatory considerations. Cross-breeding depends on natural genetic recombination and is limited by the availability of genetic variation within species or closely related species. Mutagenesis breeding induces random mutations across the genome, offering broad but unpredictable genetic changes, while transgenic breeding provides targeted introduction of desired traits but may face regulatory and public acceptance challenges due to concerns about genetic modification (Schmidt et al., 2020).

8.2.4 Modern Plant Breeding Methods

There are different modern breeding methods through which *Aloe vera* can be propagated, as discussed below:

- **Conventional *in vivo* propagation:** Propagation by seed is one of the methods of conventional *In vivo* propagation. *Aloe vera* is generally a self-incompatible plant; therefore, seeds derived only from cross-pollination are fertile, which involves a high rate of heterogeneity among the seed population (Silva et al., 2010). Using the sexual reproduction method for maintaining the species or varieties of *Aloe vera* is difficult because the natural hybridization of this species is very frequent. A significant genetic variation exists across Indian accessions for yield and quality parameters, as demonstrated by some studies. However, *Aloe vera* seeds can have variable genetic traits, so this method may result in plants with different characteristics than the plant (Supe, 2006). To propagate through seeds, collect mature seeds and plant them in a well-draining soil mix. Keep the soil consistently moist until germination occurs. Seedlessness is also another critical aspect that occurs due to environmental and genetic factors, resulting in anomalous chromosomal behavior. A low percentage (0-25%) of seed germination is observed under *in vivo* conditions as compared to seed germination under *in vitro* conditions (60-70%) (Das et al., 2016). The plants raised by seed propagation take a long gestation period before attaining maturity, 3–4 years for reaching harvestable size, and it has a life span of about 12 years. Therefore, commercial nurseries prefer propagation of *Aloe vera* agamically.

Conventionally, *Aloe vera* is also propagated through suckers (lateral shoots) and rhizome cuttings as planting material. Each mother plant produces three to four suckers in their growing season under a conventional production system.

- **Suckers and rhizome cuttings:** Conventionally, *Aloe vera* is propagated using lateral shoots and cuttings of rhizome as explants (Singh and Sood, 2009). Throughout the growing season, each mother plant produces three to four suckers in a conventional production system. Then selected suckers should be subjected to stress and shade conditions for about 5–10 days for the prevention of diseases and promotion of root suberisation, followed by transplantation in the field. This speeds up the sucker's adaptation and enhances the growth of roots when planting, and the crop will be ready for harvesting after 18 months of sowing. A large-scale production of *Aloe vera* using rhizome cuttings is impractical due to the long period of time (average two years) required for the development of a complete plant and exposure of phytopathogens in cuts (Gantait et al., 2014). There are some limiting factors, including a high probability of damaging the mother plant, production of infected seedlings, a limited number of suckers, limited availability of raw materials of high quality, seasonality, and slow vegetative growth for large-scale production of young plants commercially.
- ***In vitro* propagation/micropropagation:** This method can be used for large-scale production of uniform and healthy plantlets throughout the year and overcomes various drawbacks of *in vivo* propagation (Singh et al., 2009). *In vitro* propagation can be done by using two methods, i.e., axillary bud stimulation and de novo *in vitro* organogenesis (Almeida, 2024). Axillary bud stimulation in *Aloe vera* involves inducing the growth of new shoots from the axillary buds (located in the leaf axils), which can lead to the production of multiple shoots for propagation and multiplication of *Aloe vera* plants (Bais and Todkari, 2024). There are some methods, including pruning, use of auxin hormones, wounding, manipulation of environmental factors (temperature and light), and nutrient management, commonly used for the stimulation of axillary bud growth in *Aloe vera,* which results in increased production of shoots and overall success of plant propagation (Kanwar et al., 2015)

De novo *in vitro* organogenesis in *Aloe vera* involves the regeneration of new plant organs, such as shoots or roots, from cultured plant tissues in a laboratory setting. This process initiates with the selection of appropriate explants (small pieces of tissues taken from the parent plant, mostly leaf sections or meristematic tissues), followed by surface sterilization (bleach or ethanol as sterilizing agents) (Shibru et al., 2018). After surface sterilization, explants will be placed onto nutrient medium containing all essential components and optimized culture conditions for the induction of organogenesis (formation of new shoots or roots from explants). After initiation of organogenesis, the developing shoots or roots are subcultured onto fresh medium periodically for promotion of further growth and development (Kumari and Naseem, 2015). For successful regeneration, careful maintenance of culture conditions such as temperature, light, and nutrient levels is mandatory (Gupta et al., 2014). Once a sufficient number of shoots or roots have developed, they are transferred to soil or soil-less growing medium in a greenhouse or controlled environment. This allows the regenerated plants to adapt to natural growing conditions before they are ready for transplantation to the field.

The conventional breeding methods have been extensively used in agriculture for the development of improved crop varieties, but they also have a few limitations when applied to *Aloe vera* (Das and Bora, 2018). Some of the drawbacks of conventional breeding methods for *Aloe vera* include:

Long breeding cycles: *Aloe vera* has a relatively long breeding cycle, so it takes several years to develop and evaluate new varieties. Conventional breeding methods, which depend on sexual reproduction and repeated cycles of cross-pollination and selection, can further increase the breeding process (Gantait et al., 2014).

Limited genetic diversity: Conventional breeding methods for *Aloe vera* are frequently limited by the availability of genetic diversity within the species. This can hamper the ability to introduce novel traits or improve desirable characteristics, especially if the breeding pool is narrow or lacks genetic variation.

Genetic variation and instability: *Aloe vera* exhibits significant genetic variation within and among populations, which can complicate breeding efforts. Conventional breeding methods may lead to genetic instability and unpredictable outcomes due to the separation of traits and genetic recombination during sexual reproduction (Cristiano et al., 2016).

Incompatibility barriers: *Aloe vera* exhibits self-incompatibility, i.e., plants are often unable to self-pollinate or cross-pollinate with closely related individuals because of genetic barriers. This can curb the effectiveness of conventional breeding methods, as it may be difficult to achieve successful crosses between desired parental lines.

Labor-intensive and resource-consuming: Conventional breeding methods require significant resources, including time, labor, land, and infrastructure for field trials and evaluation of breeding populations. This can be particularly challenging for *Aloe vera*, which often grows in arid or semi-arid regions with limited agricultural resources (Gantait et al., 2023).

Limited precision and predictability: Conventional breeding methods depend on random genetic recombination and selection processes, which may lack precision and predictability in trait improvement. This can result in a high degree of variability among progeny and make it difficult to achieve consistent and targeted breeding objectives (Badar et al., 2017).

Complex trait inheritance: Many of the traits of interest in *Aloe vera*, such as gel quality, yield, and disease resistance, are controlled by multiple genes and influenced by environmental factors. Conventional breeding methods may struggle to effectively manipulate and improve complex traits due to their polygenic nature (Zakia et al., 2013).

To overcome the above-mentioned limitations of conventional breeding methods, modern biotechnological approaches such as marker-assisted selection, genetic transformation, and genomic selection may be used as alternative strategies for *Aloe vera* improvement, allowing for more precise manipulation of traits and speeding up breeding progress (Badar et al., 2017).

8.3 GENOMIC TOOLS AND TECHNOLOGIES

Genomic tools and techniques have been increasingly employed in the study of *Aloe vera* to understand its genetic makeup, improve traits, and develop new varieties (Ikram et al., 2024). Some of the genomic tools and techniques used for *Aloe vera* research are mentioned below:

DNA sequencing: DNA sequencing technologies include next-generation sequencing (NGS) platforms, such as Illumina sequencing, which has been utilized for sequencing the entire genome of *Aloe vera*. These technologies make it easy for researchers to identify genes, regulatory elements, and genetic variations within the genome (Janbandhu et al., 2024).

Transcriptomic analysis: Transcriptomic analysis, including RNA sequencing (RNA seq), is done to study gene expression patterns in *Aloe vera* tissues under different developmental stages and environmental conditions, which helps in the identification of genes involved in key biological processes and pathways (Rurek and Smolibowski, 2024).

Genetic mapping and QTL analysis: These studies are conducted for the identification of genomic regions associated with important agronomic traits in *Aloe vera,* such as drought tolerance, leaf gel yield, and disease resistance (Kuzmishyna, 2024). For this analysis, different techniques such as simple sequence repeats (SSR) or single-nucleotide polymorphisms (SNPs) are used for breeding purposes and marker-assisted selection (Yogi et al., 2023; Shah et al., 2024).

Genomic-wide association studies (GWAS): This study involves the scanning of genomes of different varieties of *Aloe vera* germplasm for the identification of genetic variations associated with traits of interest. This study helps in the elucidation of the genetic basis of complex traits and facilitates marker-assisted breeding efforts (Ninama et al., 2024).

Functional genomics: Functional genomics studies include gene expression analysis, gene knockout or overexpression studies, and gene editing using CRISPR-Cas9 technology, which are employed in the investigation of gene function and regulatory mechanisms in *Aloe vera* (Ninama et al., 2024). These approaches help in the validation of candidate genes identified through genomic analysis and elucidate their roles in plant physiology and development.

Epigenomics: These studies explore epigenetic modifications, including DNA methylation and histone modifications, in *Aloe vera* genomes. Epigenetic regulation plays a crucial role in gene expression, stress responses, and developmental processes to understand epigenetic mechanisms, which can provide insights into the biology of *Aloe vera* and crop improvement (Singh et al., 2024).

Metabolomics: Metabolomic profiling allows the comprehensive analysis of metabolites in *Aloe vera* tissues, such as secondary metabolites (aloin, polysaccharides, and phenolics), which contribute to its medicinal and nutritional properties. These studies help in the identification of metabolic pathways and regulatory mechanisms that unravel the biosynthesis and accumulation of bioactive compounds in *Aloe vera* (Ushasree et al., 2024).

Bioinformatics and genomic databases: Various bioinformatics tools and genomic databases are available, which provide valuable resources for analysis and interpretation of genomic data used in *Aloe vera* research. One of the genomic databases, such as AloeBase, facilitates data storage, retrieval, and comparative genomics analyses and helps researchers to explore the genetic diversity and evolutionary relationships among Aloe species (Singh et al., 2024).

The above-discussed genomic tools and techniques help researchers to advance their understanding of *Aloe vera* genetics, physiology, and biochemistry, which ultimately contributes to the development of improved varieties with enhanced agronomic traits and medicinal properties.

8.3.1 Marker-assisted Selection (MAS) in *Aloe vera* Breeding

MAS has become a valuable tool in plant breeding, including the breeding of *Aloe vera*, a succulent plant known for its various medicinal and cosmetic properties. This approach involves the selection of plants based on genetic markers associated with desirable traits, which helps the breeders to accelerate the breeding process compared to traditional methods (Janbandhu et al., 2024). MAS can be applied in *Aloe vera* breeding by following the procedure mentioned below (Kuzmishyna, 2024):

Identification of desirable traits: Breeders identify specific traits they want to introduce or improve in *Aloe vera,* including increased gel yield, increased production of bioactive compounds like aloin or polysaccharides, resistance to diseases or pests, or adaptation to specific environmental conditions (Kumari et al., 2024).

Identification of genetic markers: Researchers identify genetic markers linked with the desirable traits through molecular markers (such as SSRs, SNPs) or DNA sequencing. These markers could be associated with genes specific to the expression of the desired traits.

Development of molecular maps: A molecular map of the *Aloe vera* genome is created, which shows the location of genetic markers across the chromosomes. This map facilitates understanding the inheritance patterns of traits and assists in selecting the desired markers for breeding (Kumari et al., 2024).

Analysis of marker-trait association: By using statistical methods, breeders analyse the association between genetic markers and desirable traits, which helps in the identification of the most reliable markers that consistently correlate with the desired traits across different *Aloe vera* varieties or populations.

Marker-assisted selection: After the identification of reliable markers, breeders use them in the screening of *Aloe vera* plants at early growth stages (e.g., seedlings or tissue cultures) or at the DNA level. By selecting plants with the desired markers, breeders can indirectly select for the target traits without any need to wait for phenotypic expression in mature plants (Nair and Rabbani, 2024).

Validation and selection of elite lines: The selected *Aloe vera* lines with desirable marker profiles are further evaluated for their phenotypic performance in field trials or controlled environments. This validation step is used to ensure that the selected lines exhibit the desired traits in addition to genetic markers.

Optimization of breeding program: MAS facilitates the optimization of breeding programs by enabling more efficient selection of parent plants for crossing, reducing the time required for trait introgression, and increasing the chances of success in developing superior *Aloe vera* cultivars (Rani, 2024).

In conclusion, MAS holds great promise for enhancing the efficiency and precision of *Aloe vera* breeding programs, ultimately leading to the development of improved varieties with enhanced medicinal, cosmetic, and agronomic traits (Ramesh et al., 2024).

8.3.2 Genomic Insights into Medicinal Traits

Genomic insights into medicinal traits involve understanding the genetic basis of beneficial and medicinal properties found in different plants, such as *Aloe vera.* Genomic research provides valuable insights into the medicinal traits of *Aloe vera* in the following ways:

Identification of bioactive compounds: Genomic study helps in the identification of genes that are involved in the biosynthesis pathways of bioactive compounds such as polysaccharides, phenolic compounds, aloin, and different types of vitamins present in *Aloe vera.* By focusing on these genes, researchers gain knowledge about the molecular mechanisms involved in the production of medicinal compounds (Ninama et al., 2024).

Characterization of biosynthetic pathways: Genomic analysis explicates the metabolic pathways responsible for the synthesis of medicinal compounds in *Aloe vera.* An understanding of these pathways can facilitate metabolic engineering efforts aimed at enhancing the production of specific bioactive compounds or inventing novel compounds with enhanced medicinal properties (Raza et al., 2024).

Genetic variation and diversity: Genomic analysis reveals genetic variations among different *Aloe vera* accessions or cultivars, providing knowledge into the natural diversity of medicinal traits. By exploring genetic diversity, researchers can identify genetic markers associated with desirable traits or with high levels of bioactive compounds for breeding purposes (Singh et al., 2024).

Marker-assisted breeding: Genomic analysis also allows researchers to develop molecular markers associated with target genes controlling medicinal traits in *Aloe vera*. These markers are involved in marker-assisted selection (MAS) in breeding programs, allowing breeders to select plants efficiently with desirable medicinal properties at an early stage of development (Ramesh et al., 2024).

Genomic editing for trait modification: Modern technologies such as CRISPR-Cas9 provide tools for precise genome editing in *Aloe vera*. Researchers can use this technology in the modification of specific genes involved in the biosynthesis of medicinal compounds, resulting in enhancing their production or altering their composition in the improvement of therapeutic efficacy (Ushasree et al., 2024).

Understanding of environmental stress responses: A study of the genome also helps in the elucidation of molecular mechanisms responsible for *Aloe vera's* response towards environmental stresses such as drought and salinity (Raza et al., 2024). This information helps in the development of stress-tolerant varieties with enhanced medicinal properties by manipulating breeding strategies.

Comparative genomic analysis: A comparative genomic analysis with other *Aloe vera* species or related plants with known medicinal properties can provide insights into conserved pathways and genes associated with medicinal traits. This comparative approach can help in the discovery of novel medicinal compounds or pathways in *Aloe vera* (Singh et al., 2024). Overall, genomic insights into medicinal traits provide a foundation for understanding the molecular basis of *Aloe vera*'s therapeutic effects and offer opportunities for improving the quality, efficacy, and resilience of *Aloe vera*-based products through targeted breeding and biotechnological interventions.

8.4 EMERGING TECHNOLOGIES: CRISPR-CAS9 AND GENE EDITING

The development of various genome editing technologies has transformed research in the fields of crop improvement and functional genomics (Liu and Fan, 2014) (Figure 8.2). CRISPR/Cas9 (clustered regularly interspaced short palindromic repeat) is a multipurpose gene editing technology that represents a revolutionary approach to genetic manipulation that has transformative potential across various fields such as agriculture, medicine, and biotechnology (Guo et al., 2022). It is a powerful gene editing tool derived from a natural defence mechanism found in bacteria containing segments of prokaryotic DNA having short repetitive base sequences. Cas9 is an enzyme that acts as a molecular scissor, capable of cutting DNA at specific target sequences (Ceasar et al., 2016).

8.4.1 Working of CRISPR-Cas

The working of CRISPR-Cas9 depends on three steps, including Guide RNA design, DNA cleavage, and repair mechanisms (Doudna and Charpentier, 2014). A synthetic RNA molecule known as guide RNA (gRNA) is designed to complement the target DNA sequence. The guide RNA directs the Cas9 enzyme to make a cut at a precise location in the genome. Then, the Cas9 enzyme, guided by gRNA, binds to the target DNA sequence and induces a double-strand break (DSB) at that site (Barrangou,

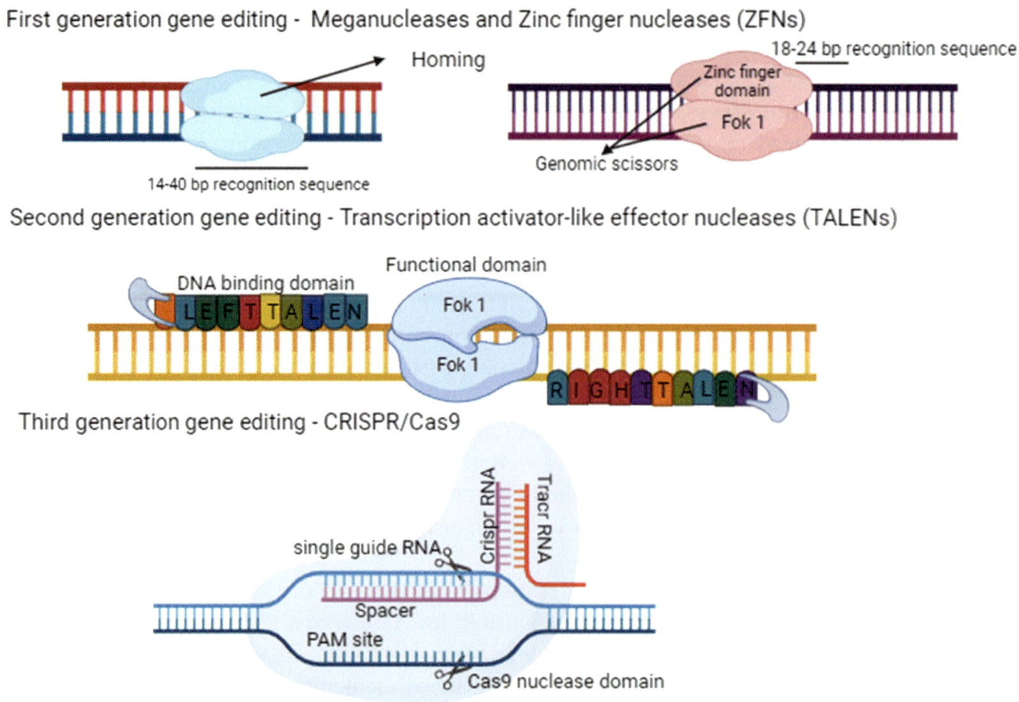

FIGURE 8.2 Different tools in genome editing used for advancement in crop improvement.

2015). Cells have natural DNA repair mechanisms that can fix the DSB through either non-homologous end joining (NHEJ) or homology-directed repair (HDR). Non-homologous end joining methods can introduce small insertions or deletions at the cutting site, leading to gene knockouts (Deltcheva et al., 2011). Homologous directed repair mechanisms can be used for insertions or replacing specific DNA sequences with high precision (Puchta, 2005; Scully et al., 2019) (Figure 8.3).

8.4.2 Applications of CRISPR-Cas9

The CRISPR-Cas9 tool can be used in different areas, as mentioned below (Xiong et al., 2021):

Gene knockout studies: CRISPR-Cas9 can be used to disrupt or knock out specific genes through the introduction of insertions and deletions (Indels) at target sites, allowing the study of gene function (Lotfi et al., 2024).

Gene editing: CRISPR-Cas9 also helps in the precise editing of DNA sequences, such as insertions, deletions, or replacements of nucleotides, which facilitates the modification of genetic traits in different species (Mehrabadi et al., 2024).

Disease modelling: CRISPR-Cas9 facilitates the generation of cellular or animal models with specific genetic mutations associated with human diseases, helping in the study of disease mechanisms and the development of novel therapies (Stadtmauer et al., 2020).

Crop improvement: The CRISPR-Cas9 tool can be used in the enhancement of crop traits such as yield, nutritional content, and stress tolerance by editing genes involved in plant development, metabolism, and defence mechanisms in the agriculture sector (Ambika et al., 2024).

Therapeutic applications: CRISPR-Cas9 is also a promising tool in the treatment of genetic disorders by correcting disease-causing mutations in patients' cells or tissues. It also has a potential for cancer immunotherapy and precision medicine (Hajian et al., 2019).

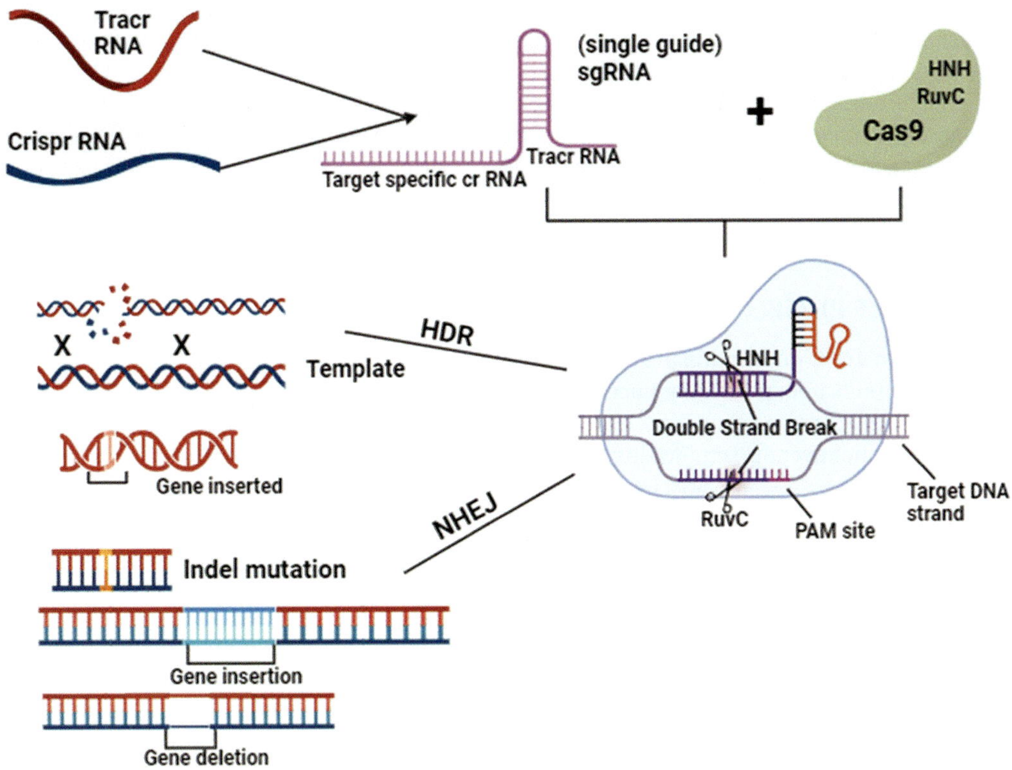

FIGURE 8.3 Mechanism of CRISPR-Cas9 genome editing tool.

Biotechnological applications: CRISPR-Cas9 is used in various biotechnological applications, including engineering of microbial strains for industrial processes, production of biofuels and pharmaceuticals, and modification of model organisms used for research purposes (Ambika et al., 2024).

The research continues in the advancement of CRISPR-Cas9 technology, mainly focusing on improving its efficiency, specificity, and delivery methods. In addition to Cas9, Cas12 and Cas13 offer expanded capabilities for gene editing, epigenome editing, and RNA manipulation (Guo et al., 2022). The CRISPR-Cas9 tool also holds a significant promise in the advancement of *Aloe vera* research and breeding programs, and can be applied in various ways, including target trait modification, developing disease-resistant and stress-tolerant varieties, and manipulating metabolic pathways (Roviqowati et al., 2024). CRISPR-Cas9 can be utilized to precisely edit genes that are involved in the production of medicinal compounds such as phenolic compounds, aloin, and polysaccharides, target genes associated with traits like leaf morphology and gel yield to improve the overall quality and productivity of *Aloe vera* plants (Doudna and Charpentier, 2014). Stress tolerance and disease-resistant varieties of *Aloe vera* can also be developed by using CRISPR-Cas9 technology by editing genes involved in plant defence mechanisms and stress response pathways, which can improve the plant's ability to thrive in diverse growing conditions. This tool can expedite traditional breeding efforts by facilitating the introgression of desirable traits from wild or exotic *Aloe vera* germplasm into elite cultivars, leading to shortened breeding cycles and enabling the rapid development of improved varieties. Gene editing techniques can also be used to introduce genetic markers or modifications for molecular authentication of *Aloe vera* cultivars to ensure product quality and authenticity in the marketplace (Janbandhu et al., 2024). CRISPR-Cas9 can also be applied to manipulate metabolic pathways involved in the biosynthesis of specific compounds in *Aloe vera,* leading to increased production of important metabolites or altering their composition to meet consumer demands (Ninama et al., 2024).

In conclusion, CRISPR-Cas9 and gene editing tools present exciting opportunities for enhancing the agronomic, medicinal, and commercial value of *Aloe vera.* By utilizing these tools, researchers and breeders can stimulate the development of novel varieties with improved traits and contribute to the sustainable cultivation and utilization of this valuable plant species (Ushasree et al., 2024).

8.5 FUTURE PERSPECTIVES AND CHALLENGES

The transition from conventional breeding methods to genome-assisted breeding methods in *Aloe vera* presents both promising opportunities and significant challenges. The future prospects associated with this transition include accelerated breeding progress, enhanced trait precision, expanded genetic diversity exploration, improved stress resilience, and optimization of the utilization of resources. Genome-assisted breeding offers the potential to significantly facilitate the breeding process in *Aloe vera* by providing precise genetic information to breeders about selection decisions, which leads to the development of improved cultivars with desirable traits in a short time period. The modern technologies also offer identification and selection of plants with specific genetic markers associated with target traits such as increased gel yield, increased secondary metabolites production, and increased stress tolerance, leading to superior performance and quality. Marker-assisted breeding techniques also facilitate the identification of novel alleles and genetic variants associated with desirable traits and enable the exploration of genetic diversity present in *Aloe vera* germplasm collections. Genome editing tools allow targeted gene modifications to alter genes to develop abiotic and abiotic stress tolerance varieties, enhanced nutritional and shelf-life of plants. CRISPR-Cas9 tool allows breeders to optimize the utilization of resources such as land, labour, and materials by focusing on the highest genetic potential for desired traits, which leads to developing efficient breeding programs and increased productivity in *Aloe vera* cultivation.

The transition from conventional breeding methods to genome-assisted breeding methods in the cultivation of *Aloe vera* is also associated with some challenges, such as genome complexity, limited genomic resources, some ethical and social considerations, and integration of omics technologies. *Aloe vera's* genome is relatively large and complex, posing challenges for genome sequencing and analysis. To overcome complexity, robust bioinformatics tools and computational resources are required for accurate assembly and annotation of the genome. Despite advancements, genomic resources for *Aloe vera* may still be limited compared to other crops. Generating comprehensive genomic datasets, including genetic maps, reference genomes, and functional annotations, requires substantial investment and collaboration among researchers. Addressing regulatory concerns and ensuring public acceptance are essential for the commercialization of genetically improved *Aloe vera* varieties. Integrating multiple omics technologies, such as genomics, transcriptomics, metabolomics, and phenomics, presents computational and analytical challenges. So, it is mandatory to develop robust pipelines for data integration and interpretation for harnessing the full potential of multi-omics approaches in *Aloe vera* breeding.

In summary, while genome-assisted breeding holds a great promise for advancing *Aloe vera* improvement programs, addressing technical, regulatory, and socio-economic challenges is mandatory to realize its full potential and ensure its responsible and sustainable application in agriculture. Collaboration between researchers, breeders, policymakers, and industry stakeholders will be crucial in overcoming these challenges and driving innovation in *Aloe vera* breeding towards improved cultivars with enhanced traits.

8.6 CONCLUSION

Plant breeding has played a vital role in food production security to feed the world. In spite of the breeding approaches used in the production of new varieties, safety assessment measures based on the product and technology used should be determined to determine whether the plant should be released into the environment. Both conventional breeding methods and genome-assisted breeding methods play a significant role in crop improvement efforts. In summary, while conventional breeding methods have been the cornerstone of crop improvement for centuries, genome-assisted methods offer unrivalled precision, efficiency, and opportunities for accelerating genetic gain. By integrating both conventional breeding and genome-assisted breeding approaches and embracing technological innovations, breeders can address current and future challenges in agriculture and contribute to global food security and sustainability.

REFERENCES

Almeida, R. O. (2024). Micropropagation of *Aloe vera*. *Micropropagation of Medicinal Plants*, *1*, 91.

Ambika, Bhati, S., & Kumar, R. (2024). Plant breeding using the CRISPR-Cas9 system for food security and facing climate change. In Jen-Tsung Chen and Sunny Ahmar (eds.) *Plant genome editing technologies: Speed breeding, crop improvement and sustainable agriculture* (pp. 149–181). Singapore: Springer Nature.

Badar, Z., Khan, S., & Rao, H. (2017). Analysis of genetic fidelity of wild type and in vitro regenerated *Aloe vera* plants through RAPD and ISSR molecular markers. *International Journal of Biotechnology and Bioengineering*, *3*(8), 259–267.

Bais, S. K., & Todkari, S. S. (2024). International Journal of Pharmacy and Herbal Technology-ISSN NO: 2583-8962 (Online). *International Journal*, *2*(1), 475–494.

Barrangou, R. (2015). Diversity of CRISPR-Cas immune systems and molecular machines. *Genome Biology*, *16*, 1–11.

Ceasar, S. A., Rajan, V., Prykhozhij, S. V., Berman, J. N., & Ignacimuthu, S. (2016). Insert, remove or replace: A highly advanced genome editing system using CRISPR/Cas9. *Biochimica et Biophysica Acta (BBA)-Molecular Cell Research*, *1863*(9), 2333–2344.

Chen, K., Wang, Y., Zhang, R., Zhang, H., & Gao, C. (2019). CRISPR/Cas genome editing and precision plant breeding in agriculture. *Annual Review Plant Biology*, *70*, 667–697. https://doi.org/10.1146/annurevarplant-050718-100049

Cristiano, G., Murillo-Amador, B., & De Lucia, B. (2016). Propagation techniques and agronomic requirements for the cultivation of Barbados aloe (*Aloe vera* (L.) Burm. F.)—A review. *Frontiers in Plant Science*, *7*, 196036.

Das, A., Haque, S. M., Ghosh, B., Nandagopal, K., & Jha, T. B. (2016). Morphological and genetic characterization of micropropagated field grown plants of *Aloe vera* L. *Plant Tissue Culture and Biotechnology*, *25*(2), 231–246.

Das, J., & Bora, S. (2018). A rapid plant production technique of *Aloe vera* L for commercialization. *SSRG International Journal of Agriculture & Environmental Science*, *5*(3), 30–37.

Deltcheva, E., Chylinski, K., Sharma, C. M., Gonzales, K., Chao, Y., Pirzada, Z. A., ... & Charpentier, E. (2011). CRISPR RNA maturation by trans-encoded small RNA and host factor RNase III. *Nature*, *471*(7340), 602–607.

Doudna, J. A., & Charpentier, E. (2014). The new frontier of genome engineering with CRISPR-Cas9. *Science*, *346*(6213), 1258096.

Enfissi, E. M., Drapal, M., Perez-Fons, L., Nogueira, M., Berry, H. M., Almeida, J., & Fraser, P. D. (2021). New plant breeding techniques and their regulatory implications: An opportunity to advance metabolomics approaches. *Journal of Plant Physiology*, *258*, 153378.

Gantait, S., Sinniah, U. R., & Das, P. K. (2014). *Aloe vera*: A review update on advancement of in vitro culture. *Acta Agriculturae Scandinavica, Section B-Soil & Plant Science*, *64*(1), 1–12.

Gantait, S., Subrahmanyeswari, T., & Mahanta, M. (2023). In vitro approaches for the production of aloin: Advancements and opportunities. *South African Journal of Botany*, *153*, 188–194.

Gao, C. (2021). Genome engineering for crop improvement and future agriculture. *Cell*, 184(6), 1621–1635. https://doi.org/10.1016/j.cell.2021.01.005

Guo, M., Chen, H., Dong, S., Zhang, Z., & Luo, H. (2022). CRISPR-Cas gene editing technology and its application prospect in medicinal plants. *Chinese Medicine*, *17*(1), 33.

Gupta, S., Sahu, P. K., Sen, D. L., & Pandey, P. (2014). In-vitro Propagation of *Aloe vera* (L.) Burm. f. *British Biotechnology Journal*, *4*(7), 806–816.

Hajian, R., Balderston, S., Tran, T., DeBoer, T., Etienne, J., Sandhu, M., ... & Aran, K. (2019). Detection of unamplified target genes via CRISPR–Cas9 immobilized on a graphene field-effect transistor. *Nature Biomedical Engineering*, *3*(6), 427–437.

Hekmatpou, D., Mehrabi, F., Rahzani, K., & Aminiyan, A. (2019). The effect of *Aloe vera* clinical trials on prevention and healing of skin wound: A systematic review. *Iranian Journal of Medical Sciences*, *44*(1), 1.

Hickey, L. T., Hafeez, N. A., Robinson, H., Jackson, S. A., Leal-Bertioli, S. C., Tester, M., ... & Wulff, B. B. (2019). Breeding crops to feed 10 billion. *Nature Biotechnology*, 37(7), 744–754. https://doi.org/10.1038/s41587-019-0152-9

Holme, I. B., Gregersen, P. L., & Brinch-Pedersen, H. (2019). Induced genetic variation in crop plants by random or targeted mutagenesis: convergence and differences. *Frontiers in Plant Science*, *10*, 488642.

Ikram, M., Rauf, A., Rao, M. J., Maqsood, M. F. K., Bakhsh, M. Z. M., Ullah, M., ... & Tahira, M. (2024). CRISPR-Cas9 based molecular breeding in crop plants: A review. *Molecular Biology Reports*, *51*(1), 227.

Janbandhu, M. S., Mehta, A., Beese, S., Pandey, S. K., Singh, B., Patel, A., & Singh, B. P. (2024). Advances and emerging trends in horticultural production and management. *Journal of Experimental Agriculture International*, *46*(3), 47–69.

Kanwar, K., Devi, V., Sharma, S., Soni, M., & Sharma, D. (2015). Effect of physiological age and growth regulators on micropropagation of *Aloe vera* followed by genetic stability assessment. *National Academy Science Letters*, *38*, 29–35.

Kumar, S., Kalita, S., Basumatary, I. B., Kumar, S., Ray, S., & Mukherjee, A. (2024). Recent advances in therapeutic and biological activities of *Aloe vera*. *Biocatalysis and Agricultural Biotechnology*, *57*, 103084.

Kumari, A., & Naseem, M. D. (2015). An efficient protocol for micropropagation of a medicinal plant *Aloe vera* L. through organ culture. *The Journal of Indian Botanical Society*, *94*(1–2), 118–125.

Kumari, M., Dubey, A. K., Kumar, R., & Kumar, A. (2024). Marker-assisted selection in plant breeding for stress tolerance. In Mohammad Abass Ahanger, Javaid Akthar Bhat, Parvaiz Ahmad, and Riffat John (eds.) *Improving stress resilience in plants* (pp. 371–387). Academic Press.

Kuzmishyna, I. I. (2024). *Medicinal plants and medicinal raw materials: Theory and practice*. Вежа-Друк (Vezha-Druk), Lutsk, Ukraine.

Kyndt, T., Quispe, D., Zhai, H., Jarret, R., Ghislain, M., Liu, Q., ... & Kreuze, J. F. (2015). The genome of cultivated sweet potato contains Agrobacterium T-DNAs with expressed genes: An example of a naturally transgenic food crop. *Proceedings of the National Academy of Sciences*, *112*(18), 5844–5849.

Liu, L., & Fan, X. D. (2014). CRISPR–Cas system: A powerful tool for genome engineering. *Plant Molecular Biology*, *85*(3), 209–218.

Lotfi, M., Butler, A. E., Sukhorukov, V. N., & Sahebkar, A. (2024). Application of CRISPR-Cas9 technology in diabetes research. *Diabetic Medicine*, *41*(1), e15240.

Mehrabadi, S., Izadi, F. S., Pasha, S., Pourali, R., Khazaei, M., Hassanian, S. M., ... & Avan, A. (2024). The potential therapeutic applications of CRISPR/Cas9 in the treatment of gastrointestinal cancers. *Current Molecular Medicine*, *25*(3), 278–288.

Misir, J., Brishti, F. H., & Hoque, M. M. (2014). *Aloe vera* gel as a novel edible coating for fresh fruits: A review. *American Journal of Food Science and Technology*, 2(3), 93–97.

Nagansurkar, S. B., Bais, S. K., & Mujawar, M. B. (2024). A review: *Aloe vera* is miracle plant. *International Journal of Pharmaceutics*, 2, 367–378.

Nair, S. K., & Rabbani, M. T. (2024). Marker-assisted selection approaches for improving quality traits. In Kailash Chandra, Subhash Chand, Ravi Prakash Saini, and Rohit Sharma (eds.) *Smart breeding* (pp. 287–328). Apple Academic Press.

Ninama, N., Gangal, L., Khayum, A., SB, H., HM, S., & Singh, A. (2024). Post-harvest biotechnology or genetic engineering solutions: Extending shelf life and reducing food waste. *Journal of Advances in Biology & Biotechnology*, *27*(4), 1–26.

Pinto-Carbó, M., Sieber, S., Dessein, S., Wicker, T., Verstraete, B., Gademann, K., ... & Carlier, A. (2016). Evidence of horizontal gene transfer between obligate leaf nodule symbionts. *The ISME Journal, 10*(9), 2092–2105.

Podevin, N., Davies, H. V., Hartung, F., Nogue, F., & Casacuberta, J. M. (2013). Site-directed nucleases: A paradigm shift in predictable, knowledge-based plant breeding. *Trends in Biotechnology, 31*(6), 375–383. https://doi.org/10.1016/j.tibtech.2013.03.004

Puchta, H. (2005). The repair of double-strand breaks in plants: Mechanisms and consequences for genome evolution. *Journal of Experimental Botany, 56*(409), 1–14.

Ramesh, D., Singh, S., Shah, P., Chakraborty, A., Satyavathi, T. C., & Thirunavukkarasu, N. (2024). Trait mapping, marker-assisted selection, and introgression breeding in pearl millet. In Vilas A Tonapi, Nepolean Thirunavukkarasu, SK Gupta, Prakash I Gangashetty, and OP Yadav (eds.) *Pearl Millet in the 21st century: Food-Nutrition-Climate resilience-Improved livelihoods* (pp. 111–140). Singapore: Springer Nature.

Rani, R. (2024). Marker-assisted selection approaches for abiotic stress tolerance in crop plants. *Smart Breeding*, 261–286.

Raza, A., Salehi, H., Bashir, S., Tabassum, J., Jamla, M., Charagh, S., ... & Varshney, R. K. (2024). Transcriptomics, proteomics, and metabolomics interventions prompt crop improvement against metal (loid) toxicity. *Plant Cell Reports, 43*(3), 1–34.

Roviqowati, F., Samanhudi, S. T., Purwanto, E., Sisharmini, A., Apriana, A., & Yunus, A. (2024). Introduction of CRISPR/Cas9 with the target genes to improve agronomic traits and leaf blight resistance in rice. *SABRAO Journal of Breeding Genetics, 56*(1), 29–44.

Rurek, M., & Smolibowski, M. (2024). Plant transcriptomic responses in a diverse organ and stress conditions.

Schmidt, C., Fransz, P., Rönspies, M., Dreissig, S., Fuchs, J., Heckmann, S., ... & Puchta, H. (2020). Changing local recombination patterns in Arabidopsis by CRISPR/Cas mediated chromosome engineering. *Nature Communications, 11*(1), 4418.

Scully, R., Panday, A., Elango, R., & Willis, N. A. (2019). DNA double-strand break repair-pathway choice in somatic mammalian cells. *Nature Reviews Molecular Cell Biology, 20*(11), 698–714.

Shah, M. R. A., Mustfa, W., Tehreem, S., Ali, K., Malik, A., Naveed, M., & Changhong, G. (2024). Advancing the production and health benefits of aloe barbadensis miller through biotechnology enabled integration of pathology and breeding approaches. *Journal of Survey in Fisheries Sciences, 11*(1), 49–56.

Shibru, S., Olani, G., & Debebe, A. (2018). In vitro propagation of *Aloe vera* Linn from shoot tip culture. *GSC Biological and Pharmaceutical Sciences, 4*(2), 01–06.

Silva, H., Sagardia, S., Seguel, O., Torres, C., Tapia, C., Franck, N., & Cardemil, L. (2010). Effect of water availability on growth and water use efficiency for biomass and gel production in *Aloe vera* (Aloe barbadensis M.). *Industrial Crops and Products, 31*(1), 20–27.

Singh, B., & Sood, N. (2009). Significance of explant preparation and sizing in *Aloe vera* L.—A highly efficient method for in vitro multiple shoot induction. *Scientia Horticulturae, 122*(1), 146–151.

Singh, D., Mittal, N., Verma, S., Singh, A., & Siddiqui, M. H. (2024). Applications of some advanced sequencing, analytical, and computational approaches in medicinal plant research: A review. *Molecular Biology Reports, 51*(1), 23.

Singh, M., Rathore, M. S., Panwar, D., Rathore, J. S., Dagla, H. R., & Shekhawat, N. S. (2009). Micropropagation of selected genotype of *Aloe vera* L.—An ancient plant for modern industry. *Journal of Sustainable Forestry, 28*(8), 935–950.

Stadtmauer, E. A., Fraietta, J. A., Davis, M. M., Cohen, A. D., Weber, K. L., Lancaster, E., ... & June, C. H. (2020). CRISPR-engineered T cells in patients with refractory cancer. *Science, 367*(6481), eaba7365.

Supe, U. (2006). In vitro propagation of Aloe barbadensis-A medicinal plant. *Indian Journal of Agricultural Biochemistry, 19*(1), 19–21.

Swaminathan, M. S. (2009). Norman E. Borlaug (1914–2009). *Nature, 461*(7266), 894–894.

Ushasree, M. V., Jia, Q., Do, S. G., & Lee, E. Y. (2024). New opportunities and perspectives on biosynthesis and bioactivities of secondary metabolites from *Aloe vera*. *Biotechnology Advances, 72*, 108325.

Van Vu, T., Das, S., Hensel, G., & Kim, J. Y. (2022). Genome editing and beyond: what does it mean for the future of plant breeding. *Planta, 255*(6), 130.

Vergauwen, D., & De Smet, I. (2017). From early farmers to Norman Borlaug—the making of modern wheat. *Current Biology, 27*(17), R858–R862.

Xiong, X., Liang, J., Li, Z., Gong, B. Q., & Li, J. F. (2021). Multiplex and optimization of dCas9-TV-mediated gene activation in plants. *Journal of Integrative Plant Biology*, *63*(4), 634–645.

Yogi, R., Kumar, N., Kumar, R., & Kumar, M. (2023). Linkage planning of QTL for agronomic and root qualities utilizing IB370 × MAS25-F_2 population developed underwater restricted conditions. *Acta Physiologiae Plantarum*, *45*, 69.

Zakia, S., Zahid, N. Y., Yaseen, M., Abbasi, N. A., Hafiz, A. A., & Mahmood, N. (2013). Standardization of micropropgation techniques for *Aloe vera*: A pharmaceutically important plant. *Pakistan Journal of Pharmacy Sciences*, *26*(6), 1083–1087.

Zsögön, A., Čermák, T., Naves, E. R., Notini, M. M., Edel, K. H., Weinl, S., & Peres, L. E. P. (2018). De novo domestication of wild tomato using genome editing. *Nature Biotechnology*, *36*(12), 1211–1216.

9 Cultivating Potential
In-vitro Propagation and Genetic Transformation of Aloe vera

Alka Jangra, Kishor Kumar, Shalu Yadav, Harsh Chaurasia, and Vinod Chhokar

9.1 INTRODUCTION

Throughout history, a multitude of both familiar and lesser-known plant species have been utilized for medicinal purposes, providing remedies for a diverse array of disorders. The preference for natural remedies stems from their well-balanced biochemical composition and the minimal risks associated with their usage. However, it is not very reassuring to note that as the twenty-first century commenced, less than 24% of newly developed pharmaceuticals exhibited potent physiological effects and featured innovative forms derived from plant constituents (Kim et al. 2015; Georgiev et al. 2018). The lack of effective technology to produce biologically active compounds from plant raw components hinders the utilization of plants. In addition, there is a dearth of progress in the development and production of therapeutic plant species in many kinds of farms, including botanical gardens. Furthermore, during the past 50 years, there has been a discernible decrease in the species variety and structural integrity of the natural plant population, especially medicinally valuable species (Bagheri et al. 2018). The damaging impact of increased anthropogenic and technogenic burdens on biogeocenosis, coupled with the rising number of "hot spots" is anticipated to result in the complete loss of eight percent of diverse plant species within the next 25 years, as indicated by scientific assessments. According to the World Resources Institute, the pervasive occurrence of global desertification has impacted a significant part in 110 countries, worldwide. Numerous species, especially the endemic ones, have disappeared, while others are declining and are at risk of extinction (Chupakhin et al. 2021). Lately, there has been a growing fascination with medicinal plants, which serve as sources of secondary metabolites that can affect microorganisms, animals, and humans in various ways. These plants not only improve overall health but are also used to treat serious ailments like diabetes, cancer, cardiovascular disease, and more. Essentially, the population's state of health emphasizes the significance of moving domestic science forward to a more modern state. This progress entails utilizing all available means to investigate the inherent qualities of therapeutic plant species to extract physiologically active substances (isolated compounds from plants), and to apply these substances in the treatment of serious ailments strategically (Babich et al. 2020).

Aloe vera is a well-known succulent plant species that is frequently used in various Ayurvedic formulations. It is a member of the Liliaceae family and stands out as one of the major therapeutic plants with a wide range of applications, including cosmetic and pharmaceutical, along with a good profit margin. *Aloe vera* exhibits exceptional properties for healing and relaxing. Pedanius Dioscorides, a Greek physician and pharmacologist, initially recorded the health benefits associated with Aloe juice in his Greek Herbal in the year 1 A.D. It is used internally as a laxative, emmenagogue, stomachic, anthelminthic, astringent, hemorrhoid cure, uterine stimulant, and disinfectant. It is also used to facilitate the healing of wounds (Gantait et al. 2014). The rising demand for medicinal plants has led to significant exploitation of Aloe, posing a threat to biodiversity.

DOI: 10.1201/9781003603108-9

Research and development programs focused on medicinal plants emphasize the importance of sustainable use and the monitoring of the genetic health of Aloe populations. This approach strengthens the emerging companies that rely on medicinal plants while simultaneously protecting biodiversity. To facilitate large-scale production of high-quality Aloe planting material, biotechnological interventions have been developed. These developments render it possible to clone and multiply a wide variety of plant species *in-vitro*, especially those superior germplasms that are difficult to propagate through conventional techniques like grafting and cutting (Campestrini et al. 2006). Plant tissue culture has become a crucial method for the various facets of *Aloe vera* metabolic engineering and *in-vitro* multiplication. Moreover, when it comes to controlling the proportion of some harmful metabolites or increasing the concentration of certain medicinally significant metabolites, metabolic engineering plays a very crucial role. It has opened new and exciting opportunities for improving plant cell culture efficiency. To achieve this goal of metabolic engineering, some molecular changes in the genome need to be executed, triggering the need for certain standard genetic transformation strategies (Jangra et al. 2023). This chapter focuses particularly on the different strategies of *in-vitro* micropropagation and the role of genetic transformation as a sustainable approach to address the perspective of metabolic engineering in *Aloe vera*.

9.2 CONVENTIONAL PROPAGATION OF *ALOE VERA*

Rhizome cuttings and suckers (lateral shoots) are the traditional planting materials used to propagate *Aloe vera*. During the growth season, each mother plant under the conventional farming method yields three to four suckers. Selected suckers should be exposed to mild moisture stress in a shaded environment for five to ten days prior to transplanting in the field to prevent infections. This encourages root suberization, which accelerates their adaptation and stimulates root growth while planting. The crop is ready for harvesting after 18 months of sowing (Cristiano et al. 2016). It is not feasible to replicate Aloe on a wide scale using rhizomes since it takes an average of two years to have a fully grown plant (Rajeswari et al. 2012). Furthermore, since the mother plant is injured when the lateral branches are topped off, a high incidence of diseases and pests is anticipated to follow. The infection manifests as yellowish brown rot at the base of mature or older leaves. Depending on how severe the infection is, this could result in either partial or total plant defoliation. Therefore, seasonality, the limited number of suckers, the high likelihood of harming the mother plant and producing infected seedlings, the lack of high-quality raw materials, and slow vegetative growth are all factors limiting the commercial production of young plants on a large scale. Aloe faces severe exploitation due to the growing demand for therapeutic herbs, endangering biodiversity. The sustainable use of medicinal herbs, along with continuous monitoring of genetic fidelity, is the primary focus of research and development initiatives on medicinal plants, ensuring uniform metabolic status. This would support the growth of the medicinal plant-based enterprises (Campestrini et al. 2006). Biotechnological interventions have been developed to provide Aloe planting material of exceptional quality on an industrial scale. These developments have made it possible to clone and multiply many plant species *in-vitro*, particularly superior germplasms that are resistant to conventional propagation techniques such as cutting and grafting (Gantait et al. 2014).

9.3 *IN-VITRO* CULTURE TECHNIQUES

It goes without saying that the medicinal plant industry's research and development requirements necessitate a steady supply of planting materials. To address this need, the plant tissue culture industry is currently paying close attention to medicinal plant species with proven pharmaceutical value. *In-vitro* morphogenesis in classical medicinal plants has been the subject of almost 6500 research publications, according to a survey of the literature published in the last 30 years. However, only 82 medicinal plants have complete micropropagation techniques accessible. The *in-vitro* regenerated plants in 42 plant species have progressed to the field evaluation stage. In most of these experiments, regeneration through somatic embryogenesis has been favored over proliferation through axillary bud or multiple shoot culture (Gantait et al. 2014). Organogenesis is a flexible regeneration process that deals with the de novo development of organs (roots or shoots). Shoots can be produced entirely via adventitious shoot formation, that is, the differentiation of non-meristematic tissues or axillary shoot formation, which is the emergence of pre-existing meristematic tissues (Seran and Ahmad 2018). Tissues lacking established meristems can undergo adventitious organogenesis, which is the creation of new structures and organizations, such as shoots or roots, that were previously absent. A synergistic interplay of physical and chemical variables is the outcome of both of

TABLE 9.1
***In-vitro* clonal micropropagation of *Aloe vera* by direct organogenesis**

Explant type	Media composition	Response of explant	References
Shoot apex	MS medium + 1.0 mg/l BAP + 0.5 mg/l NAA	Efficient micropropagation with high shoot multiplication rate	(Aggarwal and Barna 2004)
Basal shoot	MS medium + 2.5 mg/l BAP + 0.5 mg/l NAA	Prolific shoot multiplication	(Liao et al. 2004)
Shoot tip	MS medium + 2.0 mg/l BAP + 0.5 mg/l Kinetin + 0.2 mg/l NAA	High shoot proliferation rate; emergence of shoots within 2 weeks	(Ahmed et al. 2007)
Rhizomatous stem	MS medium + 1.0 mg/l BAP + 0.5 mg/l NAA	Successful induction of multiple shoots; mass propagation achieved	(Gantait et al. 2010b)
Inflorescence axis	MS medium + 6.0 mg/l 2,4-D + 100 mg/l AC (for callus induction); MS medium + 1.0 mg/l 2,4-D (for shoot regeneration)	Callus formation within 30–35 days; regenerative globular structures upon transfer to lower 2,4-D concentration	(Rathore et al. 2011)
Leaf base	MS medium + 2.0 mg/l BAP + 0.5 mg/l NAA	Direct shoot regeneration after four to six weeks	(Sahoo and Rout 2014)
Apical buds	MS medium + BAP (for shoot induction) + IBA (for root formation) + NAA (for root induction)	100% regeneration; maximum shoot and root formation (4.8±0.53 shoots, 3.5±0.35 cm shoot length)	(Gupta et al. 2014)
Meristem explants	MS medium + 1–20 mg/l $AgNO_3$ (Silver Nitrate)	Increased shoot organogenesis; optimal shoot formation at 10 mg/l $AgNO_3$	(Lee et al. 2016)
Shoot primordia	MS medium + 0.6 mg/l BAP + 0.2 mg/l IBA	High-efficiency propagation; high shoot multiplication rate	(Shibru et al. 2018)
Meristems and cauline discs	MS medium + 2.5 mg/l BAP + 0.25 mg/l IAA	Highest shoot regeneration frequency; effect of sugars and photoperiod on organogenesis	(Albarrán Ruiz et al. 2019)
Offshoot cuttings	MS medium + 0.5–2.0 mg/l BAP+ 0.5 mg/l NAA	Successful shoot proliferation; highest shoot induction at 1.5 mg/l BAP and 0.5 mg/l NAA	(Niguse et al. 2020)
Lateral shoot buds	MS medium + 4.0 mg/l BAP + 1.0 mg/l Kinetin	Shoot induction and elongation; 93.33% regeneration rate	(Singh et al. 2020)
Shoot primordia	MS medium + 4.0 mg/l BAP + 0.5 mg/l NAA	High shoot induction; 90% response; 2.7 shoots per explant; 4.7 cm shoot length	(Yasmin et al. 2022)
Leaf explants	MS medium + 4.0 mg/l BAP + 0.1 mg/l NAA	High transformation efficiency and regeneration of shoots	(Jangra et al. 2023)

these methods. An appropriate explant selection at the ideal age, exact medium formulations, certain growth regulators, a carbon supply, a gelling agent, and other physical parameters like temperature, relative humidity, and light intensity are all necessary for successful organogenesis (Gantait et al. 2010b). A number of studies have been conducted to investigate the morphogenic capacity and accomplish regeneration of *Aloe vera* (Table 9.1).

9.4 ADVENTITIOUS SHOOT REGENERATION

Adventitious Aloe shoots could be generated by both direct and indirect organogenesis. The development of a meristem occurs without the intermediate proliferation of undifferentiated callus tissue in the direct approach (Figure 9.1). After four to six weeks of growth on MS (Murashige and Skoog) media supplemented with 2.0 mg/l BAP (6-Benzylaminopurine), 0.5 mg/l NAA (Naphthaleneacetic acid), and 40 mg/l ADS (Antidiuretic hormone), leaf explants in *Aloe vera* showed direct shoot regeneration (Murashige and Skoog 1962; Sahoo and Rout 2014). Calli derived from the base of the leaf, the segments of the leaf, the leaf without sheaths, the inflorescence axis, the stem segments, the stem discs, and the rhizomatous stems were found to produce rapid and large-scale indirect adventitious shoot regeneration. Additionally, meristems and seeds have been employed as explants for callus induction (Cristiano et al. 2016). *Aloe vera* leaf explants grown on MS media supplemented with 1.0 mg/l BAP and 1.0 mg/l NAA showed the best *in-vitro* growth, based on another recent study (Wahab et al. 2020). Agar-solidified MS medium supplemented with a combination of 1.0 mg/l IAA and 4.0 mg/l BAP increased the production of *Aloe vera* plants by producing the greatest number of shoots possible. Regenerated shoots were able to root in the same MS medium that was supplemented with 0.2 mg/l IAA and 0.8 mg/l BAP (Molsaghi et al. 2014). Additionally, reports indicated that on MS medium supplemented with 0.2 mg/l IBA (Indole-3-butyric acid) and 1.0 mg/l BAP, respectively, the high shoot production was 2.90, and the multiplication factor was close to 2.87 (Shibru et al. 2018). The most crucial stage in the tissue culture procedure is sterilizing the plant material. Plant material is sterilized by removing microbial contaminants from its interior and exterior, increasing its chances of surviving *in-vitro*. Additionally, a pretreatment with a mixture of streptomycin and bavistin, followed by surface sterilization, reduced the growth of bacteria and fungi in cultures (Cristiano et al. 2016).

Auxins, such as IAA (Indole-3-acetic acid), IBA, NAA, or 2,4-D (2,4-Dichlorophenoxyacetic acid), are frequently added to the culture medium to regulate morphogenesis, either alone or in combination with cytokinins. The commencement of regenerative callus culture is reliant on the kind and concentration of plant growth regulators (PGRs).

9.5 EXPLANT SOURCES AND CULTURE ESTABLISHMENT

The types of explants, size, position, age, and physiological state employed have a considerable impact on the success of *in-vitro* plant multiplication and proliferation. Explants in Aloe include shoot tips, axillary buds, meristems, subterranean stems, and inflorescences. Traditionally, Aloe propagation is based on lateral shoots produced by the mother plant during the growth season. Shoot tips from mature plants are the most widely used explants due to their consistent ability to produce several shoots. Previously, researchers have successfully used shoot tips from high-yielding Aloe plants to promote numerous shoots (Aggarwal and Barna 2004). However, the number of lateral shoots per plant is restricted and varies over time, making it difficult to acquire homogeneous propagules for large-scale commercial production. Conventional systems produce only three to four lateral shoots per mother plant every year. This method is not only labor-intensive and time-consuming, but it also has severe limits for large-scale production of the plant. Although axillary buds have been shown to be the most efficient explants for Aloe micropropagation, some research has focused on the use of shoot tips and apical meristems (Campestrini et al. 2006; Gantait et al. 2010b). Previously, researchers demonstrated quick and extremely successful micropropagation with vegetative meristems and shoot tips, respectively. They also emphasized the potential of immature inflorescences as explant for *Aloe*

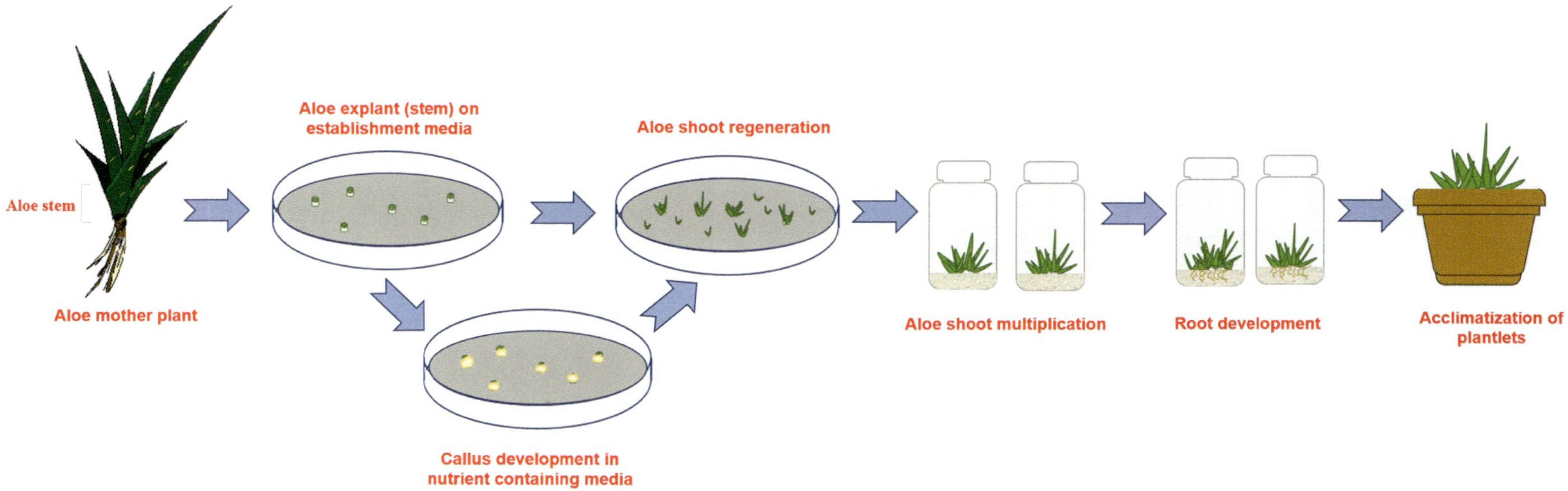

FIGURE 9.1 Overview of *in-vitro* regeneration and multiplication of *Aloe vera*.

barbadensis propagation (Velcheva et al. 2005). Other research has employed underground stems as explants for Aloe micropropagation. The previous study effectively devised a methodology for Aloe mass multiplication via direct organogenesis, resulting in many shoots from rhizomatous stems (Al-Nema and Abdullah 2023; Ahmad et al. 2024). However, a group of researchers utilize rhizomatous stem explants for *in-vitro* Aloe proliferation and conservation. However, these explants are susceptible to microbial contamination and phenolic production, which can stymie growth (Gantait et al. 2010a; Gantait et al. 2011). Seeds and meristems have also been used to induce callus and regenerate plants, thereby serving as a good source of explant. However, primary cultures derived from these explants frequently confront problems due to the excretion of phenolic substances that restrict development. Previous studies achieved fast shoot production from Aloe calli by employing polyvinylpyrrolidone (PVP) to reduce phenolic secretion in the culture media. Incorporating such phenolic adsorbents into regeneration media has been shown to overcome phenolic-induced growth retardation (Almeida 2024). Despite difficulties with callus induction, multiple investigations have successfully obtained callus cultures from various Aloe explants. These include leaf bases, stem segments, rhizomatous stems, leaf and younger stems, stem discs, and inflorescence axes (Upadhyay et al. 2007; Jayakrishna et al. 2011; Seran and Ahmad 2018; Jader and Obaid 2023). The meristem culture is extremely effective at growing virus-free plants, but it has limited success with Aloe callus induction. This emphasizes the need to choose appropriate explants based on tissue culture goals. *Aloe vera* micropropagation was carried out using a variety of explant forms, including shoot tips, leaf tips, leaf segments, axillary branches, buds, and stems for direct multiple shoot production. Among them, the majority of researchers determined that shoot tips were the best explants for micropropagation (Bhandari et al. 2010; Biswas et al. 2013).

9.6 *IN-VITRO* MULTIPLICATION AND ROOTING

Direct multiple shoot initiation and proliferation from meristem tips was initially accomplished using MS media supplemented with 0.5 mg/l BAP and 0.02 mg/l 2,4-D (Abdi et al. 2013). It has been found that IBA at a concentration of 1 mg/l can cause several axillary buds to develop into shoots and spontaneous roots from decapitated shoot explants. IBA-supplied nutritional media produced more adventitious and axillary buds than those supplemented with NAA (Das et al. 2018). According to Singh et al. (2009), multiple shoot production resulted from bud breaking that occurred in cultures after 28–32 days. The semi-solid agar-gelled MS medium comprising 13.32 μM of BAP and additives showed the most significant response. Each explant regenerated 10.3 ± 0.675 shoots that were 2.49 ± 0.345 cm long. According to a report, *in-vitro*, regenerated shoots were cultured on half-strength MS supplemented with 0.5 to 1.5 mg/l of IBA, NAA, or IAA to stimulate root development. On the tenth day of culture, roots started to appear in the medium containing 0.5 mg/l of NAA, and within 23–28 days, 95% of the roots were formed. With an average length of 3.5 ± 0.35 cm, the maximum number of roots per shoot was 4.8 ± 0.53.The quality of the roots that grew in the medium with a greater auxin content (1.0–1.5 mg/l) was lower (Baksha et al. 2005).

It was previously discovered that the presence of modest quantities of BA and NAA boosted the rooting percentage. Additionally, a negative relationship between rooting and the medium's BA concentration was noted. On MS medium enriched with 0.5 mg/l BA + 0.5 mg/l NAA and 1 mg/l BA + 0.5 mg/l NAA, the shoots demonstrated excellent rooting. The medium supplemented with 0 mg/l IBA + 1 mg/l NAA produced the greatest number of roots (9.71), whereas the medium supplemented with 1 mg/l IBA + 1 mg/l NAA produced the longest (8.75 cm) and thickest (4.3 cm) roots (Hashemabadi and Kaviani 2008). It was suggested auxin does not need to be applied externally in order to induce roots in *Aloe vera*. No roots were produced by the shoots in the original regeneration media that included BAP either by itself or in conjunction with NAA. However, after a month of culture, activated charcoal-containing media produced roots regardless of the presence of citric acid. But the roots that were started in these media were fragile and thin. Thus, the rooting capability in two additional media – one with 0.2 mg/1 NAA and the other without growth hormones – was assessed. In all rooting media, rooting took place in less than two weeks. In media containing 0.2 mg/l NAA with 1.3, 100% rooting was seen (Nayanakantha et al. 2011). Previous research observed the highest percentage of root induction (80%) in the MS medium supplemented with 0.5 mg/l IBA. Root development was robust and healthy as a result of this combination. In particular, after eight weeks of culture, over ten well-formed roots with an average length of more than six cm were seen in the same medium. This strong rooting demonstrates the potency of the selected medium and hormone levels, indicating that it may be used for effective root induction in tissue culture investigations (Dwivedi et al. 2014). These results highlight how crucial it is to optimize hormone supplements for effective plant regeneration. According to Kumar et al. (2011), within the first week after inoculation, shoots inoculated on hormone-free and IBA-supplemented media showed a rooting response. Interestingly, the hormone-free medium exhibited a stronger rooting response. The higher efficacy of the hormone-free medium was demonstrated by the 100% rooting that was seen on the fifteenth day of inoculation. In the hormone-free media, the average number of roots per shoot was 2.8 ± 0.5. In contrast to the roots grown in the IBA-supplemented media, which had thinner, less elongated roots and an average root number of 1.7 ± 1.1 per plant, the roots in this medium were thicker and more elongated. These findings emphasize the potential advantages of hormone-free media for root induction, particularly for applications requiring robust and well-formed root systems. The results also suggest that while hormonal supplementation may aid initial root initiation, hormone-free conditions could be more favorable for achieving optimal root morphology and development. This insight is valuable for researchers and practitioners seeking efficient and cost-effective methods for plant tissue culture and propagation. These findings serve as valuable insights for researchers with the aim of developing efficient and cost-effective methods for plant tissue culture and propagation.

9.7 ACCLIMATIZATION OF *IN-VITRO* REGENERATED ALOE

The *ex-vitro* acclimation period of the micropropagated plantlets is measured by the survival percentage which determines the success of a micropropagation process. Like other medicinal plants, acclimatization of plantlets is a crucial step in the efficient micropropagation of Aloe. The effects of substrate medium, light, temperature, and humidity have been the primary factors evaluated in studies on the acclimation of *in-vitro* regenerated Aloe plantlets with a well-developed rhizosphere. Regenerated Aloe plantlets transferred to a mixture of soil, sand, and pumice during greenhouse acclimatization resulted in a high (almost 100%) survival rate (Gantait et al. 2014). The proportion of survival during the micro-propagated plantlets *in-vitro* acclimation phase determines the success of a micro-propagation technique. Gupta et al. (2014) found that when regenerated plants were placed in soil obeying acclimation, they showed 85% survival. This success has been favorably connected with the ex-vivo microplants dimensions: medium (six to eight cm long, having four to five leaves) and big (13–16 cm long, having five to seven leaves) sizes have been found to differ greatly from small (three to five cm long, two to three leaves) (de Oliveira et al. 2009). Supe (2006) showed that direct sunshine is necessary for micropropagated Aloe to acclimate. Compost, loam, and clay in 1:2:2 ratios were utilized by Hosseini and Parsa (2007) as an acclimatization substrate. Transparent plastic pots were placed over the plants to maintain humidity during the procedure. Over the course of six months, five explants produced more than 2000 well-acclimatized plants (Cristiano et al. 2016). Kumar et al. (2011) disinfected newly developed Aloe plants in one percent (w/v) Bavistine to prevent any fungal infection before transplanting *in-vitro* plantlets to a mixture of soil and farmyard manure (1:1). The plants were then kept under a polyhouse and then under a shade house with 80% humidity and 31°C temperature to ensure more than 85% survival. Another research group attempted a dual-phase acclimation of micropropagated plants (Gantait et al. 2010a; Gantait et al. 2011). A 98% survival rate was attained using a combination of sand, soil, and FYM (2:1:1; v/v) out of the four growing substrates that were exploited. Additionally, intermittent water spraying and a transparent polythene sheet covering ensured high humidity, which accelerated the acclimatization process.

9.8 GENETIC TRANSFORMATION APPROACHES

Genetic engineering has become a pivotal tool for producing pharmaceutically important compounds in medicinal plants (Niazian 2019). In *Aloe vera*, genetic improvement is essential to introduce desirable traits to enhance its therapeutic properties or suppress the expression of undesirable bioactive compounds (Egbuna et al. 2022; Jangra et al. 2022). The primary objectives of genetic transformation include improving agronomic traits such as stress tolerance, growth rates, and bioactive compound production. Moreover, genetic engineering provides a powerful approach to studying gene functions, regulatory networks, and metabolic pathways (Sethi et al. 2021). Despite its importance, *Aloe vera* has seen limited attempts at genetic transformation. *Agrobacterium*-mediated transformation is preferred for producing high-quality transgenic plants due to its ability to integrate large DNA segments with minimal rearrangements and a few copies of transgenes (Tiwari 2019; Azizi-Dargahlou and Pouresmaeil 2024; Rahman et al. 2024). He et al. (2007) first reported an *Agrobacterium*-meditated transformation protocol for improvement of *Aloe* using fresh stem segments and induced shoots as explants with the pBI121 binary vector, expressing both the NPTII resistance marker and the GUS reporter. The highest transient expression of GUS was noted in the induced shoots (80%) followed by fresh stem segment (52.6%) using *Agrobacterium tumefaciens* strain EHA105. However, the high rates of transient GUS expression coupled with a low transformation efficiency (0.9%) can be attributed to inconsistencies in cell infection and regeneration, as well as potential gene rearrangements or damage within the recipient cells. These limitations underscore the need for improved protocols to enhance the efficiency of stable transformation in *Aloe vera*.

Particle bombardment is another widely used method for genetic transformation. However, this method presents several key challenges, including tissue necrosis, escapees during selection, multiple copy insertion, and browning due to phenolics that reduce viability. Velcheva et al. (2010) highlighted innovative selection and recovery strategies to address these issues, including pretreatment with compounds like dithiothreitol (DTT) and gibberellic acid (GA_3) to improve viability, stepwise selection protocols to minimize escapees, and the use of media supplements such as gallic acid and resin beads to reduce phenolic effects. Plants are becoming useful platforms for the production of desired therapeutic proteins. It offers several advantages over animal systems, such as cost-effectiveness, product safety, storage conditions, purification process, environmental concerns, as well as being cross-contamination free, and accepted by the public (Sethi et al. 2021). In this context, Lowther et al. (2012), described genetic transformation strategies for the production of transgenic *Aloe vera* plants expressing recombinant therapeutic proteins. These strategies utilized *Agrobacterium*-mediated transformation and particle bombardment methods. The recombinant proteins were cloned under suitable promoters and directed to the gel and/or pith tissues, which function as reservoirs for bioactive compounds in *Aloe vera*. Further, this method was demonstrated to develop transgenic *Aloe vera* expressing the *human interferon alpha 2 (IFNa2)* using a callus-based tissue culture system. The callus cultures initiated from *zygotic embryos* enabled the generation of transgenic plants expressing the IFNa2 with its secretion signal sequence. The biological activity and localization of IFNa2 were confirmed through RT-PCR, immunoblot analysis, and antiviral assays, demonstrating activation of the IFN signaling pathway in human A549 cells and antiviral activity against encephalomyocarditis virus.

Jangra et al. (2023) reported a more efficient *Agrobacterium*-mediated genetic transformation protocol for *Aloe vera*, achieving an 83% transient GUS expression rate with a significantly improved stable transformation efficiency (2%). This advancement was enabled by optimizing infection protocols using the *Agrobacterium tumefaciens* strain AGL1 and implementing a strict, stepwise selection process during regeneration, utilizing *hptII*. Unlike the earlier approach, the current protocol minimized chimerism by selecting only plants with consistent and stable characteristics. These improvements reduced variability and enhanced the reliability of transgenic plant recovery. Jangra et al. (2023) also introduced modifications to address common challenges such as phenolic browning and necrosis, which often compromise cell viability. Their refined protocol provides a more robust and reproducible system, ensuring higher transformation efficiency while maintaining plant viability. This upgraded and efficient approach for *Aloe vera* transformation can be employed for genetic improvement aimed at enhancing the therapeutic potential with a uniform metabolic status suitable for industrial use.

9.9 CONCLUSION

The growing global demand for *Aloe vera* and its biotic components, demands the advancement of more effective propagation and genetic transformation techniques. Conventional methods are the most popular, but they have limitations, such as a slower rate of spread and susceptibility to certain infections. *In-vitro* propagation is an adequate substitute which permits elite germplasm conservation and mass replication. Advanced micropropagation methods, including somatic embryogenesis and shoot organogenesis, have shown impressive results in producing Aloe on a vast scale. Through the introduction of traits for the synthesis of enhanced bioactive compounds and stress resistance, genetic transformation further increases the practical value of plants. However, problems like phenolic oxidation and low transformation efficiency still exist. These can be resolved by using particle bombardment, *Agrobacterium*, and other optimal genetic transformation techniques. *Aloe vera* has enormous potential for future development owing to recent advances in plant biotechnology, such as metabolic engineering and genome editing based on CRISPR (Clustered Regularly Interspaced Short Palindromic Repeats). To meet the increasing demand while maintaining genetic diversity, it will be crucial to integrate these strategies with sustainable cultivation methods in the future. Further studies in tissue culture and genetic modification will assure *Aloe vera*'s position as an essential resource in industrial and pharmaceutical applications, hence enhancing its importance as a medicinal and economically significant plant.

REFERENCES

Abdi G, Hedayat M, Modarresi M (2013) *In-vitro* micropropagation of Aloe vera-impacts of plant growth regulators, media and type of explants. *J Biol Environ Sci* 7(19).

Aggarwal D, Barna KS (2004) Tissue culture propagation of elite plant of Aloe vera Linn. *J Plant Biochem Biotechnol* 13:77–79.

Ahmad S, Tariq M, Belwal V, Shah MH, Kumar P, Sharma M, Kole C, Jain M, Kaushik P (2024) Efficient in-vitro regeneration protocol for large-scale propagation of Aloe vera (L.) *Burm f J Appl Res Med Aromat Plants* 43:100588.

Ahmed S, Kabir AH, Ahmed MB, Razvy MA, Ganesan S (2007) Development of rapid micropropagation method of Aloe vera L. *Sjemenarstvo* 24:121–128.

Al-Nema QS, Abdullah RM (2023) Propagation protocol of the medicinal plant-Aloe vera using tissue culture. *SABRAO J. Breed. Genet.* 55(1):254–259.

Albarrán Ruiz Y, Pacheco Maldonado JC, Rache Cardenal LY (2019) Organogénesis a partir de ápices meristemáticos y discos caulinares de Aloe vera L. *Cienc En Desarro* 10:9–21.

Almeida RO (2024) Micropropagation of Aloe vera. *Micropropag Med Plants* 1:91.

Azizi-Dargahlou S, Pouresmaeil M (2024) Agrobacterium tumefaciens-mediated plant transformation: a review. *Mol Biotechnol* 66:1563–1580.

Babich O, Sukhikh S, Pungin A, Ivanova S, Asyakina L, Prosekov A (2020) Modern trends in the in vitro production and use of callus, suspension cells and root cultures of medicinal plants. *Molecules* 25:5805.

Bagheri F, Tahvilian R, Karimi N, Chalabi M, Azami M (2018) Shikonin production by callus culture of Onosma bulbotrichom as an active pharmaceutical ingredient. *Iran J Pharm Res IJPR* 17:495.

Baksha R, Jahan MAA, Khatun R, Munshi JL (2005) Micropropagation of Aloe barbadensis Mill. through in vitro culture of shoot tip explants. *Plant Tissue Cult Biotechnol* 15:121–126.

Bhandari AK, Negi JS, Bisht VK, Bharti MK (2010) In vitro propagation of Aloe vera: a plant with medicinal properties. *Nat Sci* 8:174–176.

Biswas GC, Miah MA, Sohel MMH, Hossain A, Shakil SK, Howlader MS (2013) Micropropagation of Aloe indica L. through shoot tip culture. *IOSR J Agric Vet Sci* 5:30–35.

Campestrini LH, Kuhnen S, Lemos PMM, Bach DB, Dias PF, Maraschin M (2006) Cloning protocol of Aloe vera as a study-case for "Tailor-Made" biotechnology to small farmers. *J Technol Manag Innov* 1:76–79.

Chupakhin E, Babich O, Prosekov A, Asyakina L, Gureev M, Krasavin M (2021) Plants of the Russian Federation pharmacopeia: an unexhausted natural products research opportunity? *Nat Prod Res* 35:3525–3527.

Cristiano G, Murillo-Amador B, De Lucia B (2016) Propagation techniques and agronomic requirements for the cultivation of Barbados aloe (Aloe vera (L.) Burm. F.)-a review. *Front Plant Sci* 7:1410.

Das J, Bora S, Das M, Pathak P (2018) A review on in vitro culture of Aloe vera, type of explants and impact of growing media and growth regulators. *Int J Curr Microbiol Appl Sci* 7:3473–3489.

de Oliveira ET, Crocomo OJ, Farinha TB, Gallo LA (2009) Large-scale micropropagation of Aloe vera. *Hortic Sci* 44:1675–1678.

Dwivedi NK, Indiradevi A, Asha KI, Nair RA, Suma A (2014) A protocol for micropropagation of Aloe vera L.(Indian Aloe)-a miracle plant. *Res Biotechnol* 5(1):2229–791X.

Egbuna C, Gupta E, Ezzat SM, Jeevanandam J, Mishra N, Akram M, Sudharani N, Adetunji CO, Singh P, Ifemeje JC, Deepak S, Bhavana A, Walag AMP, Ansari R, Adetunji JB, Laila U, Olisah MC, Onyekere PF (2022) Correction to: aloe species as valuable sources of functional bioactives. In: Chukwuebuka Egbuna and Genevieve Dable Tupas (eds.) *Functional Foods and Nutraceuticals*. pp. C1–C1.

Gantait S, Mandal N, Bhattacharyya S, Das PK (2010a) A novel strategy for in vitro conservation of Aloe vera L. through long term shoot culture. *Faisalabad Biotechnol*. 9(3):326–331.

Gantait S, Mandal N, Bhattacharyya S, Das PK (2010b) Determination of genetic integrity in long-term micropropagated plantlets of Allium ampeloprasum L. using ISSR markers. *Faisalabad Biotechnol* 9(3): 218–223.

Gantait S, Mandal N, Das PK (2011) In vitro accelerated mass propagation and ex vitro evaluation of Aloe vera L. with aloin content and superoxide dismutase activity. *Nat Prod Res* 25:1370–1378.

Gantait S, Sinniah UR, Das PK (2014) Aloe vera: a review update on advancement of in vitro culture. *Acta Agric Scand Sect B-Soil Plant Sci* 64:1–12.

Georgiev V, Slavov A, Vasileva I, Pavlov A (2018) Plant cell culture as emerging technology for production of active cosmetic ingredients. *Eng Life Sci* 18:779–798.

Gupta S, Sahu PK, Sen DL, Pandey P (2014) In-vitro Propagation of Aloe vera (L.) Burm. f. *Biotechnol J Int* 4(7):806–816.

Hashemabadi D, Kaviani B (2008) Rapid micro-propagation of Aloe vera L. via shoot multiplication. *Int. J. Biotechnol. Biochem.* 3(2–3).

He C, Zhang J, Chen J, Ye X, Du L, Dong Y, Zhao H (2007) Genetic transformation of Aloe barbadensis Miller by Agrobacterium tumefaciens. *J Genet Genomics* 34:1053–1060.

Hosseini R, Parsa M (2007) Micropropagation of Aloe vera L. grown in South Iran. *Pak J Biol Sci* 10:1134–1137.

Jader HS, Obaid OH (2023). Effect of different concentrations of growth regulators on callus induction from different explants of Aloe vera L. in vitro. In: Ahmed Ghanim Wadday, Ali Najah Al-Shamani, Dhafer Manea Hachim AL-Hasnawi, Atheer Kadhim Ibadi, Dhurgham Hassan Abid, Faris Mohammed Ali, Ahmed Razzaq Hasan AI-Manea, Saadia H Kadhim Alsultani, Marwa Fadhil Alsafar, and Hussein Abad Gazi Jaaz (eds.) *AIP Conference Proceedings*. AIP Publishing.

Jangra A, Chaturvedi S, Sharma G, Sihag S, Tiwari S, Chhokar V (2023) Efficient Agrobacterium tumefaciens-mediated genetic transformation of Aloe vera. *Plant Cell, Tissue Organ Cult* 154:189–200.

Jangra A, Sharma G, Sihag S, Chhokar V (2022) The dark side of miracle plant-Aloe vera: a review. *Mol Biol Rep* 49:5029–5040.

Jayakrishna C, Karthik C, Barathi S, Kamalanathan D, ArulSelvi PI (2011) In vitro propagation of Aloe barbadensis Miller, a miracle herb. *Res Plant Bio* 1(5):22–26.

Kim BS, Hwang HS, Park H, Kim HH (2015) Effects of selective cleavage of high-mannose-type glycans of Maackia amurensis leukoagglutinin on sialic acid-binding activity. *Biochim Biophys Acta (BBA)-General Subj* 1850:1815–1821.

Kumar M, Singh S, Singh S (2011) In vitro morphogenesis of a medicinal plant-Aloe vera L. *Asian J Plant Sci Res* 1(1):31–40.

Lee S, Kim JK, Park SU (2016) Enhanced shoot organogenesis in Aloe saponaria following treatment with ethylene inhibitors and polyamines. *Biosci Biotechnol Res Asia* 13:17–21.

Liao Z, Chen M, Tan F, Sun X, Tang K (2004) Microprogagation of endangered Chinese aloe. *Plant Cell Tissue Organ Cult* 76:83–86.

Lowther W, Lorick K, Lawrence SD, Yeow WS (2012) Expression of biologically active human interferon alpha 2 in Aloe vera. *Transgenic Res* 21:1349–1357. https://doi.org/10.1007/s11248-012-9616-0

Molsaghi M, Moieni A, Kahrizi D (2014) Efficient protocol for rapid Aloe vera micropropagation. *Pharm Biol* 52:735–739.

Murashige T, Skoog F (1962) A revised medium for rapid growth and bio assays with tobacco tissue cultures. *Physiol Plant* 15(3):473.

Nayanakantha NMC, Singh BR, Kumar A (2011) Improved culture medium for micropropagation of Aloe vera L. *Trop Agric Res Ext* 13:87–93.

Niazian M (2019) Application of genetics and biotechnology for improving medicinal plants. *Planta* 249:953–973.

Niguse M, Sbhatu DB, Abraha HB (2020) In vitro micropropagation of Aloe adigratana Reynolds using offshoot cuttings. *Sci World J* 2020:9645316.

Rahman SU, Khan MO, Ullah R, Ahmad F, Raza G (2024) Agrobacterium-mediated transformation for the development of transgenic crops; present and future prospects. *Mol. Biotechnol.* 66:1836–1852.

Rajeswari R, Umadevi M, Rahale CS, Pushpa R, Selvavenkadesh S, Kumar KPS, Bhowmik D (2012) Aloe vera: the miracle plant its medicinal and traditional uses in India. *J Pharmacogn Phytochem* 1:118–124.

Rathore MS, Chikara J, Shekhawat NS (2011) Plantlet regeneration from callus cultures of selected genotypes of Aloe vera L.-An ancient plant for modern herbal industries. *Appl Biochem Biotechnol* 163:860–868.

Sahoo S, Rout GR (2014) Plant regeneration from leaf explants of Aloe barbadensis Mill. and genetic fidelity assessment through DNA markers. *Physiol Mol Biol Plants* 20:235–240.

Seran TH, Ahmad N (2018) Shoot organogenesis of Aloe plants with emphasis on TDZ. *Thidiazuron from urea Deriv to plant growth Regul*:359–376.

Sethi L, Kumari K, Dey N (2021) Engineering of plants for efficient production of therapeutics. *Mol Biotechnol* 63:1125–1137.

Shibru S, Olani G, Debebe A (2018) In vitro propagation of Aloe vera Linn from shoot tip culture. *GSC Biol Pharm Sci* 4:1–6.

Singh M, Rathore MS, Panwar D, Rathore JS, Dagla HR, Shekhawat NS (2009) Micropropagation of selected genotype of Aloe vera L.–an ancient plant for modern industry. *J Sustain For* 28:935–950.

Singh MK, Yadav T, Raman RK (2020) A quick method for micro-propagation of Aloe vera L. from leaf explants via callus induction. *J Entomol Zool Stud* 8:201–206.

Supe U (2006) In vitro propagation of Aloe barbadensis-a medicinal plant. *Indian J Agric Biochem* 19:19–21.

Tiwari S (2019) Enhanced Agrobacterium-mediated transformation efficiency of banana cultivar Grand Naine by reducing oxidative stress. *Sci Hortic (Amsterdam)* 246:675–685.

Upadhyay NK, Himanshu Dubey HD, Haider ZA (2007) Tissue culture propagation of Aloe vera: the medicinal wonder plant. *Ranchi J Res* 19(1):115–127.

Velcheva M, Faltin Z, Vardi A, Eshdat Y, Perl A (2005) Regeneration of Aloe arborescens via somatic organogenesis from young inflorescences. *Plant Cell Tissue Organ Cult* 83:293–301.

Velcheva M, Faltin Z, Vardi A, Hanania U, Eshdat Y, Dgani O, Sahar N, Perl A (2010) Aloe vera transformation: the role of Amberlite XAD-4 resin and antioxidants during selection and regeneration. *Vitr Cell Dev Biol – Plant* 46:477–484. https://doi.org/10.1007/s11627-010-9301-z

Wahab U, Ashfaq M, Sajjad M, Shaheen S, Sadique R, Ali M, Khan F (2020) Effects of different plant growth regulators on in vitro growth of Aloe vera L. *Bangladesh J Bot* 49:159–162.

Yasmin S, Hasan J, Hossain S, Saha S, Khatun F (2022) Auxin and cytokinin synergism in micropropagation for mass production of Aloe vera. *Biotechnol J Biotechnol Comput Biol Bionanotechnol* 103(3):301–310.

10 Physiological Responses and Adaptation Mechanisms Under Abiotic Stress in *Aloe vera*

Shalu Yadav and Vinod Chhokar

10.1 INTRODUCTION

Aloe vera is a monocotyledonous xerophytic, succulent, and crassulacean acid metabolism plant (CAM) belonging to the Aloe genus of the family *Asphodelaceae* (Solomou et al. 2020; Jangra et al. 2022). The plant originated in Africa's warm, arid climate, but owing to its extreme adaptability, it is now cultivated extensively throughout the w11111orld (Steenkamp and Stewart 2007). Of the almost 400 species identified thus far, some have significant medical qualities, such as *Aloe vera, Aloe perryi, Aloe ferox*, and *Aloe arborescens*. Amongst them, *Aloe vera* has the most therapeutic potency and is hence more prominent (Pugh et al. 2001; Babu and Noor, 2020). Owing to its pharmacological qualities, which include anti-inflammatory, anti-tumour, antiviral, anti-ulcer, and fungicidal effects, the plant is widely utilized as a traditional herb in several cosmetics and skin care industries (Gupta and Malhotra 2012; Raksha et al. 2014). Due to the presence of major chemical constituents such as anthraquinones, vitamins, enzymes, sterols, amino acids, minerals, salicylic acid, saponin, lignin, lectin, and alkylbenzenes, *Aloe vera* serves as a potent medicinal herb (Eshun and He 2004; Luta and McAnalley 2005; Joseph and Raj 2010; Choudhri et al. 2018). Twenty types of anthraquinone, the phenolic compounds, are identified in the green leaf region of *Aloe vera*. The two major anthraquinones with antibacterial and antiviral activity are Aloin and emodin, and it has been shown that the amount of these anthraquinones in the leaf exudate of 68 aloe species varies from 0.1 to 21.5% dry weight. These compounds are responsible for prolonging cell life as they are involved in the destruction of cancer cells and provide strength to the immune system (Park et al. 2009). About 98.5–99.5% of the *Aloe vera gel* is made up of water, while the remaining solids section contains over 200 distinct components, the most common of which are polysaccharides (Ogidi and Enenebeaku 2023). The polysaccharides of *Aloe vera* have been considered as the major constituents responsible for their beneficial properties (Choi and Chung 2003; Boudreau and Beland 2006; Simões et al. 2012). Previously, the whole genome transcriptome sequence database of *Aloe vera* was generated using Illumina HiSeq technology, and the progressive annotation in relation to the metabolic specificity of the plant was explored precisely (Choudhri et al. 2018). *Aloe vera* can survive in adverse hot and dry climatic conditions. Previous research has generated the transcriptome and chloroplast genome database of *Aloe vera*, and a few studies have also highlighted the resistance of *Aloe vera* toward prevailing drought stress in the environment (Ren et al. 2020). *Aloe vera* is a CAM (crassulacean acid metabolism) plant, which is characterized by nocturnal CO_2 assimilation(Silvera et al. 2010) and nocturnal stomata aperture, resulting in higher water usage efficiency (Winter et al. 2005). Since it is a CAM species and has evolved naturally to withstand dryness and high temperatures, it only needs minimal irrigation, depending on the soil's ability to retain humidity and can be cultivated widely in arid and semi-arid areas (Delatorre-Herrera et al., 2010). For better growth and reproduction, *Aloe*

DOI: 10.1201/9781003603108-10

vera requires high temperature, light soil with intense sunlight. Previous research reported that excessive water accumulation around the root of the *Aloe* plant severely hampers plant growth, resulting in root rot and fungal attack to the root of the plant (Azarfam et al. 2020). In general, *Aloe vera* is self-incompatible, possesses high heterogeneity in seed population, and the seeds are only fertile if produced through cross-pollination (Botes, Johnson, and Cowling 2009; Muñoz-Sanz et al. 2020). The prolonged climatic changes have had adverse effects on medicinal plants and increased their susceptibility to prevailing abiotic stresses. The most promising abiotic stresses such as drought, temperature, intense light and salinity stress can disturb cellular homeostasis, impair metabolic processes, and adversely affect plant growth and development. Plants adopted various defence approaches such as phenology, morphology, and modification in physiological and biochemical pathways aimed at mitigating the detrimental effects and enhancing stress tolerance (Penfield 2008). In response to drought, salt, and heat stress, plants accumulate compatible solutes like mannitol, proline, and amino acid derivatives (Mansour 2000). The information regarding the effects of environmental stresses on biochemical and phytochemical characteristics in *Aloe vera* is very limited. Currently, changing climatic conditions, global warming, and water stress have directly given rise to a different kind of abiotic stress, endangering more than 800 million hectares of land in danger and salt affected (Asseng et al. 2015). *Aloe vera,* because of its CAM member family plant, is adapted to several abiotic stresses prevalent in the environment. This book chapter encompasses several abiotic stresses and also highlights the adaptive mechanisms of *Aloe vera* to cope with these prevalent stresses. This study aims to enhance our understanding of abiotic stress tolerance in *Aloe vera* and contribute to the development of improved crop resilience strategies.

10.2 PHYSIOLOGICAL RESPONSES AGAINST ABIOTIC STRESS IN *ALOE VERA*

Any adverse conditions or components that impair or prevent a plant's metabolism, growth, or development are referred to as plant stress (Kranner et al. 2010). Many physiologists describe stress as the altered physiological condition caused by stimuli aimed to disturb an equilibrium from a functional standpoint (Kranner et al. 2010). Depending on the frequency and severity of the stressful phenomenon, as well as the plant's ontological stage, the physiological response to stress might change. Hardening or acclimation is the term used to describe a plant's physiological response to a gradually increasing stressful environment. This response protects the plant from growth inhibition or injury caused by abruptly imposed environmental stresses, without involving changes to the DNA sequence.

10.3 TEMPERATURE STRESS

The stress caused by high or low temperatures is one of the significant environmental variables that limit the global distribution of species, which is mainly controlled by latitude and altitudinal gradients that define thermal niches with specific properties. Stress imposed by both low and high temperatures directly affects the structures of molecules (DNA, lipids, and proteins) and macromolecules (membranes, chromosomes), mainly as a result of the thermodynamic effect (Ruelland and Zachowski 2010).

10.3.1 High-Temperature Stress

It exerts an important constraint on plant growth and development and is a major cause of plant mortality (McDowell et al. 2018). Due to high temperature at the cellular and subcellular level, reduction in water content was monitored with a negative effect on photosystem II functioning, rubisco activity, fluidity of the thylakoid membrane, and cell division and growth of the plant (Hasanuzzaman et al. 2013; Teskey et al. 2015). During high temperatures, the inactivation of enzymes, membrane dysfunction, microtubule de-organization, and protein denaturation and aggregation have been observed. The cells have two fates for these unfolded proteins; the heat shock protein, along with chaperones, recovers the damaged protein's structure, and the proteins that are not recovered by chaperons are degraded through ubiquitin-mediated 26S proteasomal degradation (Melnikov and Rotanova 2010). Researchers reported that in *Populus tremula,* a rise in temperature causes a rapid consumption of carbohydrates. They studied different plant responses towards low and high temperatures; it was observed that at 55ºC eight times greater rapid carbohydrate consumption was monitored than at 25ºC. After 30 minutes of continuation incubation in the dark, the content of starch in the leaves of Populus *tremula* decreased by 16% at 25ºC and 53% at 50ºC compared with the control starch content (Hasanuzzaman et al. 2013). In C_3 and C_4 plants, the elevated temperature has a significant influence on their photosynthetic capacity (Wang et al. 2008). Previous research reported that an increase in temperature results in a decline in net photosynthesis, which leads to increased photorespiration and mitochondrial respiration. The increased photorespiration has been attributed to changes in Rubisco specificity and differences in CO_2 and O_2 solubility (Busch 2020). However, the study of the enzymatic response of Rubisco toward high temperature is limited. It has only been well characterized in a few species, including spinach (Yamori et al. 2006), tobacco (Sharwood et al. 2016) and *Arabidopsis* (Galmes et al. 2019). Thus, monitoring the response to high temperatures in *Aloe vera* is an important area of research. During high-temperature inactivation of Rubisco attributed to reduced activity of Rubisco activase leads to decreased photosystem II activity and damage to the thylakoid membrane, resulting in less ATP synthesis and increased permeability of thylakoid membrane towards H^+ ions (Teskey et al. 2015; Jajoo and Allakhverdiev 2017). Previously, it was reported that the soluble sugar content increased at high temperatures (Zapata et al. 2013).

Metabolic pathway against heat stress

Heat stress has the potential to uncouple enzymes and metabolic processes that lead to the build-up of undesired and harmful reactive oxygen species (ROS). The most frequent ROS that induce oxidative stress are singlet oxygen ($^1O_2$), superoxide radical (O_2•–), hydrogen peroxide (H_2O_2), and hydroxyl radical (OH•) (Asada 2006). It has been suggested that thermal damage to photosystems leads to less photon absorption. Under these stress conditions, ROS are produced if photon intensity is absorbed by PSI and PSII, the excess of which is needed for CO_2 assimilation and can be perceived as surplus electrons (Halliwell 2006). Thermal stress may stimulate oxidative stress through peroxidation of membrane lipids and by denaturing proteins, which may disrupt the integrity of cell membranes (Rodriguez et al. 2005; Camejo et al. 2006). An increase in proline and serine content in *Aloe vera* was observed against heat stress.

10.3.2 Cold Stress

Extreme temperature stress affects medicinal plants at the biochemical, physiological, and molecular levels. It results in leaf senescence and promotes chlorophyll degradation, resulting in protein degradation (Pradhan et al. 2017). Previously, researchers in the *Arnica Montana* plant found that the ratio of quercetin to kaempferol has increased with a decrease in temperature. Researchers observed the improved accretion of artemisinin in *Artemisia annua* due to the upregulation of candidate genes involved in its biosynthesis as a result of cold stress (Vashisth et al. 2018). In response to low temperature stress, plants deploy advanced molecular mechanisms to withstand cold stress. The cold stress, which results in a metabolism shift toward vegetative storage protein production and plants' ability to withstand low temperatures, initiates several biochemical and physical changes (Welling et al. 2002; Jahed et al. 2023). Modifying the saturation level of the fatty-acid chains in membrane lipids is the key strategy plants use to maintain membrane fluidity in response to freezing tolerance (Falcone et al. 2004).

10.3.2.1 Metabolic Pathway against Cold Stress

It was reported that, in response to cold stress in apple, the R2R3 MYB transcription factors MYB88 and MYB124 were overexpressed. The induction of TFs in apples stimulates anthocyanin accumulation and ROS detoxification (Jajoo and Allakhverdiev 2017). The modulation of the expression of the C-repeat/DREB binding factor belonging to the APETELALA2/ Ethylene-Responsive Factor family plays an important role in chilling resistance and acclimation in plants (Shi, Ding, and Yang 2018).

10.4 DROUGHT STRESS

Drought stress is one of the significant abiotic stresses affecting the growth and development of plants globally. A major factor in plant death is drought stress, and plants have developed a range of coping mechanisms to deal with the various stages of water deprivation, including morphological, physiological, and biochemical changes. These fall into two categories: drought mitigation and drought avoidance (McDowell et al. 2018). In water-deficient conditions, to adjust osmotic balance, an increase in osmolyte concentration has been observed in *Aloe vera*. The increased concentration of soluble sugars, proline, and fructan will enhance the water usage efficiency of *Aloe vera* in water-deficit conditions (Delatorre-Herrera et al. 2010). The previous investigation also reported the formation of reactive oxygen species (ROS) due to the induction of oxidative stress in drought conditions (Ahmad et al. 2010). Reactive oxygen species can react with DNA, protein, lipids, and photosynthetic pigments, which eventually leads to lipid peroxidation, membrane damage, inactivation of antioxidant enzymes, and cell death. To cope with the effect of ROS, plants effectively use antioxidant defence mechanisms through several ROS detoxifying enzymes (Gill and Tuteja 2010).

Along with antioxidant enzymes, plants have developed several adaptation strategies that are associated with physiological traits, i.e., accumulation of sugars and increasing concentration of osmoprotectant (Kooyers 2015). Stress induced by water scarcity has a significant impact on *Aloe vera*'s ability to synthesize soluble sugars, proline, and secondary metabolites (Lucini et al. 2013; Ray et al. 2013). Previous research revealed that mild drought stress did not affect *Aloe vera*'s ability to produce leaves, biomass, or gel; only severe drought stress resulted in a considerable loss in biomass. They induced drought conditions for 180 days in *Aloe vera* and observed that under mild drought stress, leaf thickness was inclined to increase, while under severe drought conditions, leaf thickness and relative water content were reduced (Habibi 2018). The previous research reported the morphologic plasticity adaptation of *Aloe vera* toward water-deficit conditions. Succulent plants, in fact, can display plasticity in their photosynthesis in response to the presence of thick leaves (Habibi and Ajory 2015; Habibi 2016). It has been observed that mild drought stress increases phenolic compound production in the leaves of the *Aloe vera* plant. Previous literature reported the effect of drought stress on plants treated for 180 days; they observed that total phenolic and flavonoids were increased under mild drought stress, while, on the contrary, under severe drought conditions, their contents remained unchanged. In addition, after 120 and 180 days of exposure to severe drought stress, an increase in anthocyanin content was observed in contrast to plants grown under mild drought stress. Along with this, some researchers reported the correlation between PAL activation in the aloe leaves and phenolic compounds in treated plants. Severe drought stress suppressed PAL activity, which led to a decrease in phenolic compounds in the *Aloe* leaves (Habibi 2018). In CAM plants, sucrose helps the plant to tolerate drought stress through osmotic adjustment. Under the water deficit condition, the *Aloe vera* plants showed high soluble sugar content (Hamman 2008). It was reported that the polysaccharides improve the resistance of plants against water stress (Delatorre-Herrera et al. 2010).

10.5 SALINITY STRESS

Elevated salinity stress modifies the physiochemical characteristics of soil and the physiological, biochemical, and molecular processes of plants, which has an adverse effect on plant development and production. The effects of salinity stress on many physiological and biochemical processes, growth, development, metabolism, absorption of water and nutrients, and photosynthetic efficiency result in a significant decrease in yield (Manaf 2016; Sofy 2016). Additionally, an excessive amount of salts causes the production of reactive oxygen species (ROS), which disrupts lipids, proteins, DNA, and cellular membranes (Jiang et al. 2021). The excess of sodium and chlorine ions affects plant growth, metabolism, and photosynthetic activity; sometimes, it even causes plant death (Mahajan and Tuteja 2005; Machado and Serralheiro 2017). Previously, researchers monitored the effect of different levels of salt stress on three *Aloe* species (*Aloe vera, Aloe arborescens,* and *Aloe brevifolia*). The previous study monitored morphological, physiological, and biochemical parameters in response to different concentrations of salinity stress. A slight increase in chlorophyll content was observed in all three species when salt stress was stimulated. It was hypothesized that salinity stress could reduce photosynthetic activity and limit CO_2 incorporation through partial stomatal closure (Burman, Garg, and Kathju 2003). The proline content was high in all three *Aloe* species after salt stress exposure, which reduces the potential osmotic effect (Kovacs et al. 2019). In contrast to the control plant, total sugars, reducing sugars, and polysaccharide levels increased significantly under the salinity stress (Sedigheh 2019). Polyphenol accumulation increased in *Aloe vera* in response to salt stress, and it provides osmoprotectant, osmoregulation, carbon storage, and radical scavenging (Parida and Das 2005; Ksouri et al. 2007). Previously, researchers also monitored the increment in aloin content in *Aloe vera* plants that were subjected to salt stress (Rahimi-Dehgolan et al. 2012). The previous study concluded that the tolerance of *Aloe* plants towards salinity and dehydration could be due to the xerophytic nature and to the water reserve stored in their succulent leaves (Derouiche et al. 2023).

10.6 EVALUATING ADAPTATION RESPONSES TO ABIOTIC STRESS USING STRESS INDICATORS IN *ALOE VERA*

A thorough method is to investigate *Aloe vera*'s adaptive responses to abiotic stress using stress indicators such as osmolytes, carbon isotopes, phytochemicals, and antioxidant enzymes. Outlined below are some ways that each of these indicators could provide insights into the way *Aloe vera* responds to environmental stress.

10.6.1 Osmolyte

Osmolytes are highly soluble, low molecular weight compounds that are innocuous even in substantial amounts. Individuals may survive under extreme conditions due to the accumulation of these substances. Osmolytes are members of several groups of unique chemical compounds. These include methylamines, sugars, polyols, amino acids, and their derivatives. They are found widely in many different taxonomic groups, from higher plants to algae (Sengupta et al. 2016). Primary osmotic stresses often initiate osmolyte synthesis, although, under some circumstances, secondary stress signals such as reactive oxygen species, phytohormones, and intracellular second messengers may trigger osmolyte production. Osmolytes maintain proteins' structural stability, which prevents misfolding or aggregation of proteins during abiotic stresses (Lamitina et al. 2006). An essential defence mechanism that plants have developed to deal with the impacts of salt stress is osmotic adjustment. Osmolytes are essential for preserving the osmotic potential of ions in cells and subcellular structures under a variety of stressful circumstances (Ali et al. 2019; Ahanger et al. 2020). Plants that have higher levels of osmolyte accumulation, salt compartmentation, and antioxidant system upregulation are better able to withstand salt stress (Ahmad et al. 2017). Zinc is an important mineral that significantly increases osmolyte accumulation and improves stress tolerance via osmotic adjustments. When plants are subjected to abiotic stress, zinc treatment enhances the accumulation of soluble proteins and amino acids and significantly boosts plant development (Ma et al. 2016). Plants exposed to drought, salinity, or temperature stress have been observed to accumulate proline, and stress tolerance is mainly dependent on proline, an essential osmolyte that is greatly enhanced by zinc administration. Proline accumulation controls solute potential and water absorption (Abebe et al. 2003; Lugan et al. 2010). Due to their propensity to establish hydrogen bonds with macromolecules, osmolytes can inhibit intramolecular H-bond formation, which would otherwise irreversibly alter the three-dimensional structure of the molecules (Hosseinifard et al. 2022). By applying polyamines exogenously, it may be possible to maintain the integrity of plant cell membranes, reduce stress-induced growth inhibition, moderate the expression of osmotically responsive genes, and stimulate the activity of antioxidant enzymes (Navakoudis et al. 2003; Jangra et al. 2023).

10.6.2 Carbon Isotopes

The photosynthetic mechanisms that plants adapted during stress conditions can be better understood by examining carbon isotopes (Caemmerer et al. 2014). The carbon isotope fingerprints for various photosynthetic processes, including C_3, C_4, and CAM (Crassulacean Acid Metabolism), are different (Luttge 2004). *Aloe vera* mostly uses the CAM pathway, which works exceptionally well in dry environments. The analysis of carbon isotopes can reveal changes in the photosynthetic activity and water utilization efficiency of *Aloe vera* plants in response to abiotic stress like salinity or drought (Perron et al. 2023). The carbon isotope technique is an efficient way to screen a large number of species for the presence or absence of strong CAM (Pachon et al. 2022).

10.6.3 Phytochemical and Antioxidant Analysis

Plants produce bioactive substances called phytochemicals, frequently in response to stress.

These substances include terpenoids, alkaloids, flavonoids, and phenolics (Yeshi et al. 2022). *Aloe vera*'s phytochemical study can reveal information regarding possible adaptations and strategies to respond to stressful situations. The activation of defence mechanisms or metabolic adjustments to deal with stressors may be indicated by changes in the composition of phytochemicals (Perez-Llorca et al. 2023). Reactive oxygen species (ROS) can build up in plant tissues as a result of environmental stress, which can lead to oxidative stress (Das and Roychoudhury 2014). Plants use antioxidant defence systems, which include enzymes like peroxidase (POD), catalase (CAT), and superoxide dismutase (SOD), to counteract ROS damage (P. Sharma et al. 2012). *Aloe vera*'s ability to maintain cellular homeostasis and mitigate oxidative damage can be determined by evaluating the activity levels of these antioxidant enzymes under stress conditions (Khanam and Sharma 2013).

10.7 GENETIC TRACES OF *ALOE VERA* RESILIENCY TO CHALLENGING CONDITIONS

The roots of a plant serve as its first line of defence against abiotic stress. If the plant grows in healthy, biologically varied soil, its chances of surviving severe conditions will be excellent. The resilience of *Aloe vera* to changing climatic conditions significantly improved through its genetic makeup. Previously, through genetic analysis, researchers have identified specific genes and genetic pathways that help to withstand *Aloe vera* in highly stressful conditions. These genes could have been used to defend against pathogens, increase drought tolerance, and retain water.

10.7.1 Phytohormone Response against Stress Tolerance

Plant growth, development, and response to environmental stimuli, such as stress, are regulated by phytohormones, which are chemical messengers. *Aloe vera* apparently modulates the level of phytohormones to increase its resistance to stress conditions. For instance, the plant may increase the production of phytohormones associated with stress, including abscisic acid (ABA), which aids in drought adaptation by promoting root development and minimizing water loss in plants. *Aloe vera* reacts to stress by altering the level of phytohormones, such as abscisic acid (ABA), just like other plants. In *Aloe vera*, ABA is essential for responding to environmental stresses, especially those related to drought and salinity. In drought conditions, high ABA levels trigger various physiological and molecular responses, such as inducing stomatal closure, that help *Aloe vera* to resist water scarcity. The ABA level regulates the osmotic regulation in *Aloe vera* as it promotes the accumulation of compatible solutes, which act as osmoprotectants under water deficit conditions. ABA signaling stimulates the production of antioxidant enzymes such as superoxide dismutase, catalase, peroxidase, and nonenzymatic antioxidants, i.e., glutathione and ascorbate, to mitigate oxidative damage caused by drought stress. In *Aloe vera*, ABA regulates the expression of stress-responsive genes, especially those that are involved in drought tolerance. These genes include aquaporins, transcription factors like DREBs (dehydration-responsive element-binding proteins), and LEA (late embryogenesis abundant) proteins (Sharma 2024).

10.7.2 Role of Secondary Metabolites against Abiotic Stresses

The secondary metabolites are not directly involved in plant growth and development. Although they frequently have a significant impact on the extent to which plants respond to environmental stresses, they are synthesized through complex biochemical pathways. Secondary metabolite accumulation is a part of the plant's stress-adaptive response and improves overall survival and resilience. The phenolic acids and flavonoids are major secondary metabolites that act as antioxidants, scavenging reactive oxygen species to combat abiotic stresses. Abiotic stress conditions, such as extreme temperatures and deficiency of nutrients, have been shown to influence alkaloid biosynthesis in plants (Yeshi et al. 2022). Previous studies reported that alkaloid production in medicinal plants such as opium poppy is majorly influenced by environmental stresses. It was hypothesized that plants produce alkaloids as a dynamic and adaptive response to different stress circumstances, supporting the plant's defence mechanisms (Leontopoulos et al. 2015). The flavonoids are recognized for their antioxidant activity, which aids plants in combating oxidative stress induced by environmental factors. Quercetin has been shown to scavenge free radicals and protect plant cells from oxidative damage caused by stress (Mierziak et al. 2014).

10.7.3 Genes Involved in Resistance against Abiotic Stresses

Genetic regulations are essential processes that plants employ to manage environmental stresses. Genetic regulation entails the activation or suppression of specific genes in response to stress. Such genes could be involved in initiating signaling pathways to protect plants against environmental stress. Genetic regulation plays an important role in enabling plants to combat stressful conditions (Delavari et al. 2021). It is quite likely that *Aloe vera*'s stress stability is controlled by ABA-induced genes such as HVA22 homologues. The expression of HVA22 homologues genes was induced by ABA and environmental stresses, such as cold and drought, in the vegetative tissues (Shen et al. 2001). Previously, it was hypothesized that the glycerol kinase gene had been conserved structurally and functionally during evolution, and it could be one of the potential genes that can help *Aloe vera* to combat stress (Bahieldin et al. 2013; Cheng et al. 2014).

10.8 CONCLUSION

In summary, *Aloe vera* is considered an excellent botanical specimen because of its numerous therapeutic benefits and its remarkable ability to withstand adverse environmental conditions. Its popularity in both conventional medicine and the cosmetics industry has been established by its broad range of pharmacological properties, which include anti-inflammatory, anti-tumour, antiviral, and fungicidal properties. The therapeutic effectiveness of *Aloe vera* is highlighted by its complex biochemistry, which consists of a wide range of compounds from polysaccharides to anthraquinones. The metabolic complexities of this plant have been revealed by researchers through meticulous genomic and transcriptomic assessments, providing insight into its ability to thrive in drought-prone areas through its crassulacean acid metabolism (CAM) mechanism. Additionally, *Aloe vera*'s resilience to abiotic stresses such as drought, low or high temperature, intense light, and salinity has been a subject of further research. Its adaptability is shown by its ability to accumulate suitable solutes and alter physiological and biochemical processes. However, along with its resilience to environmental stresses, the study on *Aloe vera*'s biochemical and phytochemical properties attracts huge attention from researchers. This comprehensive book chapter serves as a beacon, illuminating the adaptive mechanisms of *Aloe vera* and providing insights into enhancing *Aloe vera* resilience strategies amidst changing climatic conditions. By deepening our understanding of *Aloe vera*'s response to abiotic stresses, we pave the way for harnessing its potential more effectively, not only in traditional medicine but also in agricultural practices geared towards sustainable crop production.

REFERENCES

Abebe, Tilahun, Arron C Guenzi, Bjorn Martin, and John C Cushman. 2003. "Tolerance of Mannitol-Accumulating Transgenic Wheat to Water Stress and Salinity." *Plant Physiology* 131(4): 1748–55.

Ahanger, Mohammad Abass, Rayees Ahmad Mir, Mohammed Nasser Alyemeni, and Parvaiz Ahmad. 2020. "Combined Effects of Brassinosteroid and Kinetin Mitigates Salinity Stress in Tomato through the Modulation of Antioxidant and Osmolyte Metabolism." *Plant Physiology and Biochemistry* 147: 31–42.

Ahmad, Parvaiz et al. 2010. "Roles of Enzymatic and Nonenzymatic Antioxidants in Plants during Abiotic Stress." *Critical Reviews in Biotechnology* 30(3): 161–75.

Ahmad, Parvaiz et al. 2017. "Zinc Application Mitigates the Adverse Effects of NaCl Stress on Mustard [Brassica Juncea (L.) Czern & Coss] through Modulating Compatible Organic Solutes, Antioxidant Enzymes, and Flavonoid Content." *Journal of Plant Interactions* 12(1): 429–37.

Ali, Shafaqat et al. 2019. "Combined Use of Biochar and Zinc Oxide Nanoparticle Foliar Spray Improved the Plant Growth and Decreased the Cadmium Accumulation in Rice (Oryza Sativa L.) Plant." *Environmental Science and Pollution Research* 26: 11288–99.

Asada, Kozi. 2006. "Production and Scavenging of Reactive Oxygen Species in Chloroplasts and Their Functions." *Plant Physiology* 141(2): 391–6.

Asseng, Senthold et al. 2015. "Rising Temperatures Reduce Global Wheat Production." *Nature Climate Change* 5(2): 143–7.

Azarfam, S P, H Nadian, A Moezzi, and A Gholami. 2020. "Effect of Silicon on Phytochemical and Medicinal Properties of Aloe Vera under Cold Stress." *Applied Ecology & Environmental Research* 18(1): 561–75.

Babu, Spoorthy N, and Ayesha Noor. 2020. "Bioactive Constituents of the Genus Aloe and Their Potential Therapeutic and Pharmacological Applications: A Review." *Journal of Applied Pharmaceutical Science* 10(11): 133–45.

Bahieldin, Ahmed et al. 2013. "Control of Glycerol Biosynthesis under High Salt Stress in Arabidopsis." *Functional Plant Biology* 41(1): 87–95.

Botes, Christo, Steven D Johnson, and Richard M Cowling. 2009. "The Birds and the Bees: Using Selective Exclusion to Identify Effective Pollinators of African Tree Aloes." *International Journal of Plant Sciences* 170(2): 151–6.

Boudreau, Mary D, and Frederick A Beland. 2006. "An Evaluation of the Biological and Toxicological Properties of Aloe Barbadensis (Miller), Aloe Vera." *Journal of Environmental Science and Health Part C* 24(1): 103–54.

Burman, U, B K Garg, and S Kathju. 2003. "Water Relations, Photosynthesis and Nitrogen Metabolism of Indian Mustard (Brassica Juncea Czern & Coss) Grown under Salt and Water Stress." *Journal of Plant Biology* 30(1): 55–60.

Busch, Florian A. 2020. "Photorespiration in the Context of Rubisco Biochemistry, CO2 Diffusion and Metabolism." *The Plant Journal* 101(4): 919–39.

Caemmerer, Susanne von, Oula Ghannoum, Jasper J L Pengelly, and Asaph B Cousins. 2014. "Carbon Isotope Discrimination as a Tool to Explore C4 Photosynthesis." *Journal of experimental botany* 65(13): 3459–70.

Camejo, Daymi et al. 2006. "Changes in Photosynthetic Parameters and Antioxidant Activities Following Heat-Shock Treatment in Tomato Plants." *Functional Plant Biology* 33(2): 177–87.

Cheng, Cheng et al. 2014. "Identification of the Glycerol Kinase Gene and Its Role in Diapause Embryo Restart and Early Embryo Development of Artemia Sinica." *Gene* 537(1): 51–62.

Choi, Seongwon, and Myung-Hee Chung. 2003. "A Review on the Relationship between Aloe Vera Components and Their Biologic Effects." *Elsevier Seminars in Integrative Medicine* 1(1): 53–62.

Choudhri, Pragati et al. 2018. "De Novo Sequencing, Assembly and Characterisation of Aloe Vera Transcriptome and Analysis of Expression Profiles of Genes Related to Saponin and Anthraquinone Metabolism." *BMC Genomics* 19(1): 1–21.

Das, Kaushik, and Aryadeep Roychoudhury. 2014. "Reactive Oxygen Species (ROS) and Response of Antioxidants as ROS-Scavengers during Environmental Stress in Plants." *Frontiers in Environmental Science* 2: 53.

Delatorre-Herrera, Jose et al. 2010. "Irrigation Restriction Effects on Water Use Efficiency and Osmotic Adjustment in Aloe Vera Plants (Aloe Barbadensis Miller)." *Agricultural Water Management* 97(10): 1564–70.

Delavari, Azar, Zahra Zinati, Sima Sazegari, and Ahmad Tahmasebi. 2021. "Integrating Expression Data and Genomic Sequences to Investigate the Transcriptional Regulation in Barley in Response to Abiotic Stress." *Biotechnologia* 102(1): 21–32.

Derouiche, M et al. 2023. "The Effect of Salt Stress on the Growth and Development of Three Aloe Species in Eastern Morocco." *Plant Stress* 9: 100187.

Eshun, Kojo, and Qian He. 2004. "Aloe Vera: A Valuable Ingredient for the Food, Pharmaceutical and Cosmetic Industries-a Review." *Critical Reviews in Food Science and Nutrition* 44(2): 91–6.

Falcone, Deane L, Joseph P Ogas, and Chris R Somerville. 2004. "Regulation of Membrane Fatty Acid Composition by Temperature in Mutants of Arabidopsis with Alterations in Membrane Lipid Composition." *BMC Plant Biology* 4: 1–15.

Galmes, Jeroni, Sebastia Capo-Bauca, Ulo Niinemets, and Concepcion Iniguez. 2019. "Potential Improvement of Photosynthetic CO2 Assimilation in Crops by Exploiting the Natural Variation in the Temperature Response of Rubisco Catalytic Traits." *Current Opinion in Plant Biology* 49: 60–7.

Gill, Sarvajeet Singh, and Narendra Tuteja. 2010. "Reactive Oxygen Species and Antioxidant Machinery in Abiotic Stress Tolerance in Crop Plants." *Plant Physiology and Biochemistry* 48(12): 909–30.

Gupta, Vinay K, and Seema Malhotra. 2012. "Pharmacological Attribute of Aloe Vera: Revalidation through Experimental and Clinical Studies." *AYU (An International Quarterly Journal of Research in Ayurveda)* 33(2): 193–96.

Habibi, Ghader. 2016. "The Role of Crassulacean Acid Metabolism Induction in Plant Adaptation to Water Deficit." *Water Stress and Crop Plants: A Sustainable Approach* 1: 12–23.

Habibi, Ghader. 2018. "Effects of Mild and Severe Drought Stress on the Biomass, Phenolic Compounds Production and Photochemical Activity of Aloe Vera (L.) Burm. F." *Acta Agriculturae Slovenica* 111(2): 463–76.

Habibi, Ghader, and Neda Ajory. 2015. "The Effect of Drought on Photosynthetic Plasticity in Marrubium Vulgare Plants Growing at Low and High Altitudes." *Journal of Plant Research* 128: 987–94.

Halliwell, Barry. 2006. "Oxidative Stress and Neurodegeneration: Where Are We Now?" *Journal of Neurochemistry* 97(6): 1634–58.

Hamman, Josias H. 2008. "Composition and Applications of Aloe Vera Leaf Gel." *Molecules* 13(8): 1599–616.

Hasanuzzaman, Mirza et al. 2013. "Physiological, Biochemical, and Molecular Mechanisms of Heat Stress Tolerance in Plants." *International Journal of Molecular Sciences* 14(5): 9643–84.

Hosseinifard, Marjanossadat et al. 2022. "Contribution of Exogenous Proline to Abiotic Stresses Tolerance in Plants: A Review." *International Journal of Molecular Sciences* 23(9): 5186.

Jahed, Khalil R, Amolpreet Kaur Saini, and Sherif M Sherif. 2023. "Coping with the Cold: Unveiling Cryoprotectants, Molecular Signaling Pathways, and Strategies for Cold Stress Resilience." *Frontiers in Plant Science* 14: 1246093.

Jajoo, Anjana, and Suleyman I Allakhverdiev. 2017. "High-Temperature Stress in Plants: Consequences and Strategies for Protecting Photosynthetic Machinery." *Plant Sstress Physiology* 2017: 138–54.

Jangra, Alka, Garima Sharma, Sonia Sihag, and Vinod Chhokar. 2022. "The Dark Side of Miracle Plant-Aloe Vera: A Review." *Molecular Biology Reports* 49(6): 5029–40.

Jangra, Alka et al. 2023. "Polyamines: The Gleam of next-Generation Plant Growth Regulators for Growth, Development, Stress Mitigation, and Hormonal Crosstalk in Plants—a Systematic Review." *Journal of Plant Growth Regulation* 42(8): 5167–91.

Jiang, Dan et al. 2021. "Exogenous Melatonin Improves the Salt Tolerance of Cotton by Removing Active Oxygen and Protecting Photosynthetic Organs." *BMC Plant Biology* 21: 1–19.

Joseph, B., & Raj, S. J. (2010). Pharmacognostic and phytochemical properties of Aloe vera linn an overview. *International Journal of Pharmaceutical Sciences Review and Research* 4(2): 106–110.

Khanam, Neelofar, and G K Sharma. 2013. "A Critical Review on Antioxidant and Antimicrobial Properties of Aloe vera L." *International Journal of Pharmaceutical Sciences and Research* 4(9): 3304.

Kooyers, Nicholas J. 2015. "The Evolution of Drought Escape and Avoidance in Natural Herbaceous Populations." *Plant Science* 234: 155–62.

Kovacs, Hajnalka, Abu Imran Baba, Laszlo Kozma-Bognar, and Laszlo Szabados. 2019. "Light Control of Salt-Induced Proline Accumulation Is Mediated by ELONGATED HYPOCOTYL 5 in Arabidopsis." *Frontiers in Plant Science* 10: 433581.

Kranner, Ilse, Farida V Minibayeva, Richard P Beckett, and Charlotte E Seal. 2010. "What Is Stress? Concepts, Definitions and Applications in Seed Science." *New Phytologist* 188(3): 655–73.

Ksouri, Riadh et al. 2007. "Salinity Effects on Polyphenol Content and Antioxidant Activities in Leaves of the Halophyte Cakile Maritima." *Plant Physiology and Biochemistry* 45(3–4): 244–9.

Lamitina, Todd, Chunyi George Huang, and Kevin Strange. 2006. "Genome-Wide RNAi Screening Identifies Protein Damage as a Regulator of Osmoprotective Gene Expression." *Proceedings of the National Academy of Sciences* 103(32): 12173–8.

Leontopoulos, S V et al. 2015. "Effect of Different Formulations of Polyphenolic Compounds Obtained from OMWW on the Growth of Several Fungal Plant and Food Borne Pathogens. Studies in Vitro and in Vivo." *Agriculture and Agricultural Science Procedia* 4: 327–37.

Lucini, Luigi, Marco Pellizzoni, and Gian Pietro Molinari. 2013. "Anthraquinones and -Polysaccharides Content and Distribution in Aloe Plants Grown under Different Light Intensities." *Biochemical Systematics and Ecology* 51: 264–68.

Lugan, Raphael et al. 2010. "Metabolome and Water Homeostasis Analysis of Thellungiella Salsuginea Suggests That Dehydration Tolerance Is a Key Response to Osmotic Stress in This Halophyte." *The Plant Journal* 64(2): 215–29.

Luta, G, and B McAnalley. 2005. "Aloe Vera: Chemical Composition and Methods Used to Determine Its Presence in Commercial Products." *GlycoScience & Nutrition* 6: 1–12.

Luttge, Ulrich. 2004. "Ecophysiology of Crassulacean Acid Metabolism (CAM)." *Annals of Botany* 93(6): 629–52.

Ma, Xiqing, Jing Zhang, and Bingru Huang. 2016. "Cytokinin-Mitigation of Salt-Induced Leaf Senescence in Perennial Ryegrass Involving the Activation of Antioxidant Systems and Ionic Balance." *Environmental and Experimental Botany* 125: 1–11.

Machado, Rui Manuel Almeida, and Ricardo Paulo Serralheiro. 2017. "Soil Salinity: Effect on Vegetable Crop Growth. Management Practices to Prevent and Mitigate Soil Salinization." *Horticulturae* 3(2): 30.

Mahajan, Shilpi, and Narendra Tuteja. 2005. "Cold, Salinity and Drought Stresses: An Overview." *Archives of Biochemistry and Biophysics* 444(2): 139–58.

Manaf, Hossam H. 2016. "Beneficial Effects of Exogenous Selenium, Glycine Betaine and Seaweed Extract on Salt Stressed Cowpea Plant." *Annals of Agricultural Sciences* 61(1): 41–8.

Mansour, Mohamed Magdy F. 2000. "Nitrogen Containing Compounds and Adaptation of Plants to Salinity Stress." *Biologia Plantarum* 43: 491–500.

McDowell, Nate et al. 2018. "Drivers and Mechanisms of Tree Mortality in Moist Tropical Forests." *New Phytologist* 219(3): 851–69.

Melnikov, E E, and T V Rotanova. 2010. "Molecular Chaperones." *Russian Journal of Bioorganic Chemistry* 36: 1–10.

Mierziak, Justyna, Kamil Kostyn, and Anna Kulma. 2014. "Flavonoids as Important Molecules of Plant Interactions with the Environment." *Molecules* 19(10): 16240–65.

Muñoz-Sanz, Juan Vicente et al. 2020. "Self-(in) Compatibility Systems: Target Traits for Crop-Production, Plant Breeding, and Biotechnology." *Frontiers in Plant Science* 11: 195.

Navakoudis, E et al. 2003. "Ozone Impact on the Photosynthetic Apparatus and the Protective Role of Polyamines." *Biochimica et Biophysica Acta (BBA)-General Subjects* 1621(2): 160–9.

Ogidi, Odangowei Inetiminebi, and Uchechi Emmanuella Enenebeaku. 2023. "Medicinal Potentials of Aloe Vera (Aloe Barbadensis Miller): Technologies for the Production of Therapeutics." In Sylvester Chibueze Izah and Matthew Chidozie Ogwu (eds.) *Sustainable Utilization and Conservation of Africa's Biological Resources and Environment*, 295–321, Springer.

Pachon, P, K Winter, and E Lasso. 2022. "Updating the Occurrence of Crassulacean Acid Metabolism (CAM) in the Genus Clusia through Carbon Isotope Analysis of Species from Colombia." *Photosynthetica* 60(2): 304.

Parida, Asish Kumar, and Anath Bandhu Das. 2005. "Salt Tolerance and Salinity Effects on Plants: A Review." *Ecotoxicology and Environmental Safety* 60(3): 324–49.

Park, Mi-Young, Hoon-Jeong Kwon, and Mi-Kyung Sung. 2009. "Evaluation of Aloin and Aloe-Emodin as Anti-Inflammatory Agents in Aloe by Using Murine Macrophages." *Bioscience, Biotechnology, and Biochemistry* 73(4): 828–32.

Penfield, Steven. 2008. "Temperature Perception and Signal Transduction in Plants." *New Phytologist* 179(3): 615–28.

Perron, Noe, Matias Kirst, and Sixue Chen. 2023. "Bringing CAM Photosynthesis to the Table: Paving the Way for Resilient and Productive Agricultural Systems in a Changing Climate." *Plant Communications* 5(3): 100772.

Pradhan, J, S K Sahoo, S Lalotra, and R S Sarma. 2017. "Positive Impact of Abiotic Stress on Medicinal and Aromatic Plants." *International Journal of Plant Sciences (Muzaffarnagar)* 12(2): 309–13.

Pugh, Nirmal, Samir A Ross, Mahmoud A ElSohly, and David S Pasco. 2001. "Characterization of Aloeride, a New High-Molecular-Weight Polysaccharide from Aloe Vera with Potent Immunostimulatory Activity." *Journal of Agricultural and Food Chemistry* 49(2): 1030–4.

Perez-Llorca, Marina, Stephan Pollmann, and Maren Muller. 2023. "Ethylene and Jasmonates Signaling Network Mediating Secondary Metabolites under Abiotic Stress." *International Journal of Molecular Sciences* 24(6): 5990.

Rahimi-Dehgolan, Ramin, Zeinolabedin Tahmasebi Sarvestani, Shams Ali Rezazadeh, and Aria Dolatabadian. 2012. "Morphological and Physiological Characters of Aloe Vera Subjected to Saline Water Irrigation." *Journal of Herbs, Spices & Medicinal Plants* 18(3): 222–30.

Raksha, Bawankar, Singh Pooja, and Subramanian Babu. 2014. "Bioactive Compounds and Medicinal Properties of Aloe Vera L.: An Update." *Journal of Plant Sciences* 2(3): 102–7.

Ray, Anirban et al. 2013. "Chemometric Studies on Mineral Distribution and Microstructure Analysis of Freeze-Dried Aloe Vera L. Gel at Different Harvesting Regimens." *Industrial Crops and Products* 51: 194–201.

Ren, Jia-Ju et al. 2020. "The Complete Chloroplast Genome of Aloe Vera from China as a Chinese Herb." *Mitochondrial DNA Part B* 5(1): 1092–93.

Rodriguez, Mayra, Eduardo Canales, and Orlando Borras-Hidalgo. 2005. "Molecular Aspects of Abiotic Stress in Plants." *Biotecnologia Aplicada* 22(1): 1–10.

Ruelland, Eric, and Alain Zachowski. 2010. "How Plants Sense Temperature." *Environmental and Experimental Botany* 69(3): 225–32.

Sedigheh, Negahdari. 2019. "Morphological, Biochemical and Soluble Sugars Characters of Aloe Vera Subjected to Saline Condition." *Egyptian Academic Journal of Biological Sciences, H. Botany* 10(1): 13–23.

Sengupta, Atreyee et al. 2016. "Polyamines: Osmoprotectants in Plant Abiotic Stress Adaptation." In Noushina Iqbal, Rahat Nazar, and Nafees A. Khan (eds.) *Osmolytes and Plants Acclimation to Changing Environment: Emerging Omics Technologies*, 97–127. Springer, New Delhi.

Sharma, Manoj Kumar. 2004. "Exploring the Biochemical Profiles of Medicinal Plants Cultivated Under Stressful Environmental Conditions." *Current Agriculture Research Journal* 12(1): 81–103.

Sharma, Pallavi, Ambuj Bhushan Jha, Rama Shanker Dubey, and Mohammad Pessarakli. 2012. "Reactive Oxygen Species, Oxidative Damage, and Antioxidative Defense Mechanism in Plants under Stressful Conditions." *Journal of Botany* 2012: 217037.

Sharwood, Robert E et al. 2016. "Temperature Responses of Rubisco from Paniceae Grasses Provide Opportunities for Improving C3 Photosynthesis." *Nature Plants* 2(12): 1–9.

Shen, Qingxi et al. 2001. "The Stress-and Abscisic Acid-Induced Barley Gene HVA22: Developmental Regulation and Homologues in Diverse Organisms." *Plant Molecular Biology* 45: 327–40.

Shi, Yiting, Yanglin Ding, and Shuhua Yang. 2018. "Molecular Regulation of CBF Signaling in Cold Acclimation." *Trends in Plant Science* 23(7): 623–37.

Silvera, Katia et al. 2010. "Evolution along the Crassulacean Acid Metabolism Continuum." *Functional Plant Biology* 37(11): 995–1010.

Simões, Joana et al. 2012. "Mass Spectrometry Characterization of an Aloe Vera Mannan Presenting Immunostimulatory Activity." *Carbohydrate Polymers* 90(1): 229–36.

Sofy, M R. 2016. "Effect of Gibberellic Acid, Paclobutrazol and Zinc on Growth, Physiological Attributes and the Antioxidant Defense System of Soybean (Glycine Max) under Salinity Stress." *International Journal of Plant Research* 6(3): 64–87.

Solomou, A D, R Germani, and M Georgakopoulou. 2020. "Ecological Value, Cultivation, Utilization and Commercialization of Aloe Vera in Greece." *JAPS: Journal of Animal & Plant Sciences* 30(4): 1013–23.

Steenkamp, Vanessa, and M J Stewart. 2007. "Medicinal Applications and Toxicological Activities of Aloe. Products." *Pharmaceutical Biology* 45(5): 411–20.

Teskey, Robert et al. 2015. "Responses of Tree Species to Heat Waves and Extreme Heat Events." *Plant, Cell & Environment* 38(9): 1699–712.

Vashisth, Divya et al. 2018. "Transcriptome Changes Induced by Abiotic Stresses in Artemisia Annua." *Scientific Reports* 8(1): 3423.

Wang, Dan et al. 2008. "Effects of Elevated CO2 on the Tolerance of Photosynthesis to Acute Heat Stress in C3, C4, and CAM Species." *American Journal of Botany* 95(2): 165–76.

Welling, Annikki, Thomas Moritz, E Tapio Palva, and Olavi Junttila. 2002. "Independent Activation of Cold Acclimation by Low Temperature and Short Photoperiod in Hybrid Aspen." *Plant Physiology* 129(4): 1633–41.

Winter, Klaus, Jorge Aranda, and Joseph A M Holtum. 2005. "Carbon Isotope Composition and Water-Use Efficiency in Plants with Crassulacean Acid Metabolism." *Functional Plant Biology* 32(5): 381–8.

Yamori, Wataru et al. 2006. "Effects of Rubisco Kinetics and Rubisco Activation State on the Temperature Dependence of the Photosynthetic Rate in Spinach Leaves from Contrasting Growth Temperatures." *Plant, Cell & Environment* 29(8): 1659–70.

Yeshi, Karma, Darren Crayn, Edita Ritmejeryt, and Phurpa Wangchuk. 2022. "Plant Secondary Metabolites Produced in Response to Abiotic Stresses Has Potential Application in Pharmaceutical Product Development." *Molecules* 27(1): 313.

Zapata, P J et al. 2013. "Characterisation of Gels from Different Aloe Spp. as Antifungal Treatment: Potential Crops for Industrial Applications." *Industrial Crops and Products* 42: 223–30.

11 Recent Advances in Molecular Markers Technology and its Applications in *Aloe vera*

Pawan Kumar and Jyoti Taunk

11.1 INTRODUCTION

Aloe vera (L.), a succulent plant member of the family Asphodelaceae, is recognized globally for its outstanding medicinal, cosmetic, and nutritional properties (Lawal et al., 2021). Over 530 species of the genus Aloe are known, and it is one of the ancient therapeutic floras acknowledged by human civilization, with chronological citations dating back over 6,000 years (Mehta, 2017; Kumar et al., 2022). Prehistoric civilizations, including Egyptians, Greeks, Romans, Indians, and Chinese, have utilized it. *Aloe vera* is thought to have emerged in the Arabian Peninsula, North Africa, and the Mediterranean region (Grace et al., 2015). It blooms well in tropical and subtropical weather conditions with nominal water needs, qualifying it as a perfect plant for drought-affected locations. It has glutinous, plump foliage that accumulates water, allowing it to endure in harsh conditions. Gel produced from its leaves, which is loaded with bioactive composites like aloin, acemannan, polysaccharides, anthraquinones, etc., has been deployed customarily in traditional medicine for centuries (Kumar et al., 2019). Its utility in healing wounds, hydration of skin, treatment of burns, eczema, psoriasis, digestive health benefits, management of diabetes, and boost in immunity have made it indispensable in both progressive healthcare and the cosmetic manufacturing (Yu et al., 2009; Svitina et al., 2019; Swaraj, 2024).

In recent years, the demand for *Aloe vera* has accomplished noteworthy growth worldwide, propelled by rising end-user awareness of its health benefits and its escalating use across a variety of sectors. In 2021, the *Aloe vera* market was evaluated to be ~ US$ 553.1 million (Majeed et al., 2024), which is anticipated to flourish at a compound annual growth rate (CAGR) of 9.40%, attaining ~ US$ 1,134.86 million by 2029 (www.databridgemarketresearch.com/reports/global-aloe-vera-market). The *Aloe vera* extract merchandise transcended US$ 3.08 billion in 2023 and is expected to witness a CAGR of more than 8.5% from 2024 to 2032 (www.openpr.com/news/3602663). With the burgeoning exigency for *Aloe vera* products globally, discerning its genetic constitution has turned out to be vital to warrant sustainable farming, upgrading product attributes, and raising its resilience to varied ecological environments.

In spite of its lucrative worth, *Aloe vera* is underexplored in terms of genetic and molecular characterization. Its genetic complexity, vegetative propagation, polyploidy, and lengthy life-cycle create obstacles for conventional breeding, restricting the scope of plant improvement efforts (Zonneveld, 2002; Nejatzadeh-Barandozi and Akbari, 2013). Contrary to chief crops and medicinal species, *Aloe vera* lacks a fully sequenced and well-annotated genome (as of September 2025, no result was observed at NCBI Genome database). This necessitates utilization of modern molecular tools to dissect its genetic architectonic, distinguish traits of interest, and assist conservation programmes. Assessment of molecular mechanisms, genetic mapping, and genome-wide classification could offer a scaffolding to tackle these problems. Various types of biomarkers are available to analyse genetic

 DOI: 10.1201/9781003603108-11

diversity in plants, *viz.* morphological, physiological, biochemical, and molecular markers. Yet, climatic conditions and developmental stages may influence these studies when the first three types (morphological, physiological, and biochemical) of markers are used. On the other hand, molecular markers (DNA sequences associated with a specific location within the genome) are not influenced by such factors. Therefore, these can be used as the most decisive tool for analysing genetic diversity and establishing strategies for speedy breeding programs.

Molecular markers include different DNA and RNA-based markers, viz. Random Amplified Polymorphic DNA (RAPD), Inter Simple Sequence Repeats (ISSR), Simple Sequence Repeats (SSR), Amplified Fragment Length Polymorphism (AFLP), Restriction Fragment Length Polymorphism (RFLP), Expressed Sequence Tags (EST), Single Nucleotide Polymorphism (SNP), cDNA Amplified fragment length polymorphism, Diversity array technology, etc. As *Aloe vera* is primarily propagated vegetatively, it leads to low genetic variability, and molecular markers can help to assess genetic diversity among wild and cultivated *Aloe* species, ensuring the conservation of unique genotypes. Further, due to lack of standardized varieties, molecular markers can help to distinguish different *Aloe vera* genotypes, which can be helpful for the authentication of high-yielding and bioactive compound-rich varieties. This is critical for ensuring the purity of commercial *Aloe vera* products and for preventing adulteration in herbal markets. Molecular marker-based tools allow researchers to explore genetic diversity, map traits of agricultural and industrial significance, and to understand the underlying genomic mechanisms that govern the plant's growth, stress tolerance, and bioactive compound production. This chapter explores the significant progress in *Aloe vera* research at molecular levels along with advancement in technologies for future progress of this commercially important species.

11.2 ADVANCEMENTS IN MARKER DEVELOPMENT IN *ALOE VERA*

11.2.1 Traditional Marker Systems

Throughout time, molecular markers have established as a significant asset for confronting challenges related to *Aloe vera* improvement, allowing meticulous classification, conservation, breeding, and enhancement of medicinal attributes. RAPD, ISSR and AFLP etc. have been utilized as traditional marker systems in plant-based research. RAPD produce random DNA fragments, which are useful in finding genetic polymorphisms without previous sequence information, while the ISSR markers make use of microsatellite regions to amplify DNA fragments, and AFLP requires restriction digestion of genomic DNA accompanied by selective polymerase chain reaction (PCR) amplification (Rana et al., 2018). Various studies exploit these conventional markers to spot and examine genetic variations. In 2017, 12 *Aloe vera* accessions from various regions of India were investigated using RAPD and ISSR-based markers (Das et al., 2017). The study showed 71.8% polymorphism, demonstrating a great intensity of genetic adaptation among the accessions. Similarly, ISSR markers were used to assess genetic diversity amongst wild and cultivated *Aloe vera* accessions, helping researchers to identify distinct genotypes and assess the genetic stability of tissue-cultured plants (Badar et al., 2017; Khatun et al., 2018). Genetic diversity and population structure of 60 *Aloe vera* genotypes were evaluated using AFLP markers, which offered insights into the genetic connection and disequilibrium within the species (Kumar et al., 2022). Nevertheless, these traditional markers have shortcomings, like low reproducibility, dominant inheritance (in RAPD and ISSR), and high methodological complexity (in AFLP). To surmount these disadvantages, advanced markers such as SSR and SNPs have been developed, which proffer better accuracy and relevance in *Aloe vera* research.

11.2.2 Simple Sequence Repeat (SSR) Markers

The SSR markers, also known as short tandem repeats (STRs) or microsatellites, consist of short, tandemly repeated DNA sequences that are characteristically stretched up to 1–6 base pairs, dispersed all through the genome, both in coding and non-coding regions. They provide broad coverage and have the capability to spot genetic variations across loci (Kalia et al., 2011). The SSR markers are meticulous, offer co-dominant inheritance, and have admirable reproducibility. The SSRs pose an efficient mode to evaluate genetic diversity in *Aloe vera*, due to their high level of polymorphism, which appears from variations in the number of repeating units while mapping genes (Rakoczy-Trojanowska and Bolibok, 2004).

In view of the fact that *Aloe vera* is not a well sequenced species, SSR markers have been developed using cross-species transfer from related monocots like *Asparagus officinalis*, *Yucca filamentosa*, and *Agave americana* (Kumar et al., 2015; Adienge et al., 2019). Information obtained from the transcriptomics experiments and genome skimming has also been used to find new SSR markers (Choudhri et al., 2018). Due to the extensive commercialization of *Aloe vera*, SSR markers have been applied for varietal fingerprinting so as to distinguish between genotypes (Kumar et al., 2015). Further, validation studies have used SSR markers to avert adulteration of *Aloe vera* products in the herbal and pharmaceutical sectors (Jha and Mughees, 2023). The SSR markers were also found to be associated with biochemical traits like aloin content, which is vital for medicinal properties (Kumar et al., 2016). However, there are also some limits of SSR markers, viz., high developmental cost, labor-intensiveness, and requirement of extensive bioinformatics analysis to identify suitable SSR loci in the genome.

11.2.3 Single-nucleotide Polymorphism (SNPs)

The SNP-based marker system has turned out to be a preference for molecular research in complex plants like *Aloe vera*. It is the most profuse kind of genetic variation, signifying a modification in a single nucleotide base at a particular position in the DNA sequence (Syvänen, 2001). This type of variation can come about in any coding, non-coding, or intergenic regions of the genome, and is often bi-allelic. The SNP polymorphisms are extremely stable and widely dispersed all over the genome, making them an indispensable technique in molecular studies (Sohi and Singh, 2021). Still, SNP marker studies in *Aloe vera* are in their primitive stages due to the unavailability of a whole genome sequence.

In plant species like *Aloe vera*, SNPs are useful in Genome-Wide Association Studies (GWAS) for connecting particular genotypic variants to phenotypic traits, like disease resistance, stress tolerance, and secondary metabolite production. After identification, SNP markers can be scrutinized using a variety of genotyping platforms, such as microarrays or NGS. In spite of their cost and technical hassle, SNPs are regarded as the benchmark for marker studies due to their ubiquitousness, reliability, and likelihood for automation, making them invaluable in modern genomics and plant breeding studies. The SNP markers have exposed low genetic variability within cultivated *Aloe vera* populations, signifying clonal propagation (Micheni, 2015). In wild populations, SNP markers have been detected to have higher genetic variation, beneficial for conservation and breeding efforts (Micheni, 2015). Population structure report using SNPs has detected sub-populations that can be beneficial for germplasm characterization (Micheni, 2015). Further, SNP markers have been correlated to vital secondary metabolites, such as aloin, polysaccharides, and antioxidants that are important for the pharmaceutical sector (Micheni, 2015). Molecular research has also elucidated SNPs linked to drought tolerance, disease resistance, and stress-responsive genes (Jaiswal et al., 2021; Choudhri et al., 2018). Recently, RNA Sequencing (RNA-Seq) based SNP discovery has bestowed insights into gene expression changes under varied environmental conditions, which laid the foundation for further amelioration in *Aloe vera* (Zou et al., 2021; Almeida et al., 2023).

11.2.4 Next-generation Sequencing (NGS) Based Markers

The introduction of NGS technologies has promoted accelerated marker discovery, promoting whole-genome sequencing, transcriptome analyses, and high-throughput SNP identification in plant species. The expansion of NGS-based markers has considerably enhanced genetic studies in plants, including *Aloe vera*, by smoothing the process of high-throughput genotyping, GWAS, and Marker-Assisted Breeding (MAB) (Prasad et al., 2024). Due to the targeted sequencing of explicit coding and non-coding regions of the genome, these markers have turned out to be crucial for genetic diversity evaluation, cultivar classification, evolutionary studies, and functional characterization investigations in *Aloe vera*. The improvements in RNA-Seq technologies have made it convenient to produce large-scale transcriptomic datasets, isolate ESTs, and identify precise markers for genetic studies and molecular breeding. Transcriptome-derived markers offer high-resolution insights into gene expression and trait-associated differences in *Aloe vera*. The linkage of transcriptome-based markers with the actively transcribing genes makes them highly applicable for examining functional attributes like biosynthesis of medicinal compounds, abiotic stress tolerance, and disease resistance. NGS has been applied to analyse the endophytic bacterial communities within *Aloe vera*. By exploring the 16S rDNA sequences (V3–V4 regions), predominant bacterial genera like Proteobacteria, Firmicutes, Actinobacteria, and Bacteroidetes have been identified. This meta-genomic methodology has provided insights into the microbial multiplicity linked with *Aloe vera*, which can impact plant health and secondary metabolite production (Akinsanya et al., 2015; Marokane-Radebe et al., 2024).

11.2.5 Epigenetic Markers

Epigenetic markers are alterations to the DNA that do not alter the primary genetic sequence but impact gene expression and the phenotype of the organism. This has the potential to regulate important traits of *Aloe vera*, such as biosynthesis of medicinal compounds, disease resilience, and growth under adverse environmental conditions (Chen et al., 2017; Hao and Xiao, 2018). The epigenetic process can also affect seed germination, flowering, and other metabolic processes in plants via mechanisms such as DNA methylation, histone modification, and chromatin remodelling. DNA methylation controls the expression of genes involved in secondary metabolite production, while histone modifications may impact the plant's adaptation to abiotic stresses (Shahrajabian et al., 2019). These epigenetic markers can be applied in breeding programs designed to ameliorate medicinal characteristics of *Aloe vera*, as they can be inherited and affect the expression of traits across generations without altering the DNA sequence itself. Research in other medicinal plants has found that methylation levels can impact biosynthesis of phenolic and flavonoid compounds, which may also be applicable to *Aloe vera* (Yang et al., 2018; Koirala et al., 2016).

Future prospects of using Chromatin Immunoprecipitation Sequencing (ChIP-Seq) could help categorize histone modification configurations in *Aloe vera*. This method involves cross-linking of DNA and histone proteins *in vivo*, which is followed by chromatin shearing and immunoprecipitation using specific antibodies targeting modified histones. The enriched DNA fragments are then sequenced to create high-resolution histone modification maps across the genome (Park, 2009). In case of *Aloe vera*, ChIP-Seq can help in the identification of regulatory elements associated with stress responses, secondary metabolite biosynthesis, and developmental processes by identifying active (e.g., H3K4me3, H3K27ac) or repressive (e.g., H3K9me3, H3K27me3) histone marks (Du et al., 2013). Knowledge of these modifications can help illuminate gene expression dynamics and probable mechanisms influencing *Aloe vera*'s medicinal properties and resilience towards environmental conditions.

11.2.6 Reduced-representation Sequencing Methods for Marker Discovery

Approaches such as Restriction-site Associated DNA Sequencing (RAD-Seq) and Genotyping-by-Sequencing (GBS) are reduced representation sequencing methodologies that allow genome-wide SNP discovery and precise genotyping in non-model organisms like *Aloe vera*, which is short of a fully sequenced reference genome. These techniques have been exploited in many species for genetic diversity analyses, population structure examination, and marker development (He et al., 2014; Andrews et al., 2016). The RAD-Seq data have been used to deduce the phylogenetic connection of *Aloe vera* within the Aloaceae family. Genotyping-by-Sequencing is a sophisticated NGS-based technique that resourcefully detects genome-wide SNPs by dropping genome complexity using restriction enzymes. It facilitates high-resolution genetic mapping, GWAS, and marker-aided selection in diverse organisms, including non-model plants (He et al., 2014). GBS is scalable, cost-efficient, and applicable even in species with complex or unsequenced genomes. Recent progression incorporates machine learning and pan-genomics to augment variant calling precision and genotype imputation (Upadhyaya et al., 2024). Amalgamation of multi-omics approaches, including genomics, transcriptomics, proteomics, and metabolomics, will cultivate more wide-ranging marker systems in the genus Aloe.

11.2.7 Marker Development Using Bioinformatics Tools and Machine Learning Algorithms

Molecular marker designing using bioinformatics and machine learning (ML) has revolutionized genomics by enabling efficient identification, selection, and validation of genetic markers for various applications, including genetic mapping, trait association studies, breeding programs, and evolutionary analysis. The traditional methods of marker discovery relied on labour-intensive experimental techniques, but advancements in bioinformatics and ML have significantly improved the speed, accuracy, and scalability of this process. Computational tactics, including genome annotation, phylogenetics, and marker-trait associations, have enhanced the exactness of genetic research (Chavhan et al., 2024). Moreover, machine learning algorithms are now being applied to envisage gene functions, catalogue genetic variants, and optimize marker assortment (Okser et al., 2014).

Bioinformatics plays a crucial role in identifying different types of molecular markers, including SNPs, SSRs, Insertions/Deletions (Indels), Copy Number Variations (CNVs), and Structural Variants (SVs) (Bhat et al., 2022). Various bioinformatics tools aid in detecting these markers from genomic data. Variant calling tools such as GATK, FreeBayes, and SAMtools help identify SNPs and Indels, while SSR detection tools like MISA and TRF allow for the discovery of repeat motifs in DNA sequences (Yao et al., 2020). Additionally, structural variants and CNVs can be identified using DELLY, LUMPY, and CNVnator. These tools help researchers extract valuable information from raw genomic data to identify markers relevant to their study objectives.

The process of marker development begins with genomic data acquisition, where researchers obtain sequence data from public repositories such as NCBI GenBank, Ensembl, the 1,000 Genomes Project, Phytozome, and Aloe Genome Project, or raw reads generated from Illumina, PacBio, and Oxford Nanopore. Generally, reads are aligned against the reference genome using bioinformatics tools such as BWA, Bowtie2, and Minimap2, but in the case of Aloe vera, no reference genome is available at NCBI; hence, the raw reads are assembled first using the de novo assembly approach. This generated assembly will work as a reference genome and will further be used to align raw reads. Once the sequences are aligned, variant calling algorithms detect genetic variations across different individuals or populations. However, identifying variations alone is not sufficient – annotation tools such as ANNOVAR, SnpEff, and Variant Effect Predictor (VEP) help determine the functional impact of these variations by mapping them to genes and regulatory elements.

Once potential markers are identified and annotated, the next step is their validation and experimental testing. Bioinformatics tools assist in designing primers and probes for molecular assays to confirm marker presence in the genome. Primer3 (Untergasser et al., 2012) and Primer-BLAST (Ye et al., 2012) are widely used for designing primers for PCR, qPCR, and high-throughput genotyping. In the case of SSR markers, tools such as BatchPrimer3 are employed for designing flanking primers. These markers are then experimentally validated in PCR-based genotyping assays, sequencing-based methods, or hybridization techniques to ensure their accuracy and reproducibility.

Machine learning has further enhanced the marker discovery process by enabling predictive modeling, feature selection, and functional annotation of genetic markers. Supervised learning models such as Random Forest (RF), Support Vector Machines (SVM), and XGBoost are frequently used to predict marker-trait associations in GWAS and Genomic Selection. These models analyze large-scale genomic datasets to determine which markers are most strongly associated with traits of interest, such as disease resistance, drought tolerance, or yield improvement. Unsupervised learning techniques, including K-means clustering and Principal Component Analysis (PCA), help classify genetic populations and identify subgroups based on genetic similarity. This approach is valuable in breeding programs and population genetics, where understanding genetic structure is essential for selecting diverse parent lines.

Deep learning techniques, particularly Convolutional Neural Networks (CNNs) and Recurrent Neural Networks (RNNs), are also gaining prominence in molecular marker research. These models analyze DNA sequence motifs, chromatin accessibility, and regulatory regions to predict the functional importance of genetic variations. For example, CNNs can detect SNPs that alter transcription factor binding sites, while RNNs can predict long-range genomic interactions affecting gene expression. Tools such as DeepSEA and RegulomeDB leverage deep learning to predict the regulatory effects of SNPs, helping researchers prioritize functionally significant markers.

In the case of *Aloe vera*, little progress has been made in this direction, with a huge area of possibilities to apply such models for marker development and genetic prediction. The deep learning-based attempts for disease prediction in *Aloe vera* plants have achieved an accuracy of 96.59% (Saini and Agarwal, 2024), which guarantees quick detection of leaf diseases, prospectively reducing crop losses. Scientific investigations have discovered the use of Federated Learning combined with Convolutional Neural Networks (CNNs) to detect and categorize *Aloe vera* leaf diseases (Sharma et al., 2023; Upadhyay et al., 2024(b); Thapliyal et al., 2024). This novel approach tackles multiple leaf diseases by utilizing datasets from various resources, thereby enhancing collaborative learning without relying on centralized data storage. Another ML-based model has exploited the EfficientNetB7 architecture to catalogue *Aloe vera* leaf diseases, like leaf spot and aloe rust, with a classification correctness of 94.11% (Malek et al., 2023). The dataset encompassed 2,770 images of diseased leaves, and the model's performance highlights the ability of deep learning in plant disease diagnostics. These tools augment the effectiveness of *Aloe vera* breeding, conservation, and commercial value by incorporating a data-driven process of decision-making in genetic improvement programs to ensure the selection of superior genotypes with high medicinal value, better stress tolerance, and disease resistance. As marker systems resume to progress, the future of *Aloe vera* research will profit from more accurate and economical genetic tools, eventually contributing to its worldwide mercantile and therapeutic applications.

AFLP: Amplified Fragment Length Polymorphism, ITS: Internal Transcribed Spacer, ISSR: Inter Simple Sequence Repeats, matK: Maturase K, RAPD: Random Amplified Polymorphic DNA, rbcL: Ribulose-1,5-bisphosphate carboxylase/oxygenase large subunit, rDNA: Ribosomal DNA, SCAR: Sequence Characterized Amplified Region, SSR: Simple Sequence Repeats

11.3 APPLICATIONS OF MOLECULAR MARKERS IN *ALOE VERA*

Molecular marker systems have modernized the study of *Aloe vera*, delivering visions into its genetic makeup, refining breeding efficacy, and certifying the sustainability of its curative and business-related applications. Besides, their amalgamation with high-throughput sequencing and bioinformatics empowers GWAS and genomic selection, flagging the way for precision-driven *Aloe vera* advancement programs. Progressive marker technologies like NGS can further augment the potential of *Aloe vera* in the biotechnology and pharmaceutical sectors. An impression of applications of molecular markers in *Aloe vera* is presented in the following section along with a brief representation in Figure 11.1.

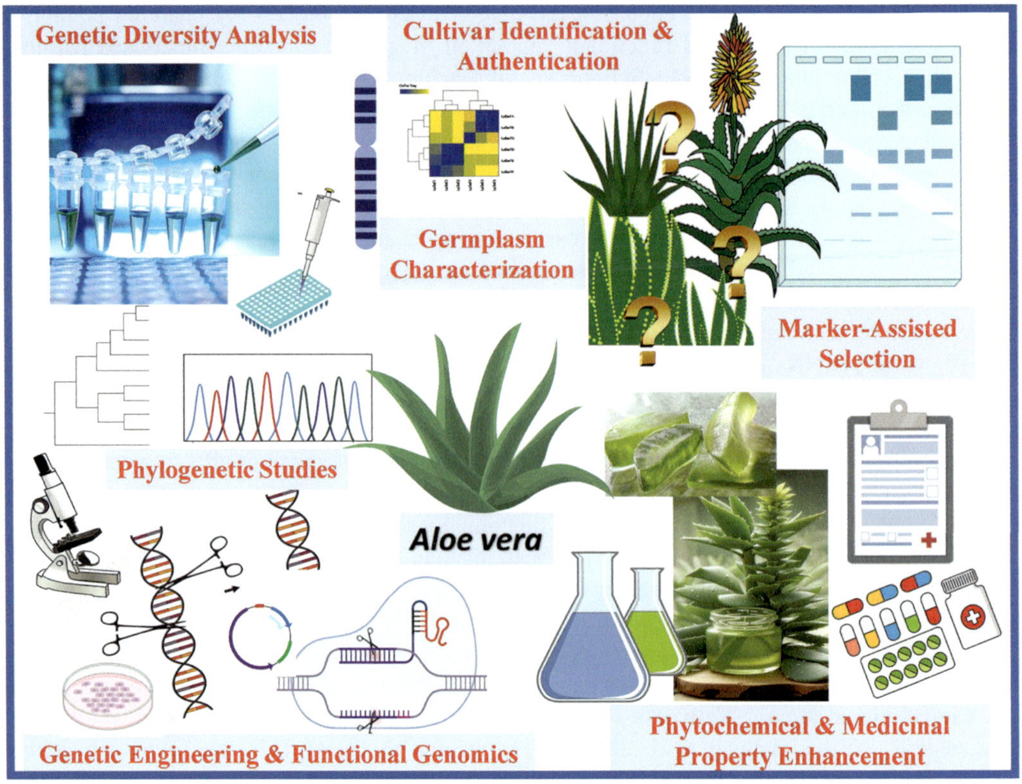

FIGURE 11.1 Applications of molecular markers in *Aloe vera* improvement.

11.3.1 Genetic Diversity and Germplasm Characterization

Genetic diversity is the basis of plant breeding, conservation, and sustainable utilization, and its evaluation in *Aloe vera* is critical for warranting the long-standing practicability of this economically cherished species. Molecular markers have been extensively employed to assess genetic variation within and among different *Aloe vera* populations, providing valuable insights into the genetic makeup of germplasm collections, allowing researchers to differentiate between closely allied genotypes, classify unique genotypes, and ascertain phylogenetic relationships. The SSR markers established rich genetic diversity among *Aloe vera* genotypes from varied agro-climatic zones in India, with similarity indices ranging from 0.53 to 0.93 (Kumar et al., 2015).

Characterizing germplasm at the molecular level will help to categorize genetic resources into discrete groups, which delivers operative conservation strategies such as *in situ* and *ex situ* preservation. The AFLP markers were employed to assess the genetic diversity and population structure of 60 *Aloe vera* genotypes, which provided insights into the genetic interactions, helped to cluster them into distinct groups, and discovered significant genetic variability within the population (Kumar et al., 2022). Furthermore, molecular markers helped in the identification of elite cultivars with important traits like high gel yield, superior medicinal properties, and stress tolerance, which are central for commercial production. For example, Ganesh and Alagukannan (2006) evaluated 21 *Aloe vera* ecotypes across South India, illuminating substantial variations in growth and yield parameters. The ecotype "TCY" demonstrated superior attributes, including greater plant height, number of leaves, and leaf weight, while "TVM" recorded the highest leaf thickness, leaf volume, and gel yield per leaf (Ganesh and Alagukannan 2006). These findings accentuate the importance of molecular characterization in selecting elite *Aloe vera* cultivars with desirable traits.

TABLE 11.1
Molecular markers used for genetic/genomic studies in *Aloe vera*

Marker Type	Type of Study	No. of Markers	Type and No. of Genotypes	Reference
RAPD and ITS	Molecular characterization	20 RAPD primers; ITS regions	10 *Aloe vera* genotypes from different regions in India	Das et al., 2017
RAPD and ISSR	Genetic fidelity assessment	10 RAPD primers and 10 ISSR primers	In vitro-propagated *Aloe vera* plantlets; Wild type and in vitro regenerated *Aloe vera* plants	Khatun et al., 2018; Badar et al., 2017
RAPD and ISSR	Genetic diversity assessment	64 RAPD, 25 ISSR	12 elite accessions of *Aloe barbadensis* Mill. collected from different geographical regions of India; 12 accessions from different regions of India	Deore, 2007; Bhaludra et al., 2014; Kumar et al., 2016
RAPD and ISSR	Genetic stability assessment	10 RAPD primers and 10 ISSR primers	*In vitro* propagated *Aloe vera* plantlets; number not specified	Rathore et al., 2011
RAPD and ISSR	Genetic diversity and population structure assessment	10 RAPD primers and 7 ISSR primers	15 populations of three *Aloe* species (*A. secundiflora*, *A. turkanensis*, and *A. scabrifolia*) from selected arid and semi-arid lands (ASALs) of Kenya; 30 samples per population	Adienge et al., 2019
RAPD and ISSR	Genetic diversity assessment	24 RAPD primers and 6 ISSR primers	*Aloe vera* genotypes from different provinces of Himachal Pradesh	Rana, S. and Kanwar, 2017
RAPD	Genetic variation analysis	19	15 Aloe species from Karachi, Pakistan	Rao et al., 2023
RAPD	Genetic diversity assessment	10	10 *Aloe vera* accessions from different regions of Iran	Nejatzadeh-Barandozi et al., 2012; Nejatzadeh-Barandozi, 2013
RAPD	Morphological and genetic characterization	10	Micropropagated *Aloe vera* plantlets	Das et al., 2015
RAPD	Genetic fidelity assessment	10	*In vitro* regenerated *Aloe barbadensis* plantlets	Sahoo, S. and Rout, 2014
RAPD	Genetic diversity assessment	10	8 Aloe species	Ezzat et al., 2016
RAPD	Genetic diversity assessment	14	38 *Aloe barbadensis* accessions from various regions of India, including Rajasthan	Chandra and Choudhary, 2014
RAPD	Genetic diversity assessment	19	15 Aloe accessions from various regions of India	Panwar et al., 2013
RAPD	Genetic diversity assessment	10	5 Aloe species from Kenya, including *A. secundiflora*, *A. lateritia*, *A. turkanensis*, *A. vera*, and *A. kedongensis*	Micheni, 2015

(*continued*)

TABLE 11.1 (Continued)
Molecular markers used for genetic/genomic studies in *Aloe vera*

Marker Type	Type of Study	No. of Markers	Type and No. of Genotypes	Reference
RAPD and Morphological	Genetic diversity assessment	20 RAPD primers	11 Aloe accessions from India, including *A. vera*, *A. perryi*, *A. lotus*, *A. zeylanicum*, and seven strains of *A. vera*	Nayanakantha et al., 2010
AFLP	Genetic diversity assessment	8 AFLP primer combinations	12 *Aloe vera* accessions from different locations in Madhya Pradesh, India	Tripathi et al., 2011
AFLP and Morphological	Genetic diversity assessment	-	Aloe species	Saljooghianpour, 2013
SCAR	Molecular authentication and quality control	-	Crocus sativus and Aloe barbadensis; number of samples not specified	Gul et al., 2016
SSR	Genetic diversity analysis	20	12 accessions from different agro-climatic zones of India	Kumar et al., 2015
SSR	Method development	5	18 *Aloe vera* genotypes from different regions of India	Kumar et al., 2018
matK, rbcL, and ITS	Molecular identification and phylogenetic analysis	3 markers types	*Aloe shadensis*	Algarni, 2022
Ribosomal DNA (rDNA) sequences	Genetic variation analysis	Internal Transcribed Spacer 1 (ITS1) region	*Aloe vera* plants from Japan, USA, and Egypt	Yagi et al., 2006
Target Capture Sequencing	Phylogenetic analysis tool development	-	Various Aloe species	Woudstra et al., 2021

The statistics obtained from genetic diversity studies is further applied to direct breeding programs, safeguarding the selection of superior genotypes, while preserving genetic variability. Rana et al. (2018) discussed the deployment of molecular markers in assessing genetic diversity in *Aloe vera*, emphasizing their potential in ascertaining superior genotypes with suitable medicinal traits. Another study by Singh et al. (2013) scrutinized the molecular properties and bioactivity scores of antioxidant compounds isolated from *Aloe vera*. The lead compounds were identified through molecular marker-based prediction, which contributed towards the selection of *Aloe vera* varieties with enhanced medicinal properties. Molecular studies have also identified genes associated with stress responses in *Aloe vera*; for example, the DREB (Dehydration-Responsive Element-Binding) gene isolated from *Aloe vera* plays a role in enhancing cold stress tolerance (Wang and He, 2007).

Furthermore, understanding genetic diversity in *Aloe vera* aids in the protection of traditional landraces and wild relatives, which are essential reservoirs of unique alleles that may offer resistance to diseases, pests, and environmental stresses. Progression in DNA barcoding has facilitated the accurate identification of *Aloe vera* and its relatives (Algarni, 2022; Woudstra et al., 2024). A customized target capture sequencing tool has been established to enhance molecular identification, thereby assisting conservation efforts by precisely distinguishing between species and averting the loss of genetic diversity (Woudstra et al., 2021). As a result, molecular marker-based germplasm characterization plays a prominent role in the sustainable management and genetic improvement of *Aloe vera*, augmenting its utility in pharmaceutical, cosmetic, and nutraceutical sectors.

11.3.2 Cultivar Identification and Authentication

Cultivar identification and verification are critical in *Aloe vera* due to its widespread commercial use in pharmaceuticals, cosmetics, and health products. The incidence of adulterants and morphologically analogous species in the marketplace imposes accurate detection means to confirm product purity and value. Customarily, morpho-chemical analysis methods are time-consuming and occasionally fail to differentiate closely related cultivars, making molecular markers a critical tool for cultivar authentication (Haider, 2011). On the other hand, molecular markers help in ascertaining unique genetic fingerprints, permitting science and industry professionals to confirm the authenticity of plant material at the genetic level. Numerous DNA-based molecular markers have been utilized to segregate *Aloe vera* cultivars from related species such as *Aloe ferox* and *Aloe arborescens* (Treutlein et al., 2003; Nejatzadeh-Barandozi et al., 2012; Ezzat et al., 2016; Rao et al., 2023).

DNA barcoding, mainly through the use of conserved gene regions such as ITS (Internal Transcribed Spacer) and matK (Maturase K), has surfaced as a reliable approach for rapid species identification (Algarni, 2023). The barcoding of chloroplast DNA has been explored for species identification within the Aloe genus. A study by Das and Joshi (2023) evaluated two chloroplast genes, matK and rbcL, to obtain specific barcodes for various *Aloe* species, including *Aloe vera*, *Aloe ferox*, and *Aloe arborescens*. Evolutionary analyses based on these sequences have allowed the clustering of various Aloe species into distinct clades, enabling accurate identification and delineation.

Molecular markers also aid in distinguishing high-yielding or medicinally superior cultivars, certifying that only the best varieties are chosen for large-scale cultivation and commercialization. Validation of *Aloe vera* cultivars is vital in the pharmaceutical industry, where contamination with other species can compromise the medicinal efficacy of the products. The International Aloe Science Council (IASC) has issued statements on the significance of correct identification and evaluation of *Aloe vera* (Pandey and Singh, 2016). The IASC governs a third-party certification program intended to authenticate the purity and quality of *Aloe vera* products, accentuating the need for correct authentication in the industry. Regulating genetic authentication methods not only boosts consumer trust but also supports regulatory compliance by delivering verifiable proof of cultivar distinctiveness. Furthermore, the precise identification of elite cultivars allows the establishment of licensed propagation programs, warranting genetic fidelity in tissue culture-based mass propagation (Rohela et al., 2022). By integrating molecular marker methods, researchers and industry stakeholders can safeguard the genetic integrity of *Aloe vera* cultivars, ultimately enhancing their commercial value and sustainable utilization.

11.3.3 Identification of Genes and Markers for Marker-assisted Selection (MAS)

Marker-assisted selection is a powerful tool in plant breeding that exploits molecular markers to fast-track the assortment of desirable traits in crops, including medicinal plants like *Aloe vera*. Since *Aloe vera* is primarily propagated vegetatively rather than through seeds, conventional breeding approaches are often slow and inefficient. Hence, MAS offers a precise, rapid, and cost-effective method for selecting superior genotypes based on genetic markers linked to key traits such as high gel yield, enhanced bioactive compound production, disease resistance, and abiotic stress tolerance (e.g., drought or salinity resistance) (Collard and Mackill, 2008; Nair and Rabbani, 2024).

The use of MAS in *Aloe vera* breeding can allow breeders to identify superior plants even at the early growth stages, lowering the time and labor needed for field trials. By screening a large number of genotypes at the molecular level, MAS assists in identifying and propagating elite cultivars with the highest medicinal and commercial value. Therefore, understanding the reproductive biology of *Aloe vera* is vital for developing effective breeding strategies. Research on the floral biology and breeding behaviour of *Aloe barbadensis* has provided foundational knowledge for breeding programs (Tapre et al., 2020). Identification of markers linked to genes responsible for the biosynthesis of bioactive compounds like aloin, acemannan, and polysaccharides is essential to enable the selection of genotypes with enriched therapeutic properties. A comprehensive transcriptome analysis of *Aloe vera* was piloted to explore the genetic basis of its medicinal properties (Choudhri et al., 2018). This study utilized *de novo* sequencing and assembly to describe the transcriptome, pinpointing genes involved in the biosynthesis of secondary metabolites, including aloin and saponins. The research highlighted the upregulation of aloin biosynthesis pathway-related genes upon methyl jasmonate induction, signifying their role in the plant's defence processes. In another study, the expression of a glucomannan mannosyl transferase gene (GMMT) from *Aloe barbadensis* Miller was checked (Salinas et al., 2018). This gene is involved in the biosynthesis of acemannan, a key acetylated glucomannan polysaccharide contributing to the plant's succulence and water retention capabilities (González-Delgado et al., 2023). The results offered insights into the molecular mechanisms underlying acemannan synthesis.

Additionally, MAS facilitates the introgression of beneficial traits from wild or related Aloe species, refining the adaptability of *Aloe vera* to diverse environmental conditions. Likewise, the concept of "introgressiomics" has been proposed as a comprehensive approach to utilize crop wild relatives in breeding programs (Prohens et al., 2017). This strategy involves the systematic introgression of beneficial genes from wild species into cultivated varieties, enhancing traits such as stress tolerance and adaptability.

Another noteworthy benefit of MAS in *Aloe vera* breeding is its function in maintaining genetic purity in tissue culture propagation programs. Since micropropagation is widely used for large-scale production, molecular markers help in confirming the genetic stability of regenerated plants, ensuring that selected elite traits are preserved. Also, MAS supports the conservation and sustainable utilization of *Aloe vera* genetic resources by allowing the identification and protection of rare or superior genotypes. Further, GWAS and quantitative trait loci (QTL) mapping enhance our understanding of genetic variations associated with desirable traits. Altaf et al. (2024) discussed the position of molecular markers in modern plant breeding, highlighting the importance of QTL mapping and GWAS in categorising genetic loci linked with complex traits such as stress tolerance and yield improvement. The review emphasizes how these methodologies have been effectively exploited in various crops to dissect complex traits, offering a framework that could be tailored for *Aloe vera* improvement. A meta-analysis protocol described by Kumari et al. (2024) concentrates on integrating QTL data from multiple studies to detect consistent genetic regions associated with specific traits. This approach augments the robustness of QTL identification and could be implemented in *Aloe vera* breeding programs once sufficient data becomes available. With advancements in genome screening, the future of *Aloe vera* breeding will increasingly depend on MAS to develop improved varieties with optimized growth, medicinal properties, and resilience to environmental stresses. Thus, MAS is a priceless approach for boosting the efficacy, accuracy, and success rate of *Aloe vera* breeding programs.

11.3.4 Phytochemical and Medicinal Property Enhancement

Molecular markers play a fundamental role in enriching the phytochemical and medicinal properties of *Aloe vera* by expediting the identification and selection of genotypes with superior bioactive compound production. *Aloe vera* is well-known for its therapeutic benefits, primarily due to the presence of bioactive compounds such as aloin, acemannan, polysaccharides, flavonoids, anthraquinones, and amino acids, which exhibit antibacterial, antiviral, anti-inflammatory, and wound-healing properties (Babu and Noor, 2020). The concentration of these compounds differs considerably among different *Aloe vera* genotypes, making it vital to employ molecular marker techniques to identify and propagate high-yielding medicinal cultivars. By associating specific markers to genes involved in phytochemical biosynthesis pathways, researchers can proficiently screen *Aloe vera* germplasm and hand-pick varieties with improved therapeutical potential. For example, markers associated with the *PAL* (Phenylalanine Ammonia-Lyase) gene, a key enzyme in flavonoid biosynthesis, can be used to spot genotypes with higher antioxidant properties (Thiyagarajan et al., 2016; Yang et al., 2022). Likewise, genes controlling the production of acemannan, a polysaccharide known for its immune-stimulatory and wound-healing effects, can be targeted using marker-assisted selection to enhance these therapeutic properties in cultivated varieties (Quezada et al., 2017; Liu et al., 2019).

Molecular tools also aid in optimizing plant growth conditions to maximize bioactive compound synthesis. Studies using transcriptomic and metabolomic approaches have found key regulatory genes involved in secondary metabolite production in *Aloe vera*, paving the way for the genetic improvement of elite cultivars through breeding or genetic engineering techniques (Ushasree et al., 2024; Batra and Dey, 2025). Besides, the application of biotechnological interventions such as tissue culture and genetic transformation, guided by molecular markers, enables the large-scale propagation of high phytochemical-yielding *Aloe vera* plants with coherent medicinal properties (Saljooghianpour and Javaran, 2013; Amoo et al., 2013; Sharma et al., 2014).

Another significant application of molecular markers is to ensure the genetic purity and stability of these high-value medicinal cultivars. Since adulteration or genetic drift during vegetative propagation can alter phytochemical profiles, markers help maintain the authenticity and efficacy of *Aloe vera* products in the pharmaceutical and cosmetic industries. With advancements in genomics and transcriptomics, future research can further refine the identification of functional genes responsible for medicinal compound biosynthesis, directing the development of genetically improved *Aloe vera* varieties with optimized medicinal potential.

11.3.5 Genetic Engineering and Functional Genomics

The bioactive compounds of *Aloe vera* are widely used in pharmaceuticals, cosmetics, and nutraceuticals. Hence, genetic engineering-based targeted modification is favoured to augment the medicinal, agricultural, and industrial capability of *Aloe vera*. The transfer of specific genes through techniques such as *Agrobacterium*-mediated transformation or CRISPR-Cas9 genome editing offers precise control over gene function (Li et al., 2000; Arora and Narula, 2017; Zlobin et al., 2020). However, the genetic improvement of *Aloe vera* has been tricky due to its predominant vegetative propagation and complex genome structure. A study by Jangra et al. (2023) reported a successful *Agrobacterium tumefaciens*-mediated genetic transformation method for *Aloe vera*. This protocol attained a transformation efficiency of 83.33% using stem explants, commemorating a significant improvement over previous such attempts. The study utilized the *gusA* reporter gene to confirm successful integration and expression, setting the groundwork for future genetic modifications aimed at enhancing desirable traits in *Aloe vera*.

Functional genomics aims to understand gene expression and regulation by employing techniques such as transcriptomics, proteomics, and metabolomics. In *Aloe vera*, transcriptome sequencing (RNA-Seq) has been used to detect genes accountable for secondary metabolite biosynthesis, including those involved in the production of bioactive compounds acemannan (a polysaccharide with immunostimulatory properties) and aloin (a potent laxative compound). Evidently, a study by Choudhri et al. (2018) utilized *de novo* sequencing and assembly to characterize the *Aloe vera* transcriptome. This research identified genes involved in the biosynthesis of secondary metabolites, such as saponins and anthraquinones, which are key to the plant's therapeutic effects. The study also analyzed gene expression profiles, providing insights into the metabolic pathways active in various tissues. In another vital study, the draft genome sequence of *Aloe vera* was reported, comprising approximately 12.93 Gbp and harboring 86,177 protein-coding genes (Jaiswal et al., 2021). This research provided insights into the genetic basis of *Aloe vera's* unique characteristics, such as drought resistance and medicinal properties.

Additionally, a study focused on the isolation and functional characterization of a novel aldo-ketoreductase (AvAKR) from *Aloe vera* leaves through heterologous gene expression in *Escherichia coli* (Jangra et al., 2023). High-throughput sequencing technologies and bioinformatics tools help in determining gene regulatory networks, allowing the identification of transcription factors and signaling pathways involved in stress responses and phytochemical biosynthesis. A study by Choudhri et al. (2018) highlighted the upregulation of genes associated with the aloin biosynthesis pathway upon methyl jasmonate induction, suggesting their role in the plant's defense mechanisms during stress conditions in *Aloe vera*. Also, various transcription factors, such as bHLH, NAC, and MYB, were identified, which may play crucial roles in controlling these metabolic processes.

While specific studies on *Aloe vera* are limited, research in other plant species provides insights into the regulatory networks governing stress responses and secondary metabolite production. For instance, a review by Hu et al. (2022) discusses how plants establish intricate hierarchical regulatory networks involving transcription factors to survive drought stress. These networks conduct stress signals, leading to the stimulation of specific genes that mitigate the undesirable effects of environmental stresses. Furthermore, a study by Yadav et al. (2013) underscores the role of transcription factors (bHLH, NAC, and MYB) from major families in regulating plant responses to abiotic stresses, including drought and salinity. These transcription factors are integral constituents of the gene regulatory networks that modulate the expression of stress-responsive genes and the biosynthesis of secondary metabolites. Jointly, these studies underline the significance of transcription factors and signaling pathways in arbitrating stress responses and phytochemical biosynthesis. The integration of high-throughput sequencing and bioinformatics continues to unravel the complex gene regulatory networks in *Aloe vera* and other plants, offering a groundwork for future research aimed at augmenting stress tolerance and optimizing the production of valuable secondary metabolites.

11.3.6 Phylogenetic Studies

Molecular markers have become indispensable tools in phylogenetic studies, helping researchers understand their genetic background, evolutionary processes, and taxonomic positioning between various species and cultivars within the Aloe genus, including *Aloe vera*. The phylogenetic analysis involves the use of molecular data, such as DNA sequences or molecular markers, to construct evolutionary trees or phylogenies that portray the genetic relatedness among different *Aloe* species or cultivars (Deore, 2007). These markers permit researchers to track the origins and diversification of *Aloe vera*, linking its closest relatives and clarifying the genetic basis of key traits like medicinal properties, growth habits, and environmental adaptability. A comprehensive analysis by Grace et al. (2015) utilized nuclear and plastid DNA sequences from 197 Aloe species to construct a phylogenetic framework. This study traced the origins of *Aloe vera* to the Arabian Peninsula, advocating that its widespread medicinal use is linked to its evolutionary history and leaf succulence. The research also signified that *Aloe vera* is not phylogenetically distant from other aloes, implying that factors other than genetic divergence contribute to its prominence in traditional medicine.

By matching genetic sequences from *Aloe vera* with those of other species in the Aloe genus, molecular markers provide a sharper picture of its evolutionary history, helping to catalogue new or unexplored species, simplify taxonomic doubts, and develop our understanding of how *Aloe vera* has adjusted to diverse ecological conditions (Odeleye et al., 2009; Woudstra et al., 2024). This molecular approach also aids in conservation efforts by identifying genetically distinct populations or rare wild relatives of *Aloe vera*, guiding *ex-situ* preservation and breeding programs. Lee et al. (2016) scanned the phylogenetic relationships among 40 *Aloe* species, including a putative hybrid between *A. vera* and *A. arborescens*. By evaluating chloroplast intergenic spacers and the psbK–psbI region, they found that the genus Aloe is polyphyletic, leading to the reclassification of several species into distinct genera such as *Kumara*, *Aloidendron*, and *Aloiampelos*. This reorganization imitates the complex evolutionary relationships within the group. Eventually, molecular markers in phylogenetic studies extend a robust framework for understanding the evolutionary dynamics of *Aloe vera* and its genetic resources.

11.4 CHALLENGES AND LIMITATIONS

In the face of the noteworthy progress made in the expansion and application of molecular markers in *Aloe vera*, numerous challenges linger that need to be tackled to exploit their potential in breeding, conservation, and functional genomics. *Aloe vera* has a large and complex genome with high levels of heterozygosity, repetitive sequences, and polyploidy (Brandham, 2004). This complexity obscures the identification and interpretation of molecular markers, particularly those derived from whole-genome sequencing (WGS). Additionally, the absence of a fully sequenced and well-annotated reference genome limits the scope of marker discovery and functional genomic studies. Compared to model plants, *Aloe vera* lacks comprehensive genomic information such as high-quality reference genomes, gene expression data, and functional annotations. This gap thwarts the progress of robust molecular markers for many applications, including MAS, GWAS, and functional genomics.

The effectiveness of molecular markers derived from one population or geographic region may not always extend to other populations of *Aloe vera* due to genetic diversity. Validation across different cultivars and populations is crucial to warrant marker robustness, accuracy, and transferability. While NGS-based approaches have appreciably lowered the cost of sequencing, high-depth sequencing needed for comprehensive marker discovery can still be exorbitantly expensive, mainly for large-scale projects. Furthermore, the bioinformatics infrastructure and expertise required to manage and analyze large sets of genomic data can be a barrier for researchers in low-resource settings. Epigenetic markers, such as DNA methylation and histone modifications, are increasingly recognized for their role in regulating gene expression in response to environmental stimuli. Nonetheless, the complexity of epigenetic regulation in *Aloe vera*, conjoined with the deficit of high-throughput techniques to investigate these modifications, poses challenges in applying epigenetic markers for breeding or functional genomics.

11.5 FUTURE PROSPECTS

One of the most unrelenting requirements for *Aloe vera* research is the generation of a high-quality, fully sequenced, and annotated reference genome. This would offer a critical reserve for discovering genetic variants, developing precise molecular markers, and combining genomic data with transcriptomics and metabolomics to comprehend trait inheritance and regulatory mechanisms. Future research should emphasize integrating genomics, transcriptomics, proteomics, and metabolomics to secure a holistic view of *Aloe vera's* biology. This multi-omics approach would permit researchers to associate molecular markers with gene expression profiles, protein functions, and metabolite production, authorizing the expansion of more accurate breeding strategies and a deeper interpretation of the molecular basis of medicinal properties.

Advancements in third-generation sequencing technologies, such as PacBio and Oxford Nanopore, offer more precise genome assembly, notably for complex and large genomes like *Aloe vera*. These methodologies can also provide insights into the structural variations, long-range haplotypes, and epigenetic effects, which are puzzling to identify with traditional methods. The rising understanding of epigenetic mechanisms in plants offers an opportunity to develop epigenetic markers for *Aloe vera*. These markers could be used to examine the regulation of important characters, such as medicinal compound production and stress tolerance. They could be integrated into breeding programs to select for advantageous epigenetic modifications. The elaboration of high-throughput techniques for epigenomic profiling will be necessary for advancing research in this area. The development of cost-effective, high-throughput technologies for genotyping, such as SNP arrays, genotyping-by-sequencing (GBS), and low-cost sequencing platforms, will make molecular markers more approachable to breeders and researchers.

Diminishing costs will democratize molecular marker-based research, especially in regions where assets are limited. The incorporation of molecular markers with gene-editing technologies like CRISPR-Cas9 will revolutionize *Aloe vera* breeding. By distinguishing markers associated with valuable traits, breeders can apply precision breeding techniques to target specific genes for modification, leading to more effective development of *Aloe vera* cultivars with heightened medicinal and agronomic attributes. The creation of specialized databases and collaborative platforms for *Aloe vera* genomic data will simplify the sharing of molecular markers, sequencing data, and research findings among global research communities. This will fast-track progress in marker development, functional genomics, and breeding applications.

11.6 CONCLUSION

Molecular marker technology has made momentous strides in augmenting our understanding of *Aloe vera's* genetic makeup, medicinal properties, and breeding potential. With the arrival of high-resolution tools like SNPs, SSRs, and NGS, researchers can now investigate the genetic variability of *Aloe vera* with unmatched precision. These markers have empowered the identification of traits associated with stress tolerance, medicinal compound production, and disease resistance, paving the way for more effective breeding programs. Molecular markers also expedite cultivar verification, conservation attempts, and GWAS, ultimately supplementing the sustainable utilization of *Aloe vera* in various sectors. Despite challenges such as genomic complexity, limited resources, high costs, sophisticated epigenetic studies, and the lack of a fully annotated reference genome, the continued innovation of sequencing technologies and the amalgamation of multi-omics methodologies promise to overcome these obstacles. As molecular marker technology progresses, it will unlock pristine prospects for improving *Aloe vera* cultivars with enriched medicinal and agronomic attributes, certifying its continual applicability in the pharmaceutical, cosmetic, and agricultural sectors.

ACKNOWLEDGMENT

The authors acknowledge the assistance of OpenAI for language refinement, with final oversight and revisions made by the authors themselves.

REFERENCES

Adienge, A., Muturi, G., Nadir, S., Gicheru, J., Kinyua, J. and Ngaira, J., 2019. Genetic diversity and population structure of three commercial indigenous Aloe species in selected ASALs of Kenya. *Journal of Plant Science and Molecular Breeding*, *8*, pp.1–10.

Akinsanya, M.A., Goh, J.K., Lim, S.P. and Ting, A.S.Y., 2015. Metagenomics study of endophytic bacteria in Aloe vera using next-generation technology. *Genomics Data*, *6*, pp.159–163.

Algarni, A., 2023. Evaluation of plastid and nuclear DNA markers in barcoding of Aloe saudiarabica, KSA. *Cellular and Molecular Biology*, *69*(2), pp.126–132.

Algarni, A.A., 2022. Molecular identification and phylogenetic analysis of Aloe shadensis from Saudi Arabia based on matK, rbcL and ITS DNA barcode sequence. *Saudi Journal of Biological Sciences*, *29*(2), pp.1125–1133.

Almeida, A., Raadam, M.H., Lopez-Villanueva, A., Lira, I.O.H.D. and Khakimov, B., 2023. In-depth investigation on Triterpenoid production from the desert plants Aloe vera (L.) Burm. f. and Opuntia robusta JC Wendl. prompted by their low specific methane production. *BioEnergy Research*, *16*(2), pp.898–911.

Altaf, M.T., Liaqat, W., Jamil, A., Mohamed, H.I., Fahad, M., Jan, M.F. and Baloch, F.S., 2024. A critical review: Breeding objectives, genomic resources, and marker-assisted methods in Sorghum (Sorghum bicolor L.). *Journal of Soil Science and Plant Nutrition*, *24*(3), pp.4597–4623.

Amoo, S.O., Aremu, A.O. and Van Staden, J., 2013. Shoot proliferation and rooting treatments influence secondary metabolite production and antioxidant activity in tissue culture-derived Aloe arborescens grown ex vitro. *Plant Growth Regulation*, *70*, pp.115–122.

Andrews, K.R., Good, J.M., Miller, M.R., Luikart, G. and Hohenlohe, P.A., 2016. Harnessing the power of RADseq for ecological and evolutionary genomics. *Nature Reviews Genetics*, *17*(2), pp.81–92.

Arora, L. and Narula, A., 2017. Gene editing and crop improvement using CRISPR-Cas9 system. *Frontiers in Plant Science*, *8*, p.1932.

Babu, S.N. and Noor, A., 2020. Bioactive constituents of the genus Aloe and their potential therapeutic and pharmacological applications: A review. *Journal of Applied Pharmaceutical Science*, *10*(11), pp.133–145.

Badar, Z., Khan, S. and Rao, H., 2017. Analysis of genetic fidelity of wild type and in vitro regenerated aloe vera plants through RAPD and ISSR molecular markers. *International Journal of Biotechnology & Bioengineering*, *3*(8), pp.259–267.

Batra, N. and Dey, P., 2025. Untargeted metabolomics of Aloe volatiles: Implications in pathway enrichments for improved bioactivities. *Heliyon, 11*, pp.1–12.

Bhaludra, C.S.S., Yadla, H., Cyprian, F.S., Bethapudi, R.R., Basha, S.D. and Anupalli, R.R., 2014. Genetic diversity analysis in the genus Aloe vera (L.) using RAPD and ISSR markers. *International Journal of Pharmacology*, 10(8), pp.479–86.

Bhat, G.R., Sethi, I., Rah, B., Kumar, R. and Afroze, D., 2022. Innovative in silico approaches for characterization of genes and proteins. *Frontiers in Genetics*, *13*, p.865182.

Brandham, P., 2004.The chromosomes of Aloe–variation on a theme. In Tom Reynolds (ed.) *Aloes* (pp. 373–378). Florida: CRC Press.

Chandra, D. and Choudhary, P., 2014. Diversity analysis of different accessions of Aloe barbadensis Mill. (syn. *Aloe vera* L.) collected from Rajasthan using RAPD marker system. *The Bioscan*, *9*(1), pp.7–10.

Chavhan, R.L., Hinge, V.R., Wankhade, D.J., Deshmukh, A.S., Mahajan, N. and Kadam, U.S., 2024. Bioinformatics for molecular breeding and enhanced crop performance: Applications and perspectives. In Jen-Tsung Chen (ed.) *Bioinformatics for Plant Research and Crop Breeding* (pp. 21–74). Hoboken, New Jersey, USA: John Wiley & Sons.

Chen, Y., Hong, T., Wang, S., Mo, J., Tian, T. and Zhou, X., 2017. Epigenetic modification of nucleic acids: From basic studies to medical applications. *Chemical Society Reviews*, *46*(10), pp.2844–2872.

Choudhri, P., Rani, M., Sangwan, R.S., Kumar, R., Kumar, A. and Chhokar, V., 2018. De novo sequencing, assembly and characterisation of Aloe vera transcriptome and analysis of expression profiles of genes related to saponin and anthraquinone metabolism. *BMC Genomics*, *19*, pp.1–21.

Collard, B.C. and Mackill, D.J., 2008. Marker-assisted selection: An approach for precision plant breeding in the twenty-first century. *Philosophical Transactions of the Royal Society B: Biological Sciences*, *363*(1491), pp.557–572.

Das, A., Haque, S.M., Ghosh, B., Nandagopal, K. and Jha, T.B., 2015. Morphological and genetic characterization of micropropagated field grown plants of Aloe vera L. *Plant Tissue Culture and Biotechnology*, *25*(2), pp.231–246.

Das, A., Nandagopal, K. and Jha, T.B., 2017. Molecular characterization of some Indian Aloe vera populations through RAPD and ITS markers. *Plant Biosystems-An International Journal Dealing with all Aspects of Plant Biology*, *151*(4), pp.695–703.

Das, S.K. and Joshi, A., 2023. Development of specific barcodes for identification of Aloe species based on chloroplast DNA barcoding. *bioRxiv*, pp.2023–07.

Deore, P.M., 2007. *Assessment of genetic diversity in Aloe barbadensis Mill. using molecular markers* (Doctoral dissertation, Anand Agricultural University).

Du, Z., Li, H., Wei, Q., Zhao, X., Wang, C., Zhu, Q., Yi, X., Xu, W., Liu, X.S., Jin, W. and Su, Z., 2013. Genome-wide analysis of histone modifications: H3K4me2, H3K4me3, H3K9ac, and H3K27ac in Oryza sativa L. Japonica. *Molecular Plant*, *6*(5), pp.1463–1472.

Ezzat, S.M., El Sayed, A.M. and Salama, M.M., 2016. Use of random amplified polymorphic DNA (RAPD) technique to study the genetic diversity of eight aloe species. *Planta Medica*, *82*(15), pp.1381–1386.

Ganesh, S. and Alagukannan, G. 2006. Variation and correlation studies in Aloe vera L. ecotypes. *Madras Agricultural Journal*, *93*(Jul-Dec), p.1.

González-Delgado, M., Minjares-Fuentes, R., Mota-Ituarte, M., Pedroza-Sandoval, A., Comas-Serra, F., Quezada-Rivera, J.J., Sáenz-Esqueda, Á. and Femenia, A., 2023. Joint water and salinity stresses increase the bioactive compounds of Aloe vera (Aloe barbadensis Miller) gel enhancing its related functional properties. *Agricultural Water Management*, *285*, p.108374.

Grace, O.M., Buerki, S., Symonds, M.R., Forest, F., van Wyk, A.E., Smith, G.F., Klopper, R.R., Bjorå, C.S., Neale, S., Demissew, S. and Simmonds, M.S., 2015. Evolutionary history and leaf succulence as explanations for medicinal use in aloes and the global popularity of Aloe vera. *BMC Evolutionary Biology*, *15*, pp.1–12.

Gul, I., Nissar, U. and Abdin, M.Z., 2016. Molecular authentication and quality control of Crocus sativus and Aloe barbadensis in raw material source and polyherbal medicine employing scar markers. *Pakistan Journal of Botany*, *48*(3), pp.1091–8.

Haider, N., 2011. Identification of plant species using traditional and molecular-based methods. In R.E. Davis (ed.) *Wild Plants: Identification, Uses and Conservation* (pp. 1–62). New York, USA: Nova Science Publishers, Inc.

Hao, D.C. and Xiao, P.G., 2018. Deep in shadows: Epigenetic and epigenomic regulations of medicinal plants. *Chinese Herbal Medicines*, *10*(3), pp.239–248.

He, J., Zhao, X., Laroche, A., Lu, Z.X., Liu, H. and Li, Z., 2014. Genotyping-by-sequencing (GBS), an ultimate marker-assisted selection (MAS) tool to accelerate plant breeding. *Frontiers in Plant Science*, *5*, p.484.

Hu, Y., Chen, X. and Shen, X., 2022. Regulatory network established by transcription factors transmits drought stress signals in plant. *Stress Biology*, *2*(1), p.26.

Jaiswal, S.K., Mahajan, S., Chakraborty, A., Kumar, S. and Sharma, V.K., 2021. The genome sequence of Aloe vera reveals adaptive evolution of drought tolerance mechanisms. *Iscience*, *24*(2), p.102079.

Jangra, A., Chaturvedi, S., Sharma, G., Sihag, S., Tiwari, S. and Chhokar, V., 2023. Efficient Agrobacterium tumefaciens-mediated genetic transformation of *Aloe vera. Plant Cell, Tissue and Organ Culture (PCTOC)*, *154*(1), pp.189–200.

Jangra, A., Chaturvedi, S., Sihag, S., Sharma, G., Tiwari, S. and Chhokar, V., 2023. Identification and functional characterization of a novel aldo–keto reductase from *Aloe vera. Planta*, *258*(6), p.107.

Jha, S.R. and Mughees, M., 2023. Tools for safety evaluation and conservation of medicinal plants. In Altaf Ahmad and Ambreen Asif (eds.) *Omics Studies of Medicinal Plants* (pp. 263–278). Boca Raton, Florida: CRC Press.

Kalia, R.K., Rai, M.K., Kalia, S., Singh, R. and Dhawan, A.K., 2011. Microsatellite markers: An overview of the recent progress in plants. *Euphytica*, *177*(3), pp.309–334.

Khatun, M.M., Tanny, T., Yesmin, S., Salimullah, M.D. and Alam, I., 2018. Evaluation of genetic fidelity of in vitro-propagated Aloe vera plants using DNA-based markers. *Sci Asia*, *44*, pp.87–91.

Koirala, N., Thuan, N.H., Ghimire, G.P., Van Thang, D. and Sohng, J.K., 2016. Methylation of flavonoids: Chemical structures, bioactivities, progress and perspectives for biotechnological production. *Enzyme and Microbial Technology*, *86*, pp.103–116.

Kumar, R., Naik, P.K., Kumar, A., Aggarwal, H., Kumar, A. and Chhokar, V., 2016. A combined approach using RAPD, ISSR and bioactive compound for the assessment of genetic diversity in Aloe vera (L.) Burm. f. *Indian Journal of Biotechnology*, *15*, pp.538–549.

Kumar, R., Salar, R.K., Kumar, A., Kumar, A. and Chhokar, V., 2018. A low cost, high throughput gel electrophoresis method for separation of SSR markers in Aloe vera. *The Pharma Innovation Journal*, *7*(3), pp.622–627.

Kumar, R., Salar, R.K., Naik, P.K., Yadav, M., Kumar, A., Kumar, A., Yogi, R., Kumar, M. and Chhokar, V., 2022. Elucidation of genetic diversity and population structure of sixty genotypes of Aloe vera using AFLP markers. *South African Journal of Botany*, *147*, pp.1146–1155.

Kumar, R., Singh, A.K., Gupta, A., Bishayee, A. and Pandey, A.K., 2019. Therapeutic potential of Aloe vera—A miracle gift of nature. *Phytomedicine*, *60*, p.152996.

Kumar, S., Yadav, M., Yadav, A. and Yadav, J.P., 2015. Molecular assessment of genetic diversity in Indian accessions of Aloe vera using SSR marker. *International Journal of Pharmacy and Pharmaceutial Science*, *7*(12), pp.149–54.

Kumari, A., Sharma, D., Sahil, Kumar, K., Sevanthi, A.M. and Agarwal, M., 2024. Meta-analysis of mapping studies: Integrating QTLs towards candidate gene discovery. In Priyanka Anjoy, Kuldeep Kumar, Girish Chandra, and Kishor Gaikwad (eds.) *Genomics Data Analysis for Crop Improvement* (pp. 191–216). Singapore: Springer Nature Singapore.

Lee, Y.S., Park, H.M., Kim, N.H., Waminal, N.E., Kim, Y.J., Lim, K.B., Baek, J.H., Kim, H.H. and Yang, T.J., 2016. Phylogenetic relationship of 40 species of genus Aloe L. and the origin of an allodiploid species revealed by nucleotide sequence variation in chloroplast intergenic space and cytogenetic in situ hybridization. *Genetic Resources and Crop Evolution*, *63*, pp.235–242.

Li, W., Guo, G. and Zheng, G., 2000. Agrobacterium-mediated transformation: State of the art and future prospect. *Chinese Science Bulletin*, *45*, pp.1537–1546.

Liu, C., Cui, Y., Pi, F., Cheng, Y., Guo, Y. and Qian, H., 2019. Extraction, purification, structural characteristics, biological activities and pharmacological applications of acemannan, a polysaccharide from aloe vera: A review. *Molecules*, *24*(8), p.1554.

Majeed, H., Iftikhar, T. and Manzoor, R., 2024. Extraction and characterization of novel alternative cellulosic fiber for sustainable textiles from Aloe barbadensis Miller stems (agricultural waste). *Heliyon*, *10*(18), pp.e37428–e37428.

Malek, M.A., Debnath, A. and Reya, S.S., 2023, September. Classification of Aloe Vera leaf diseases using deep learning. In Mohammad Shamsul Arefin, M. Shamim Kaiser, Touhid Bhuiyan, Nilanjan Dey, and Mufti Mahmud (eds.) *International Conference on Big Data, IoT and Machine Learning* (pp. 591–604). Singapore: Springer Nature Singapore.

Marokane-Radebe, C., Raimi, A., Amoo, S. and Adeleke, R., 2024. Metabolomic profiling and 16 S rRNA metabarcoding of endophytes of two Aloe species revealed diverse metabolites. *AMB Express*, *14*(1), p.122.

Mehta, I., 2017.'History of Aloe Vera'-(A Magical Plant). *IOSR Journal of Humanities and Social Science*, *22*, pp.21–4.

Micheni, C.M., 2015. *Genetic Diversity of Aloe Species in Kenya and the Efficacy of Aloe Secundiflora, Aloe Lateritia and Aloe Turkanesis on Fusarium Oxysporum and Pythium Ultimum* (Doctoral dissertation, University of Nairobi).

Nair, S.K. and Rabbani, M.T., 2024. Marker-assisted selection approaches for improving quality traits. In Kailash Chandra, Subhash Chand, Ravi Prakash, and Saini; Rohit Sharma (eds.) *Smart Breeding* (pp. 287–328). Apple Academic Press.

Nayanakantha, N.M.C., Singh, B.R. and Gupta, A.K., 2010. Assessment of genetic diversity in Aloe germplasm accessions from India using RAPD and morphological markers. *Ceylon Journal of Science*, *39*, pp.1–9.

Nejatzadeh-Barandozi, F., 2013. Genetic diversity in Aloe vera accessions from Iran based on agro-morphological, phytochemical and random amplified polymorphic DNA (RAPD) markers. *Journal of Medicinal Plants Research*, *7*, pp.1869–1877.

Nejatzadeh-Barandozi, F. and Akbari, L., 2013. Karyotypic variation of the Aloe vera L. and Aloe littoralis Baker in Iran. *Iranian Journal of Biotechnology*, *11*(4), pp.233–237.

Nejatzadeh-Barandozi, F., Naghavi, M.R., Enferadi, S.T., Mousavi, A., Mostofi, Y. and Hassani, M.E., 2012. Genetic diversity of accessions of Iranian Aloe vera based on horticultural traits and RAPD markers. *Industrial Crops and Products*, *37*(1), pp.347–351.

Lawal, I., I Oyediran, R., B Olaniyi, M., F Okanlawon, T., O Oyeleye, A., O Akanni, F., O Rafiu, B., O Omogbene, T., A Adam, A., Y Babalola, O. and A Idowu, O., 2021. A review on the diverse uses, conservation measures and agronomic aspect of Aloe vera (L.) Burm. f. *European Journal of Medicinal Plants*, *32*(9), pp.39–51.

Odeleye, O.M., Elujoba, A.A. and Gbolade, A.A., 2009. Comparative pharmacognostic studies on Aloe schweinfurthii and Aloe vera (Aloeceae) Leaves. *Planta Medica*, *75*(04), pp.P-23.

Okser, S., Pahikkala, T., Airola, A., Salakoski, T., Ripatti, S. and Aittokallio, T., 2014. Regularized machine learning in the genetic prediction of complex traits. *PLoS Genetics*, *10*(11), p.e1004754.

Pandey, A. and Singh, S., 2016. Aloe vera: A systematic review of its industrial and ethno-medicinal efficacy. *International Journal of Pharmaceutical Research & Allied Sciences*, *5*(1), pp. 21–33.

Panwar, B.S., Rakesh Singh, R.S., Dwivedi, V.K., Ashok Kumar, A.K. and Pummy Kumari, P.K., 2013. Genetic diversity among Indian Aloe accessions based on RAPD analysis. *International Journal of Medicinal and Aromatic Plants*, *3*(3), pp.326–333.

Park, P.J., 2009. ChIP–seq: Advantages and challenges of a maturing technology. *Nature Reviews Genetics*, *10*(10), pp.669–680.

Prasad, S.S., Singh, A., Patil, K., Ramteke, P.W., Kumar, N., Shukla, P.K., Yashwanth, V., Aakanksha, K. and Srija, M.S.L., 2024. Next-Generation Sequencing (NGS) application in plant breeding. In Jameel M. Al-Khayri, Krishnananda Pralhad Ingle, Shri Mohan Jain, and Suprasanna Penna (eds.) *Plant Molecular Breeding in Genomics Era: Concepts and Tools* (pp. 149–171). Cham: Springer Nature.

Prohens, J., Gramazio, P., Plazas, M., Dempewolf, H., Kilian, B., Díez, M.J., Fita, A., Herraiz, F.J., Rodríguez-Burruezo, A., Soler, S. and Knapp, S., 2017. Introgressiomics: A new approach for using crop wild relatives in breeding for adaptation to climate change. *Euphytica*, *213*, pp.1–19.

Quezada, M.P., Salinas, C., Gotteland, M. and Cardemil, L., 2017. Acemannan and fructans from Aloe vera (Aloe barbadensis Miller) plants as novel prebiotics. *Journal of Agricultural and Food Chemistry*, *65*(46), pp.10029–10039.

Rakoczy-Trojanowska, M. and Bolibok, H., 2004. Characteristics and a comparison of three classes of microsatellite-based markers and their application in plants. *Cellular and Molecular Biology Letters*, *9*(2), pp.221–238.

Rana, S. and Kanwar, K., 2017. Assessment of genetic diversity in Aloe vera L. among different provinces of HP. *Journal of Medicinal Plants Research*, *5*(3), pp.348–354.

Rana, S., Sharma, D. and Bakshi, N., 2018. A mini review on morphological, biochemical and molecular characterization of Aloe vera L. *International Journal of Chemical Studies*, *4*, pp.3109–3115.

Rao, H., Khan, S., Azhar, A., Badar, Z., Qureshi, S.T. and Saifullah, 2023. Molecular assesment of Aloe species revealed by RAPAD markers. In Muhanad M. El Hag (ed.) *Plants as Medicine and Aromatics*, 31. Oxfordshire, United Kingdom: Taylor & Francis Group.

Rathore, M.S., Chikara, J., Mastan, S.G., Rahman, H., Anand, K.G.V. and Shekhawat, N.S., 2011. Assessment of genetic stability and instability of tissue culture-propagated plantlets of Aloe vera L. by RAPD and ISSR markers. *Applied Biochemistry and Biotechnology*, *165*, pp.1356–1365.

Rohela, G.K., Jogam, P., Saini, P., Sandhya, D., Peddaboina, V. and Shekhawat, M.S., 2022. Assessing the genetic stability of in vitro raised plants. In Shubhpriya Gupta and Preeti Chaturvedi (eds.) *Commercial Scale Tissue Culture for Horticulture and Plantation Crops* (pp. 245–276). Singapore: Springer Nature Singapore.

Sahoo, S. and Rout, G.R., 2014. Plant regeneration from leaf explants of Aloe barbadensis Mill. and genetic fidelity assessment through DNA markers. *Physiology and Molecular Biology of Plants*, *20*, pp.235–240.

Saini, K. and Agarwal, P., 2024, September. Aloe vera disease prediction using machine learning techniques. In *2024 5th International Conference on Smart Electronics and Communication (ICOSEC)* (pp. 1117–1122). IEEE.

Salinas, P., Salinas, C., Contreras, R.A., Zuñiga, G.E., Dupree, P. and Cardemil, L., 2018. Expression of a glucomannan mannosyltransferase gene (GMMT) from Aloe vera is induced by water deficit and abscisic acid. *bioRxiv*, p.306829.

Saljooghianpour, M., 2013. Genetic diversity in different accessions of Aloe sp. using Morphological and AFLP markers. *Indian Journal of Agricultural Sciences*, *83*(12), pp.1501–1504.

Saljooghianpour, M. and Javaran, T.A., 2013. Identification of phytochemical components of aloe plantlets by gas chromatography-mass spectrometry. *African Journal of Biotechnology*, *12*(49), pp.6876–6880.

Singh, S.S., Gupta, A.K. and Verma, A.V., 2013. Molecular properties and bioactivity score of the Aloe vera antioxidant compounds-in order to lead finding. *Research Journal of Pharmaceutical, Biological and Chemical Sciences*, pp. 876–881.

Shahrajabian, M.H., Sun, W. and Cheng, Q., 2019. DNA methylation as the most important content of epigenetics in traditional Chinese herbal medicine. *Journal of Medicinal Plants Research*, *13*(16), pp.357–369.

Sharma, P., Kharkwal, A.C., Abdin, M.Z. and Varma, A., 2014. Piriformospora indica improves micropropagation, growth and phytochemical content of Aloe vera L. plants. *Symbiosis*, *64*, pp.11–23.

Sharma, V., Kukreja, V., Mehta, S. and Garg, A., 2023, December. Precision agriculture: Federated learning CNNs Aloe vera leaf disease analysis. In *2023 Fourth International Conference on Smart Technologies in Computing, Electrical and Electronics (ICSTCEE)* (pp. 1–6). IEEE.

Singh, S.S., Gupta, A.K. and Verma, A.V., 2013. Molecular properties and bioactivity score of the Aloe vera antioxidant compounds-in order to lead finding. *Research Journal of Pharmaceutical, Biological and Chemical Sciences*, pp.876–881.

Sohi, N. and Singh, A., 2021. Bio-computing approaches and tools for identification of single nucleotide polymorphisms: A comprehensive review. *International Journal of Computer Aided Engineering and Technology*, *15*(4), pp.427–457.

Svitina, H., Swanepoel, R., Rossouw, J., Netshimbupfe, H., Gouws, C. and Hamman, J., 2019. Treatment of skin disorders with Aloe materials. *Current Pharmaceutical Design*, *25*(20), pp.2208–2240.

Swaraj, J.C.S., 2024. Medicinal and nutritional importance of Aloe barbadensis Miller in human health. In Mohammad Azam Ansari, Shoaib Shoaib, and Najmul Islam (eds.) *Medicinal Plants and their Bioactive Compounds in Human Health: Volume 1* (pp. 91–105). Singapore: Springer Nature Singapore.

Syvänen, A.C., 2001. Accessing genetic variation: Genotyping single nucleotide polymorphisms. *Nature Reviews Genetics*, *2*(12), pp.930–942.

Tapre, V.V., Deshmukh, S.D., Deshmukh, A.G., Pawar, A.R., Muradi, B.M. and Sarode R.B., 2020. Studies on floral biology and breeding behavior of Aloe (*Aloe barbadensis*). *International Journal of Chemical Studies, 8*(3), pp.756–758. DOI: 10.22271/chemi.2020.v8.i3i.9296

Thapliyal, N., Thapliyal, S., Singh, A.N., Kukreja, V. and Mehta, S., 2024, March. Aloe Vera leaf diseases pathology: Harnessing federated learning CNNs for enhanced detection. In *2024 11th International*

Conference on Reliability, Infocom Technologies and Optimization (Trends and Future Directions) (ICRITO) (pp. 1–6). IEEE.

Thiyagarajan, K., Vitali, F., Tolaini, V., Galeffi, P., Cantale, C., Vikram, P., Singh, S., De Rossi, P., Nobili, C., Procacci, S. and Del Fiore, A., 2016. Genomic characterization of phenylalanine ammonia lyase gene in buckwheat. *PLoS One*, *11*(3), p.e0151187.

Treutlein, J., Smith, G.F., Van Wyk, B.E. and Wink, M., 2003. Phylogenetic relationships in Asphodelaceae (subfamily Alooideae) inferred from chloroplast DNA sequences (rbcL, matK) and from genomic fingerprinting (ISSR). *Taxon*, *52*(2), pp.193–207.

Tripathi, N., Saini, N. and Tiwari, S., 2011. Assessment of genetic diversity among Aloe vera accessions using amplified fragment length polymorphism. *International Journal of Medicinal and Aromatic Plants*, *1*, pp.115–121.

Untergasser, A., Cutcutache, I., Koressaar, T., Ye, J., Faircloth, B.C., Remm, M. and Rozen, S.G., 2012. Primer3—new capabilities and interfaces. *Nucleic Acids Research*, *40*(15), pp.e115–e115.

Upadhyay, D., Verma, A., Kukreja, V. and Mehta, S., 2024(b), March. Tech-Driven agronomy: Federated learning CNN's for Aloe Vera leaf disease diagnosis. In *2024 11th International Conference on Reliability, Infocom Technologies and Optimization (Trends and Future Directions)(ICRITO)* (pp. 1–6). IEEE.

Upadhyaya, S.R., Danilevicz, M.F., Dolatabadian, A., Neik, T.X., Zhang, F., Al-Mamun, H.A., Bennamoun, M., Batley, J. and Edwards, D., 2024. Genomics-based plant disease resistance prediction using machine learning. *Plant Pathology*, *73*(9), pp.2298–2309.

Ushasree, M.V., Jia, Q., Do, S.G. and Lee, E.Y., 2024. New opportunities and perspectives on biosynthesis and bioactivities of secondary metabolites from Aloe vera. *Biotechnology Advances*, *72*, p.108325.

Wang, Y.M. and He, C.F., 2007. Isolation and characterization of a cold-induced DREB gene from Aloe vera L. *Plant Molecular Biology Reporter*, *25*, pp.121–132.

Woudstra, Y., Rees, P., Rakotoarisoa, S.E., Roensted, N., Howard, C. and Grace, O.M., 2024. An updated DNA barcoding tool for Aloe vera and related CITES-regulated species. *bioRxiv*, pp.2024–07.

Woudstra, Y., Viruel, J., Fritzsche, M., Bleazard, T., Mate, R., Howard, C., Rønsted, N. and Grace, O.M., 2021.A customised target capture sequencing tool for molecular identification of Aloe vera and relatives. *Scientific Reports*, *11*(1), p.24347.

Yadav, N.R., Taunk, J., Rani, A., Aneja, B. and Yadav, R.C., 2013. Role of transcription factors in abiotic stress tolerance in crop plants. In Tuteja N. and Gill S. S (eds.) *Climate change and plant abiotic stress tolerance* (pp. 605–640). Weinheim, Germany: Wiley-VCH Verlag GmbH & Co. KGaA.

Yagi, A., Sato, Y., Miwa, Y., Kabbash, A., Moustafa, S., Shimomura, K. and El-Bassuony, A., 2006. Ribosomal DNA sequence analysis of different geographically distributed Aloe vera plants: Comparison with clonally regenerated plants. *Saudi Pharmaceutical Journal*, 14(3–4), pp.208–211.

Yang, D., Huang, Z., Jin, W., Xia, P., Jia, Q., Yang, Z., Hou, Z., Zhang, H., Ji, W. and Han, R., 2018. DNA methylation: A new regulator of phenolic acids biosynthesis in Salvia miltiorrhiza. *Industrial Crops and Products*, *124*, pp.402–411.

Yang, L., Li, W.C., Fu, F.L., Qu, J., Sun, F., Yu, H. and Zhang, J., 2022. Characterization of phenylalanine ammonia-lyase genes facilitating flavonoid biosynthesis from two species of medicinal plant Anoectochilus. *PeerJ*, *10*, p.e13614.

Yao, Z., You, F.M., N'Diaye, A., Knox, R.E., McCartney, C., Hiebert, C.W., Pozniak, C. and Xu, W., 2020. Evaluation of variant calling tools for large plant genome re-sequencing. *BMC Bioinformatics, 21*, pp.1–16.

Ye, J., Coulouris, G., Zaretskaya, I., Cutcutache, I., Rozen, S. and Madden, T.L., 2012. Primer-BLAST: A tool to design target-specific primers for polymerase chain reaction. *BMC Bioinformatics*, *13*, pp.1–11.

Yu, Z., Jin, C., Xin, M. and JianMin, H., 2009. Effect of Aloe vera polysaccharides on immunity and antioxidant activities in oral ulcer animal models. *Carbohydrate Polymers*, *75*(2), pp.307–311.

Zlobin, N.E., Lebedeva, M.V. and Taranov, V.V., 2020. CRISPR/Cas9 genome editing through in planta transformation. *Critical Reviews in Biotechnology*, *40*(2), pp.153–168.

Zonneveld, B.J., 2002. Genome size analysis of selected species of Aloe (Aloaceae) reveals the most primitive species and results in some new combinations. *Bradleya*, *2002*(20), pp.5–12.

Zou, Y., Han, C., Wang, F., Tan, Y., Yang, S., Huang, C., Xie, S. and Xiao, X., 2021. Integrated metabolome and transcriptome analysis reveal complex molecular mechanisms underlying selenium response of Aloe vera L. *Journal of Plant Biology*, *64*, pp.135–143.

12 Harnessing Genomics for Aloe Improvement

A New Era in Plant Breeding

Pooja Garg, Dalveer Singh, Priya Upadhyay, Anamika Kashyap, Ranjeet Kushwaha, and Mahesh Rao*

12.1 INTRODUCTION

Agri genomics plays a crucial role in modern agriculture by revolutionizing crop improvement techniques. Traditional breeding methods rely on diversity and selection (Begna et al. 2022), while agri-genomics leverages advancements in genetic engineering and molecular biology to enhance phenotypic characterization and utilize molecular markers for selection. Bioinformatics, a key component of agri-genomics, integrates biological science with information technology to handle vast genomic data effectively, aiding in sequencing, annotation, and data analysis for crop enhancement (Mishra and Pandey 2021). Additionally, genome editing technologies like CRISPR/Cas offer precision breeding opportunities, enabling the creation of designer crops with desired traits to combat climate change challenges and boost global food production (Bhattacharya, Parkhi, and Char 2020). Genetic engineering, through the creation of genetically modified organisms (GMOs), further enhances crop yield, quality, pest resistance, and food security in agriculture.

12.2 GENOMIC TOOLS IN PLANT BREEDING: GENOMICS-ASSISTED BREEDING

With advancements in the field of breeding, the focus has diverted from phenotype-based selection to genotype-based selection. A whole new aspect of breeding aided by genomic approaches, known as genomics-assisted breeding (GAB), has proven to be of much assistance in the development of new cultivars that are better than the varieties developed by traditional breeding approaches. GAB allows breeders to begin with a large population of only those offspring that are genotypically characterized and then use only a selected subset for more expensive phenotypic evaluation (Leng et al. 2017). Thus, it saves time, resources, and energy. In the context of Aloe, genomics-assisted breeding holds the potential to speed up the selection of cultivars with desired traits like high medicinal content, drought tolerance, and disease resistance. For example, developing Aloe cultivars with increased aloin content can be accelerated through GAB approaches. Resistance to shattering has been produced in canola by the application of genomics-assisted breeding, and Bayer CropScience has released some new cultivars that are shattering-resistant (Braatz et al. 2017). In addition to this, various other multinational companies have developed and released soybean, sunflower, and maize varieties with the help of marker-assisted breeding (Eathington et al. 2007). Traditional plant breeding relies on phenotypic selection, where desirable traits are observed in mature plants. However, this approach can be slow and inefficient. Agri-genomics offers powerful tools like Marker-Assisted Selection (MAS) and Genomic Selection (GS) to revolutionize breeding programs.

* The authors contributed equally to this work.

DOI: 10.1201/9781003603108-12

12.2.1 Marker-assisted Selection

MAS utilizes DNA markers linked to genes controlling desired traits. Breeders can identify these markers early in the development stage, allowing the selection of plants with beneficial genes even before the trait becomes visible (Chen 2025). This restructure breeding cycles and enhances the selection accuracy for specific traits (Jannink, Lorenz, and Iwata 2010). Traditional plant breeding relies on selecting plants based on their observable traits. This approach, while effective, has limitations. It can be slow, as desirable traits often take time to manifest, and environmental factors can influence the observed phenotype. MAS emerges as a powerful tool in agri-genomics, offering breeders a more efficient and targeted way to select plants with desired characteristics. It utilizes DNA markers, which are specific variations in a plant's genome, linked to genes controlling desired traits. These markers, unlike the desired trait itself, can be identified early in a plant's development stage. By analyzing these markers, breeders can identify plants with the genes responsible for the desired trait, even before the physical expression of the trait becomes evident. In the case of Aloe, MAS could be used to select for high aloin content or other medicinal compounds in early generations, reducing the time to release high-value cultivars.

The first step of MAS involves identifying DNA markers linked to the desired trait. This can be achieved through various methods, including genetic mapping and QTL (Quantitative Trait Loci) analysis. Recent advancements in genotyping technologies, like next-generation sequencing (NGS), are facilitating the identification of a wider range of markers with higher precision (Kumar et al. 2021). Once the markers are identified, breeders develop assays to detect these markers in plant DNA samples. Techniques like PCR (Polymerase Chain Reaction) are commonly used. Plant DNA is extracted and analyzed using the developed assays. Plants with the desired markers, indicative of the beneficial genes, are selected for further breeding.

Advantages of MAS

- By focusing on the presence of beneficial genes, MAS allows for early selection, significantly shortening breeding cycles. Breeders can discard plants lacking the desired genes early on, allowing them to focus resources on promising candidates.
- Environmental factors can often mask the expression of a desired trait. MAS relies on genetic markers, not influenced by the environment, offering a more reliable method for selecting plants with the desired characteristics.

Limitations of MAS

- The effectiveness of MAS relies on the strength of the linkage between the marker and the desired gene. Weak linkage can lead to inaccurate selection.
- Developing marker assays and genotyping a large number of plants can be expensive.
- MAS is most effective for well-characterized traits controlled by single genes. Complex traits influenced by multiple genes are more challenging to address using MAS alone.

12.2.2 Genomic Selection

GS takes a more comprehensive approach. It analyses the entire genome of an individual for markers associated with multiple traits. This allows breeders to predict the overall genetic merit of a plant at a young stage, significantly accelerating breeding cycles and improving the selection of plants with a combination of desirable traits (Bhat et al. 2016; Budhlakoti et al. 2022). Traditional plant breeding and MAS have significantly improved crop varieties. However, due to certain limitations associated with these methods, GS emerges as a revolutionary approach in agri-genomics, offering breeders unprecedented power and speed in selecting plants with desired traits. By examining a vast array of genetic markers (Single Nucleotide Polymorphisms or SNPs), GS aims to predict the overall genetic merit of a plant for a particular trait, even at a young stage (Grace et al. 2015). This comprehensive analysis allows breeders to make informed selections much earlier in the breeding cycle. In Aloe, GS can be applied to predict desirable traits such as drought resistance, medicinal compound content, and yield potential, ensuring that only the most suitable plants are selected for breeding programs.

Advantages of GS

- GS considers a much larger number of genetic markers compared to MAS, leading to more accurate predictions of the genetic potential a plant holds for a specific trait.
- GS is particularly effective for complex traits influenced by multiple genes and environmental factors. By analyzing the entire genome, GS can capture the intricate interplay of genes contributing to the desired trait.
- Early and accurate prediction of the genetic merit of a plant allows breeders to discard individuals unlikely to possess the desired traits. This significantly accelerates breeding cycles by focusing resources on promising candidates.

A diverse population of plants with known phenotypic data for the target trait is established. This population serves as the training set for the GS model. The genotype of each plant in the training population is analyzed using high-throughput genotyping technologies. Additionally, the desired trait or phenotype is measured for each plant. Statistical models are developed using the combined genotypic and phenotypic data from the training population. These models learn to associate specific genetic variations (SNPs) with the observed phenotypic values. The developed GS model is used to predict the genetic merit for the target trait in new, untested individuals based solely on their genotypic data. Breeders can then select individuals predicted to have superior genetic potential for further breeding.

Limitations

- The accuracy of GS heavily relies on the quality and diversity of the training population. A well-established training population with comprehensive phenotypic data is crucial for building robust predictive models.
- Developing and interpreting complex statistical models used in GS can be challenging. Expertise in bioinformatics and statistical analysis is often required.
- High-throughput genotyping technologies can be expensive, particularly when dealing with large breeding populations.

12.2.3 Genome-wide Association Studies (GWAS)

Genome-wide association studies (GWAS) serve as a powerful discovery tool in agri-genomics. It involves analyzing the genomes of a large population of plants with varying phenotypic traits. By statistically examining correlations between specific genetic variations and the observed traits, GWAS helps identify regions of the genome associated with those traits. GWAS can be applied to Aloe to identify genomic regions associated with medicinal compound content, drought tolerance, and pest resistance. These regions can then be targeted for future breeding efforts.

12.2.4 Genomic Editing Techniques (CRISPR/Cas9, TALENs)

Agri-genomics has introduced revolutionary genome editing techniques like CRISPR/Cas9 and TALENs. These tools allow for precise manipulation of the genome at specific locations. This can help breeders introduce desirable genes for traits such as disease resistance or improved nutritional content, inactivate genes responsible for undesirable traits like inefficient nutrient uptake, and fine-tune gene expression for optimized plant growth and development (Kaul et al., 2020; Sathee et al., 2022). For Aloe, CRISPR/Cas9 could be used to introduce or enhance genes responsible for the production of valuable compounds like aloin or to increase drought resistance, which is crucial for cultivation in arid regions.

These techniques offer unprecedented precision and speed compared to traditional breeding methods, enabling the creation of novel and improved crop varieties with a wider range of beneficial characteristics.

12.3 GENOMIC DIVERSITY AND PLANT BREEDING

Plant breeding has played a crucial role in ensuring food security for millennia. However, traditional breeding methods face challenges in the face of a growing global population and changing environmental conditions. Agri-genomics offers a powerful suite of tools to revolutionize plant breeding by harnessing the immense potential of genetic diversity.

12.3.1 Harnessing Genetic Diversity for Crop Improvement

Crop varieties are the product of domestication and selection processes over generations. This has resulted in a narrowing of the genetic base of many crops, making them vulnerable to pests, diseases, and environmental stresses. Maintaining and utilizing genetic diversity within and between crop species is vital for sustainable crop improvement.

- **Enhanced disease and pest resistance:** Diverse germplasm collections can harbour genes for resistance to various pathogens and pests. Breeders can introgress these resistance genes into elite cultivars to create crops with broader resistance profiles.
- **Improved stress tolerance:** Plants possess genetic variations influencing their tolerance to abiotic stresses like drought, salinity, and extreme temperatures. Utilizing diverse germplasm allows breeders to develop crops adapted to a wider range of environmental conditions.
- **Superior yield and quality:** Diverse genetic backgrounds can lead to the identification of novel genes influencing yield, nutritional content, and other desirable traits. Breeders can leverage this diversity to create crops with improved agronomic performance and consumer preferences.

12.3.2 Genomic Approaches to Characterize Genetic Resources

The vast genetic diversity within and between crop species can be effectively characterized using genomic approaches. These approaches provide breeders with valuable information for selecting germplasm with desired traits.

- **Genotyping by sequencing (GBS):** This high-throughput sequencing technique allows for cost-effective genotyping of large numbers of individuals, revealing the genetic makeup of diverse germplasm collections.
- **Genome-wide association studies (GWAS):** By analyzing the genomes of diverse populations along with phenotypic data, GWAS helps identify regions of the genome associated with specific traits. This information can guide breeders in selecting germplasm with desirable genetic variations.
- **Diversity arrays (DArTseq):** This genotyping technique utilizes targeted sequencing of specific regions of the genome known to be informative for diversity analysis. This allows for rapid and cost-effective assessment of genetic diversity within germplasm collections.

12.3.3 Genomic Signatures of Domestication and Adaptive Evolution

Understanding the genetic changes that occurred during domestication and subsequent adaptation to different environments is crucial for utilizing genetic diversity effectively.

- **Domestication genes:** By comparing the genomes of wild ancestors and domesticated crop varieties, researchers can identify genes that were targeted by selection during domestication. These genes often control traits like seed size, plant architecture, and flowering time, and can be valuable for breeding programs.
- **Adaptation genes:** Analyzing the genomes of geographically diverse crop populations can reveal genes associated with adaptation to specific environmental conditions. Breeders can leverage these genes to develop crops with improved stress tolerance for specific regions.

12.4 ACCELERATING CROP IMPROVEMENT

Crop improvement has always been a priority for farmers and breeders as they strive to develop more resilient varieties to changing environmental conditions and pests. The advancement of genomic technologies has completely transformed the breeding process, making it possible to select desirable characteristics rapidly and precisely.

12.4.1 Reducing Breeding Cycle Time with Genomic Tools

Crop generation time now limits crop yield and production, and breeders and researchers are under pressure to expedite varietal development (Pandey et al. 2022). A breeder can increase the number of generations in a year by complementing it with improving targeted quality traits with a shorter crop cycle. By minimizing time, field space, and total resources, plant breeding and genomics-assisted breeding tactics have collectively supported quick progression towards producing innovative crop varieties with higher agronomic performance (Budhlakoti et al. 2022). Because this method can potentially shorten crop generation times, it has drawn interest from the breeding community. Reducing breeding cycle time enables breeders to generate novel varieties in less time than it would have taken with old methods, which is one of the most significant advantages of leveraging genomics in crop improvement (Budhlakoti et al. 2022). In addition to enabling farmers to receive enhanced crops earlier, this high breeding speed also assists with addressing critical problems like food security and climate change. For agriculture to become more robust and sustainable, the power of genomics must be fully utilized (Bohra et al. 2020).

Using genomic techniques to shorten the breeding cycle has become essential to current breeding programs and agriculture. Breeders may select which individuals are ideal for breeding based on their genetic diversity with a better understanding of genomic techniques like genomic prediction, GS, and MAS (Heffner et al. 2010). Plant breeders may reap great advantages from MAS, yet it has not significantly boosted complex trait enhancement. The rationale for this is that it cannot identify QTLs with minor implications (Xu and Crouch 2008; Bernardo 2008; Heffner et al. 2010). In an effort to circumvent this issue, a promising technique known as genomic selection looks for big- and small-effect QTLs. It uses many molecular markers to cover the entire genome to estimate the values of complex traits (Meuwissen, Hayes, and Goddard 2001). GS is an intriguing method to increase the effectiveness of plant breeding because of the precision of predictions discovered in GS studies and the advancements made in high-throughput genotyping technology (Heffner et al. 2010). Lorenzana and Bernardo (2009) discovered that each cycle's GS response would be at least half that of phenotypic selection. GS can reduce breeding cycles and boost gains per unit of time by substituting genomics estimates of breeding values for labor-intensive phenotypic assessment of extremely complicated features. This is particularly true for perennial crops, as it takes many years to conduct phenotypic assessments (Wong and Bernardo 2008). In a scenario where three cycles of GS could be performed for each cycle of phenotypic selection, Lorenzana and Bernardo (2009) believe GS gains per year would reach 1.5 times those of phenotypic selection based on genomics estimates of breeding value accuracy discovered in their work on biparental plant populations.

In crop development programs, it can be used with genetic engineering and contemporary breeding techniques with tremendous potential; for instance, it can be used with CRISPR-Cas9 for gene insertion and MAS to create elite hybrid lines (Wolter, Schindele, and Puchta 2019). The industrial potential of reducing the breeding cycle has also received attention lately. This initiative aims to improve the crop and standardize the early blossoming of these crucial subsistence crops in Asia and Africa (Chiurugwi et al. 2019).

12.4.2 Precision Breeding for Stress Tolerance

Precision breeding requires a greater understanding of genomic resources that are now accessible and more effectively utilized in plant breeding programs (Gray, Li, and Dhekney 2014). In the era of omics, it is relatively easy to determine the genetic elements and genes from several databases. The control of major agronomic traits usually involves the unknown involvement of many genes and regulatory machinery. Precision breeding aligns biologically with conventionally bred crops and the entire plant lifecycle, which signifies a technical enhancement of existing breeding methodologies. Individual lines can be tested, as has always been done for conventionally bred crops, to accelerate functional analyses and produce improved cultivars (Gray, Li, and Dhekney 2014). It is necessary to have a more comprehensive understanding of the organization of the genome's structure and the sequence-function correlations to build functional markers, which are highly utilized in precision breeding programs (Gray, Li, and Dhekney 2014).

Agronomic traits are regulated by a complex network of regulatory sequences and factors, often influenced by dynamic sequence alterations like gene duplication and mutations (Alvarez-Buylla et al. 2007). Functional markers are helpful in marker-assisted recurrent selection (MARS), GS, MAS, marker-assisted backcross breeding (MABB), genetic diversity analysis, and germplasm evaluation in precision breeding (Xu, McCouch, and Zhang 2005; Yi et al. 2009; Salgotra, Gupta, and Stewart 2014). In order to provide superior wheat cultivars with long-lasting resistance to various biotic stresses, functional markers for these diseases, such as Lr 34 (leaf rust), Yr 18 (yellow rust), and Pm 38 (powdery mildew), have been developed and are being used (Tommasini et al. 2006; Lagudah et al. 2009). Various diseases and insect pests, such as bacterial leaf blight, blast, brown spot, and brown hopper, may decrease rice crop yield. The whole genome sequences of the rice subspecies indica and japonica are now publicly available, which has led to the development of several functional markers for bacterial leaf blight, including Xa3, Xa5, Xa13, Xa21, and Xa38 (Iyer-Pascuzzi and McCouch 2007; Hur et al. 2013; Hajira et al. 2016).

Compared to biotic stresses, limited functional markers have been produced for abiotic stresses. In wheat, a transcription factor gene TaMYB2 contains an SNP associated with dehydration tolerance, based on which allele-specific functional markers have been created for use in MABB (Garg, Lata, and Prasad 2012). Other SNP-based functional markers were created for the TaAQP gene and TtASR1, which codes for drought and salt tolerance, respectively, and have been verified (Pandey et al. 2013; Hamdi et al. 2020). In rice, a flooding tolerance gene has been identified, which has been utilized in the development of a new variety, "Swarna", providing tolerance in the rainy season, specifically in the majority of Asian countries that experience floods during the rice-growing periods (Neeraja et al. 2007). Several other SNP-based functional markers have been developed in foxtail millet for the SiDREB2 gene, which controls the dehydration response; in sorghum for the SbMATE gene, which gives resistance to aluminium stress; in cowpea for the CPRD12 gene, which confers tolerance to salt and drought (Contour-Ansel et al. 2006; Lata and Prasad 2013; Too et al. 2018).

12.4.3 Improving Yield and Quality Traits through Genomic Interventions

In order to achieve higher yields in crop improvement programs, plants must be strengthened to be tolerant to both biotic and abiotic stresses and enhance nutritional quality. Traditional breeding techniques have been instrumental in selecting desirable traits across species through cross-breeding. However, chemically inducing mutations or through radioactive irradiation, alongside naturally occurring mutations, have significantly broadened the pool of selectable traits (Ku and Ha 2020). Regardless of the efficiency, the conventional methods are habitually laborious and demanding, with the drawback of significant mutations hindering the development of new commercial varieties (Pacher and Puchta 2017).

Addressing these challenges, tools constructed for genome editing have emerged as a promising breeding technology, allowing for modifying crop genomes without introducing foreign genes, thus circumventing concerns associated with transgenesis (Chen et al. 2019). These tools have swiftly revolutionized crop improvement, offering higher fidelity and proficiency in genetic alteration than traditional breeding methods. As agricultural challenges intensify due to climate change, including decreased fertile land and water scarcity, innovative crop breeding technologies are urgently needed to bolster agricultural productivity and foster sustainable development (Chen et al. 2019). Stress resilience and high yields are crucial in adapting to a dynamic environment. Opportunities for improvement lie in harnessing natural genetic variation, functionally characterizing genes, manipulating endogenous or transferred genes with precise regulatory control, developing cost-effective and safe small molecules for pre-stress treatment or post-recovery, and enhancing plant health through symbiotic interactions with microorganisms (Bailey-Serres et al. 2019).

The evolution of breeding techniques from conventional to molecular and then to genomics-assisted breeding has been driven by the need to address agricultural challenges effectively (Singh et al. 2020). This progression, from gene cloning and manipulation, is now diversified into various approaches like gene silencing and CRISPR/Cas9 to generate lines with improved agronomic and climate-resilient traits. Genome sequencing has offered unprecedented avenues to the structural and functional elements of protein-coding genes organized within the chromosomes of various species, facilitating more profound insights into genetic mechanisms underlying desirable traits (Singh et al. 2020). Zhang et al. (2012) analyzed the foxtail millet genome and discovered that among 1,500 genes, 580 genes were marked for their response to water. The aquaporin-encoding genes in foxtail millet were characterized by Singh et al. (2019), who also demonstrated tolerance to salt stress and dehydration through overexpressing SiPIP3 and SiSIP1 in heterologous yeast systems.

Furthermore, pangenomics has considerable promise for uncovering more extensive structural variations (SVs) that significantly contribute to phenotypic variety, covering copy number variation (CNV) and presence/absence variation (PAV). Otherwise, it is still challenging to identify these SVs using reference-based resequencing studies or single reference genome analysis (Luo et al. 2019; Bohra et al. 2020). Due to NGS technology advancements and concurrent cost reductions, large-scale SNP markers have been found and used for crop improvement. In *Brassica napus*, 588 accessions from 21 different nations were resequenced, yielding approximately 5,294,158 SNPs and 1,307,151 indels, among which 60 loci were found to be strongly linked with agronomic variables like seed quality, oil content, and stress resilience through a genome-wide association study (GWAS) may be helpful for genetic enhancement (Singh et al. 2020).

The statistical correlation among phenotypic observations and genotypic data for understanding the genetic basis of disparities between complex traits is estimated by studying quantitative trait loci (QTL). In soybeans, a wide array of agronomic traits in plants, such as both biotic and abiotic stress resilience, yield, factors controlling production, days to flowering, root structure, nutrient absorption, and nitrogen fixation, are associated with quantitative trait loci (QTLs) (Ku and Ha 2020). In a study by Zhao et al. (2016), the genomes of over 10,000 F2 individuals from 17 representative hybrid rice crosses were sequenced, uncovering numerous QTLs, with the majority located within 300 kilobases of the genome. In order to improve the genetic resolution obtained by QTL-seq, Zhang et al. (2019) have developed a novel technique termed quantitative trait gene sequencing (QTG-seq). Target QTL selection in the first backcross generation (BC1F1) is combined with comparatively high coverage sequencing of particular BC1F2 pools in the QTG-seq. This makes it possible to analyze a quantitative trait in a "near qualitative" way. By employing this method, the scientists could identify a maize plant height QTL (qPH7) within a 150 kb genomic region home to a causative gene that produces an NF-YC transcription factor (Zhang et al. 2019).

12.5 FUTURE PROSPECTS AND CHALLENGES

Agri-genomics holds immense potential for revolutionizing plant breeding and addressing critical challenges facing global agriculture. In the case of Aloe species, genomic advancements can help improve medicinal compound biosynthesis, stress tolerance, and agronomic traits. However, translating this potential into real-world solutions requires navigating some key prospects and challenges.

12.5.1 Addressing Socioeconomic and Environmental Challenges

The application of agri-genomics offers solutions to some of the most pressing challenges in Aloe breeding and cultivation:

- By developing superior Aloe varieties with increased aloin and bioactive compound content, improved stress tolerance, and optimized growth characteristics, agri-genomics can contribute significantly to the pharmaceutical, cosmetic, and nutraceutical industries.
- Enhancing nutrient and water use efficiency in Aloe species can reduce the environmental footprint of its large-scale cultivation, particularly in arid and semi-arid regions where Aloe is commonly grown.
- Genomic tools can help develop drought-resistant and disease-tolerant Aloe varieties, ensuring sustainable production in the face of climate change and emerging pathogens.

12.5.2 Collaborative Research and Global Initiatives

To fully realize the potential of genomics-driven Aloe improvement, fostering collaboration and coordinated global efforts is crucial:

- **Public-private partnerships:** Collaboration between research institutions, pharmaceutical companies, and Aloe growers is essential for developing and commercializing improved Aloe cultivars with high medicinal value and resilience.
- **Capacity building:** Building scientific capacity in developing countries through training programs and technology transfer initiatives is vital for equitable access to the benefits of agri-genomics.
- **Global regulatory frameworks:** Harmonized international regulations governing genetic modifications in medicinal plants, including Aloe, are essential for addressing biosafety concerns, intellectual property rights, and consumer acceptance.

12.6 CONCLUSION

While these genomic tools are powerful, they are best utilized when integrated with conventional breeding programs. Traditional breeding methods provide a strong foundation for selecting superior parental lines. Integrating genomic tools like MAS and GS allows breeders to make more informed selections within these lines, accelerating the breeding process and achieving targeted improvements.

Furthermore, information gained through GWAS can guide the choice of parental lines with desirable genetic variations, increasing the efficiency of crossing programs. Finally, genomic editing techniques, while powerful, are best used to introduce specific modifications within well-established and adapted crop varieties. Most of the advancements in genomic-assisted breeding have been made in staple and economically significant crops, while relatively little work has been done in Aloe. Given its medicinal and commercial importance, greater emphasis should be placed on applying these advanced breeding tools to Aloe improvement. By combining the strengths of traditional breeding with the precision of agri-genomics tools, breeders can develop superior Aloe varieties with improved yields, enhanced resistance to biotic and abiotic stresses, and better phytochemical compositions.

REFERENCES

Alvarez-Buylla, E.R., M. Benítez, E.B. Dávila, Á. Chaos, C. Espinosa-Soto, and P. Padilla-Longoria. 2007. Gene Regulatory Network Models for Plant Development. *Current Opinion in Plant Biology* 10, no. 1: 83–91.

Bailey-Serres, J., J.E. Parker, E.A. Ainsworth, G.E.D. Oldroyd, and J.I. Schroeder. 2019. Genetic Strategies for Improving Crop Yields. *Nature* 575, no. 7781: 109–118.

Begna, T., H. Gichile, and W. Yali. 2022. Genetic Diversity and its Impact in Enhancement. *Global Journal of Agricultural Research* 10: 13–25.

Bernardo, R. 2008. Molecular Markers and Selection for Complex Traits in Plants: Learning from the Last 20 Years. *Crop Science* 48, no. 5: 1649–1664.

Bhat, J.A., S. Ali, R.K. Salgotra, Z.A. Mir, S. Dutta, V. Jadon, A. Tyagi, et al. 2016. Genomic Selection in the Era of Next Generation Sequencing for Complex Traits in Plant Breeding. *Frontiers in Genetics* 7 (December 27): 221.

Bhattacharya, A., V. Parkhi, and B. Char (eds.) 2020. Introduction to Genome Editing Techniques: Implications in Modern Agriculture. In *CRISPR/Cas Genome Editing: Strategies And Potential For Crop Improvement*. Cham: Springer International Publishing, 1–30.

Bohra, A., U. Chand Jha, I.D. Godwin, and R. Kumar Varshney. 2020. Genomic Interventions for Sustainable Agriculture. *Plant Biotechnology Journal* 18, no. 12: 2388–2405.

Budhlakoti, N., A.K. Kushwaha, A. Rai, K.K. Chaturvedi, A. Kumar, A.K. Pradhan, U. Kumar, et al. 2022. Genomic Selection: A Tool for Accelerating the Efficiency of Molecular Breeding for Development of Climate-Resilient Crops. *Frontiers in Genetics* 13, February: 832153.

Braatz, J. 2017. Production of Oilseed Rape with Increased Seed Shattering Resistance. Doctoral dissertation.

Chen, J.-T. 2025. *Biotechnology, Multiple Omics, and Precision Breeding in Medicinal Plants*. Boca Raton: CRC Press.

Chen, K., Y. Wang, R. Zhang, H. Zhang, and C. Gao. 2019. CRISPR/Cas Genome Editing and Precision Plant Breeding in Agriculture. *Annual Review of Plant Biology* 70: 667–697.

Chiurugwi, T., S. Kemp, W. Powell, and L.T. Hickey. 2019. Speed Breeding Orphan Crops. *Theoretical and Applied Genetics* 132, no. 3: 607–616.

Contour-Ansel, D., M.L. Torres-Franklin, M.H. Cruz De Carvalho, A. D'Arcy-Lameta, and Y. Zuily-Fodil. 2006. Glutathione Reductase in Leaves of Cowpea: Cloning of Two CDNAs, Expression and Enzymatic Activity under Progressive Drought Stress, Desiccation and Abscisic Acid Treatment. *Annals of Botany* 98, no. 6: 1279–1287.

Eathington, S.R., T.M. Crosbie, M.D. Edwards, R.S. Reiter, and J.K. Bull. 2007. Molecular Markers in a Commercial Breeding Program. *Crop Science* 47, no. S3: S154–S163.

Garg, B., C. Lata, and M. Prasad. 2012. A Study of the Role of Gene TaMYB2 and an Associated SNP in Dehydration Tolerance in Common Wheat. *Molecular Biology Reports* 39, no. 12 (December 1): 10865–10871.

Grace, O.M., S. Buerki, M.R. Symonds, F. Forest, A.E. van Wyk, G.F. Smith, R.R. Klopper, et al. 2015. Evolutionary History and Leaf Succulence as Explanations for Medicinal Use in Aloes and the Global Popularity of Aloe Vera. *BMC Evolutionary Biology* 15, no. 1 (December 26): 29.

Gray, D.J., Z.T. Li, and S.A. Dhekney. 2014. Precision Breeding of Grapevine (Vitis Vinifera L.) for Improved Traits. *Plant Science* 228: 3–10.

Hajira, S.K., R.M. Sundaram, G.S. Laha, A. Yugander, S.M. Balachandran, B.C. Viraktamath, K. Sujatha, et al. 2016. A Single-Tube, Functional Marker-Based Multiplex PCR Assay for Simultaneous Detection of Major Bacterial Blight Resistance Genes Xa21, Xa13 and Xa5 in Rice. *Rice Science* 23, no. 3: 144–151.

Hamdi, K., F. Brini, N. Kharrat, K. Masmoudi, and I. Yakoubi. 2020. Abscisic Acid, Stress, and Ripening (Tt ASR1) Gene as a Functional Marker for Salt Tolerance in Durum Wheat. *BioMed Research International* 2020: 7876357.

Heffner, E.L., A.J. Lorenz, J.L. Jannink, and M.E. Sorrells. 2010. Plant Breeding with Genomic Selection: Gain per Unit Time and Cost. *Crop Science* 50, no. 5: 1681–1690.

Hur, Y.J., J.U. Jeung, S.Y. Kim, H.S. Park, J.H. Cho, J.Y. Lee, Y.B. Sohn, et al. 2013. Functional Markers for Bacterial Blight Resistance Gene Xa3 in Rice. *Molecular Breeding* 31, no. 4 (April 1): 981–985.

Iyer-Pascuzzi, A.S., and S.R. McCouch. 2007. Functional Markers for Xa5-Mediated Resistance in Rice (Oryza Sativa, L.). *Molecular Breeding* 19, no. 4: 291–296.

Jannink, J.-L., A.J. Lorenz, and H. Iwata. 2010. Genomic Selection in Plant Breeding: From Theory to Practice. *Briefings in Functional Genomics* 9, no. 2 (March 1): 166–177.

Kaul, T., S.K. Sony, N.M. Raman, M. Eswaran, R. Verma, A.T. Prakash, J. Bharti, K.F.A. Motelb, and R. Kaul. 2020. How Crisp Is CRISPR? CRISPR-Cas-Mediated Crop Improvement with Special Focus on Nutritional Traits. In N. Tuteja, R. Tuteja, N. Passricha, and S. K. Saifi (eds.) *Advancement in Crop Improvement Techniques*, 159–197. Elsevier.

Ku, H.K., and S.H. Ha. 2020. Improving Nutritional and Functional Quality by Genome Editing of Crops: Status and Perspectives. *Frontiers in Plant Science* 11: 577313.

Kumar, A., A. Bohra, R.R. Mir, R. Sharma, A. Tiwari, M.A. Khan, and R.K. Varshney. 2021. Next generation breeding in pulses: Present Status and Future Directions. *Crop Breeding and Applied Biotechnology* 21(spe): e394221S13.

Lagudah, E.S., S.G. Krattinger, S. Herrera-Foessel, R.P. Singh, J. Huerta-Espino, W. Spielmeyer, G. Brown-Guedira, L.L. Selter, and B. Keller. 2009. Gene-Specific Markers for the Wheat Gene Lr34/Yr18/Pm38 Which Confers Resistance to Multiple Fungal Pathogens. *Theoretical and Applied Genetics* 119, no. 5: 889–898.

Lata, C., and M. Prasad. 2013. Validation of an Allele-Specific Marker Associated with Dehydration Stress Tolerance in a Core Set of Foxtail Millet Accessions. *Plant Breeding* 132, no. 5: 496–499.

Leng, P., T. Lübberstedt, and M. Xu. 2017. Genomics-Assisted Breeding: A Revolutionary Strategy for Crop Improvement. *Journal of Integrative Agriculture* 16, no. 12: 2674-2685

Lorenzana, R.E., and R. Bernardo. 2009. Accuracy of Genotypic Value Predictions for Marker-Based Selection in Biparental Plant Populations. *Theoretical and Applied Genetics* 120, no. 1: 151–161.

Luo, T., M. Xian, C. Zhang, C. Zhang, L. Hu, and Z. Xu. 2019. Associating Transcriptional Regulation for Rapid Germination of Rapeseed (Brassica Napus L.) under Low Temperature Stress through Weighted Gene Co-Expression Network Analysis. *Scientific Reports* 9, no. 1: 1–16.

Meuwissen, T.H.E., B.J. Hayes, and M.E. Goddard. 2001. Prediction of Total Genetic Value Using Genome-Wide Dense Marker Maps. *Genetics* 157, no. 4: 1819–1829.

Mishra, R., and D.K. Pandey. 2021. Agri/Bioinformatics: Shaping Next-Generation Agriculture. In Upadhyay, A.K., R. Sowdhamini, V.U. Patil. (eds.) *Bioinformatics for Agriculture: High-Throughput Approaches*, 111–134. Singapore: Springer Singapore.

Neeraja, C.N., R. Maghirang-Rodriguez, A. Pamplona, S. Heuer, B.C.Y. Collard, E.M. Septiningsih, G. Vergara, et al. 2007. A Marker-Assisted Backcross Approach for Developing Submergence-Tolerant Rice Cultivars. *Theoretical and Applied Genetics* 115, no. 6: 767–776.

Pacher, M., and H. Puchta. 2017. From Classical Mutagenesis to Nuclease-Based Breeding – Directing Natural DNA Repair for a Natural End-Product. *Plant Journal* 90, no. 4 (May 1): 819–833.

Pandey, B., P. Sharma, D.M. Pandey, I. Sharma, and R. Chatrath. 2013. Identification of New Aquaporin Genes and Single Nucleotide Polymorphism in Bread Wheat. *Evolutionary Bioinformatics* 2013, no. 9 (November 3): 437–452.

Pandey, S., A. Singh, S.K. Parida, and M. Prasad. 2022. Combining Speed Breeding with Traditional and Genomics-Assisted Breeding for Crop Improvement. *Plant Breeding* 141, no. 3: 301–313.

Salgotra, R.K., B.B. Gupta, and C.N. Stewart. 2014. From Genomics to Functional Markers in the Era of Next-Generation Sequencing. *Biotechnology Letters* 36, no. 3: 417–426.

Sathee, L., B. Jagadhesan, P.H. Pandesha, D. Barman, S. Adavi B, S. Nagar, G.K. Krishna, S. Tripathi, S.K. Jha, and V. Chinnusamy. 2022. Genome Editing Targets for Improving Nutrient Use Efficiency and Nutrient Stress Adaptation. *Frontiers in Genetics* 13 (June 14): 900897.

Singh, R.K., A. Prasad, M. Muthamilarasan, S.K. Parida, and M. Prasad. 2020. Breeding and Biotechnological Interventions for Trait Improvement: Status and Prospects. *Planta* 252, no. 4: 1–18.

Singh, R.K., S. Shweta, M. Muthamilarasan, R. Rani, and M. Prasad. 2019. Study on Aquaporins of Setaria Italica Suggests the Involvement of SiPIP3;1 and SiSIP1;1 in Abiotic Stress Response. *Functional and Integrative Genomics* 19, no. 4: 587–596.

Tommasini, L., N. Yahiaoui, P. Srichumpa, and B. Keller. 2006. Development of Functional Markers Specific for Seven Pm3 Resistance Alleles and Their Validation in the Bread Wheat Gene Pool. *Theoretical and Applied Genetics* 114, no. 1: 165–175.

Too, E.J., A.O. Onkware, B.A. Iyo Were, S. Gudu, A. Carlsson, and M. Geleta. 2018. Molecular Markers Associated with Aluminium Tolerance in Sorghum Bicolor. *Hereditas* 155: 20.

Wolter, F., P. Schindele, and H. Puchta. 2019. Plant Breeding at the Speed of Light: The Power of CRISPR/Cas to Generate Directed Genetic Diversity at Multiple Sites. *BMC Plant Biology* 19, no. 1: 1–8.

Wong, C.K., and R. Bernardo. 2008. Genomewide Selection in Oil Palm: Increasing Selection Gain per Unit Time and Cost with Small Populations. *Theoretical and Applied Genetics* 116, no. 6: 815–824.

Xu, Y., and J.H. Crouch. 2008. Marker-Assisted Selection in Plant Breeding: From Publications to Practice. *Crop Science* 48, no. 2: 391–407.

Xu, Y., S.R. McCouch, and Q. Zhang. 2005. How Can We Use Genomics to Improve Cereals with Rice as a Reference Genome? *Plant Molecular Biology* 59, no. 1: 7–26.

Yi, M., K.T. Nwe, A. Vanavichit, W. Chai-arree, and T. Toojinda. 2009. Marker Assisted Backcross Breeding to Improve Cooking Quality Traits in Myanmar Rice Cultivar Manawthukha. *Field Crops Research* 113, no. 2: 178–186.

Zhang, H., X. Wang, Q. Pan, P. Li, Y. Liu, X. Lu, W. Zhong, et al. 2019. QTG-Seq Accelerates QTL Fine Mapping through QTL Partitioning and Whole-Genome Sequencing of Bulked Segregant Samples. *Molecular Plant* 12, no. 3: 426–437.

Zhang, W., H. Zhi, B.-H. Liu, J. Xie, J. Li, W. Li, G. Jia, et al. 2012. Screening of Indexes for Drought Tolerance Test at Booting Stage in Foxtail Millet. *Journal of Plant Genetic Resources* 13, no. 5: 765–772.

Zhao, Y., J. Huang, Z. Wang, S. Jing, Y. Wang, Y. Ouyang, B. Cai, et al. 2016. Allelic Diversity in an NLR Gene BPH9 Enables Rice to Combat Planthopper Variation. *Proceedings of the National Academy of Sciences of the United States of America* 113, no. 45: 12850–12855.

13 Genetic Enhancement and Future Prospects

Aloe vera

Ananya Mukherjee, Trivedi Nandan Kishorchandra, Aayushee Thakur, and Vishal Sharma

13.1 INTRODUCTION

Aloe vera is a drought-resistant, succulent plant known for its thick, fleshy leaves that contain a gel-like substance with various medicinal and commercial applications (Figure 13.1). *Aloe vera*, or *Aloe barbadensis* Miller in botanical terminology, is a member of the Lily (*Liliaceae*) family. Medically, the gel extracted from *Aloe vera* leaves is renowned for its healing properties, particularly in soothing burns, wounds, and sunburns. Its anti-inflammatory and antioxidant qualities contribute to managing skin conditions and promoting faster recovery. *Aloe vera* is also utilized in oral care products, cosmetics, and hair care items, reflecting its impact on personal care and beauty industries. Its presence in skincare products, such as lotions and creams, emphasizes its moisturizing benefits (Figure 13.2). Additionally, *Aloe vera* has found a place in the beverage industry, with *Aloe vera* drinks marketed for their potential digestive support. *Aloe vera* is known by several names due to its therapeutic value, including "the silent healer," "heaven's blessing," and "the wand of heaven" (Gupta and Rawat, 2017).

The first mention of *Aloe vera* in 'Dioscorides Medical Treatise De Materia Medica' was translated into English by John Goodyew. A.D.1655; additionally, it was employed as a laxative in the early 1800s. During the mid-1930s, effective treatment was first used to treat severe and chronic radiation dermatitis. Chromone and anthraquinone, along with their glycoside derivatives, are the two primary classes of active constituents of *Aloe vera* plant extract. Other constituents include flavonoids, phenylpropanoids, coumarins, phytosterols, naphthalene analogs, lipids, and vitamins. Plant extracts containing *Aloe vera* possess antimicrobial properties that either destroy or inhibit the growth of microorganisms, such as bacteria, fungi, and viruses (Kahramanoğlu et al., 2019). *Aloe vera* gel coatings are a secure and efficient substitute for chemical postharvest treatments. Some important compounds that are found in *Aloe vera* are listed in Table 13.1.

Genetic enhancement can be employed to increase the concentrations of beneficial compounds in *Aloe vera* gel, such as polysaccharides, antioxidants, and anti-inflammatory substances. This could enhance the plant's medicinal properties, making it even more effective in applications related to wound healing, skin care, and other health benefits. *Aloe vera* is often cultivated in arid regions, and improving its drought resistance through genetic enhancement can enhance its adaptability to challenging environmental conditions.

This could contribute to increased yield and sustainability in regions prone to water scarcity. Genetic modification can be utilized to enhance *Aloe vera*'s resistance to pests and diseases. This would reduce the need for chemical pesticides, promoting environmentally friendly and sustainable cultivation practices. Genetic engineering may be employed to enhance the nutritional content of *Aloe vera*, making it a more valuable dietary supplement. This could involve increasing the levels of vitamins, minerals, or other bioactive compounds with potential health benefits.

DOI: 10.1201/9781003603108-13

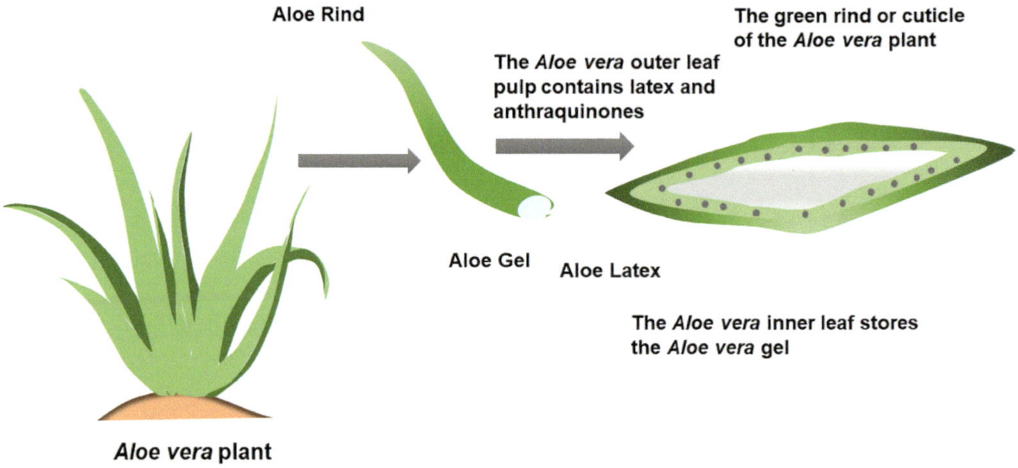

FIGURE 13.1 Graphical representation of *Aloe vera* plant and its parts.

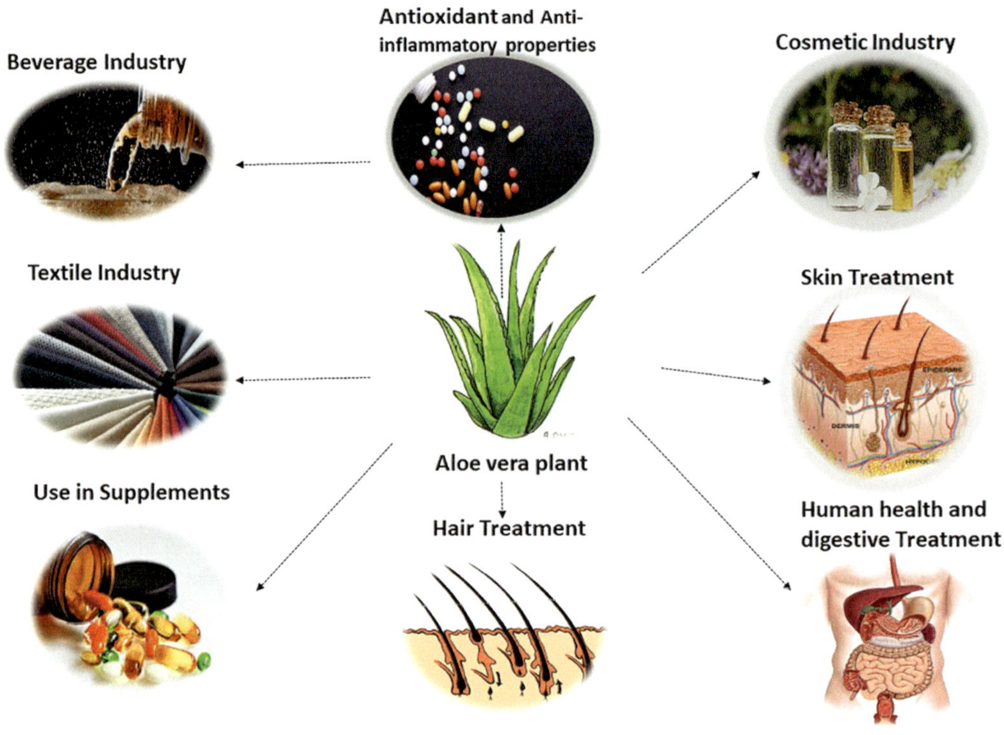

FIGURE 13.2 Various medicinal potential and industrial applications of *Aloe vera.*

13.2 GENETIC ENHANCEMENT OF *ALOE VERA*

Genetic enhancement of *Aloe vera* is a promising new area of research that has the potential to improve the plant's many medicinal and commercial properties. *Aloe vera* is a succulent plant that has been used for centuries for its healing and soothing properties. The plant's gel contains a variety of active compounds, including anthraquinones, sterols, amino acids, salicylic acids, carbohydrates, polysaccharides, vitamins, minerals, and enzymes (Choudhri et al., 2018). These compounds have

TABLE 13.1
The essential compounds that are present in *Aloe vera*

Sr. No.	Type	Compounds
1.	Anthraquinones/ anthrones	Aloe-emodin, aloetic-acid, anthranol, aloin A and B (collectively known as barbaloin), isobarbaloin, emodin, ester of cinnamic acid
2.	Carbohydrates	Pure mannan, acetylated mannan, acetylated glucomannan, glucogalactomannan, galactan, pectic substance, arabinogalactan, galactoglucoarabinomannan, galactogalacturan, xylan, cellulose
3.	Enzymes	Alkaline phosphatase, amylase, carboxypeptidase, carboxylase, catalase, cyclooxidase, phosphoenolpyruvate, cyclooxygenase, superoxide dismutase, lipase, oxidase
4.	Inorganic Compounds	Calcium, chlorine, phosphorous, chromium, copper, magnesium, iron, manganese, potassium, sodium, zinc
5.	Non-essential and essential amino acids	Alanine, arginine, aspartic acid, glutamic acid, glycine, histidine, hydroxyproline, isoleucine, leucine, lysine, methionine, proline, threonine, tyrosine, valine, phenylalanine
6.	Proteins	Lectins, lectin-like substance
7.	Saccharides	Mannose, glucose, L-rhamnose, aldopentose
8.	Vitamins	B1, B2, B6, C, β-carotene, choline, folic acid, α-tocopherol
9.	Miscellaneous	Arachidonic acid, γ-linolenic acid, potassium sorbate, steroids (campestrol, cholesterol, β-sitosterol), triglicerides, triterpenoid, gibberlin, lignins, salicylic acid, uric acid.
10.	Sterols	Campesterol, Cholesterol, β-sitosterol

Sources: (Surjushe et al., 2008), (Reynolds et al., 1999), (Baruah et al., 2016), (Reynolds and Dweck, 1999), (Radha and Laxmipriya, 2015), (Hamman, 2008), (Collard and Mackill, 2008), (Raksha et al., 2014)

been shown to have a variety of health benefits, including anti-inflammatory, antitumor, antiviral, antiulcer, fungicidal, etc (Gupta and Malhotra, 2012). Genetic enhancement can be used to improve the production of these active compounds in *Aloe vera* plants. For example, scientists have developed transgenic *Aloe vera* plants that produce higher levels of Acemannan, a polysaccharide that is effective in wound healing and immune modulation (Sierra-García et al., 2014). Other researchers are working on developing *Aloe vera* plants that produce higher levels of different active compounds, such as anthraquinones, which have anti-inflammatory and anticancer properties (Bai et al., 2023).

In addition to increasing the production of active compounds, genetic enhancement can also be used to improve other characteristics of *Aloe vera* plants, such as drought tolerance, disease resistance, and yield. For example, transgenic *Aloe vera* plants can be developed that are resistant to the fungus *Fusarium oxysporum*, which is a major cause of disease in *Aloe vera* plants, by increasing the content of phenolic compounds, as they are antifungal agents (Danish et al., 2020). Other researchers are working on developing *Aloe vera* plants that are more tolerant to drought, which would make them more suitable for cultivation in arid regions. Although genetic enhancement of *Aloe vera* is still in its early stages of development, it has the potential to revolutionize the production and use of this valuable plant. By improving the plant's medicinal properties and other characteristics, genetic enhancement can make *Aloe vera* even more beneficial to human health.

13.3 CURRENT STATUS OF GENETIC ENGINEERING IN *ALOE VERA*

The current state of genetic engineering in *Aloe vera* is still in its early stages, but there has been significant progress in recent years. The most common method of *Aloe vera* transformation is *Agrobacterium tumefaciens*-mediated gene transfer. This method involves infecting *Aloe vera* explants with a strain of *Agrobacterium tumefaciens* that has been engineered to carry the desired gene (He et al., 2007). A variety of genes have been successfully transferred to *Aloe vera*, including genes for resistance to pests and diseases, genes for tolerance to abiotic stresses, and genes for enhanced production of bioactive compounds. However, there are still some challenges and limitations to the current genetic engineering techniques for *Aloe vera*. One challenge is the relatively low transformation efficiency. This means that only a small percentage (0.9%) of *Aloe vera* explants that are infected with *Agrobacterium tumefaciens* will actually become transformed. Another challenge is the difficulty in regenerating whole plants from transformed explants (He et al., 2007). This is due to the potential benefits of these plants, such as increased production of bioactive compounds, enhanced tolerance to abiotic stresses, and new or improved medicinal or cosmetic properties.

13.4 POTENTIAL APPLICATIONS OF GENETICALLY ENHANCED *ALOE VERA*

13.4.1 Increased Production of Bioactive Compounds

Increasing the plant's ability to produce its active compounds is one of the key objectives of *Aloe vera* genetic engineering. These compounds have a variety of health benefits, including wound healing, anti-inflammatory, and antioxidant effects (Sánchez et al., 2020). One way to increase the production of active compounds is to overexpress the genes that encode the enzymes involved in their biosynthesis. Another way to increase the production of active compounds is to introduce genes from other plants that are known to produce these compounds.

13.4.2 Enhanced Tolerance to Abiotic Stresses

Genetically engineered *Aloe vera* plants could be developed to be more tolerant to abiotic stresses, such as drought, salinity, and extreme temperatures. This would make it possible to grow *Aloe vera* in a broader range of climates and to increase the plant's productivity. *Aloe vera* is a succulent plant, but it is still susceptible to drought stress. Genetic enhancement could be used to develop *Aloe vera* plants that are more tolerant to drought, which would make them more suitable for cultivation in arid regions (Silva et al., 2010). One way to improve drought tolerance is to overexpress genes that are involved in water conservation. For example, scientists have developed transgenic (Frutos et al., 1999). *Aloe vera* plants that overexpress the gene for aquaporin, a protein that helps to transport water across cell membranes. This resulted in an increase in the plant's water use efficiency and drought tolerance (Kapoor et al., 2020). Another way to improve drought tolerance is to introduce genes from other plants that are known to be drought-tolerant.

13.4.3 Increased Resistance to Pests and Diseases

Aloe vera plants are susceptible to a number of pests and diseases. Genetic enhancement could be used to develop *Aloe vera* plants that are resistant to these pests and diseases, which would reduce the need for pesticides and herbicides. One way to increase resistance to pests and diseases is to introduce genes from other plants that are known to be resistant to these pests and diseases. For example, the introduction of the gene for Bt toxin, a protein that is toxic to certain insects, from the bacterium *Bacillus thuringiensis* into *Aloe vera* plants. This increased the plant's resistance to insect pests. Another way to increase resistance to pests and diseases is to overexpress genes that are involved in the plant's immune response. For example, developing transgenic *Aloe vera* plants that overexpress the gene for pathogenesis-related protein 1 (PR1), a protein that is involved in the plant's defense against pathogens. This increased the plant's resistance to fungal diseases (Stintzi et al., 1993).

13.3.4 Improved Yield

Genetic enhancement could also be used to improve the yield of *Aloe vera* plants. This would make it possible to produce more *Aloe vera*-based products at a lower cost. One way to improve yield is to overexpress genes that are involved in plant growth and development. For example, by developing transgenic *Aloe vera* plants that overexpress the gene for isopentenyl transferase (IPT), an enzyme that is involved in CK biosynthesis and activation, which are plant hormones that promote growth. This resulted in an increase in the size and yield of the plants (Frébortová and Frébort, 2021). Another way to improve yield is to introduce genes from other plants that are known to produce high yields. For example, introducing genes for the transcription factors OsDREBs, which are involved in regulating rice grain yield and enhancing drought tolerance, from rice into *Aloe vera* plants (Chen et al., 2008). This might increase the yield of aloe gel. Overall, genetic enhancement has the potential to make *Aloe vera* an even more valuable plant for human health and industry. Thus, by increasing the production of active compounds, improving drought tolerance and disease resistance, and improving yield, genetic enhancement can make *Aloe vera* even more beneficial to society. It is important to note that genetic enhancement is a controversial topic, and there are concerns about the potential negative impacts of genetically modified organisms (GMOs). However, genetic enhancement of *Aloe vera* is still in its early stages of development, and there is a lot of potential for this technology to be used to improve the plant safely and responsibly.

13.4 METHODS FOR GENETIC GAIN

Since ancient times, *Aloe vera* has been acknowledged as a species with a variety of medicinal uses. Thanks to developments in science, it has also become clear how valuable this plant can be as an active ingredient in functional foods, edible coatings, cosmetics, and pharmaceuticals (Martínez-Burgos et al., 2022). Various reports on genetic enhancement in *Aloe vera* are reported in Table 13.2.

13.4.1 Conventional Breeding Techniques

Applying a selective breeding process, conventional breeding chooses crops based on their superior performance. The most well-known traditional breeding techniques were hybridization, mass selection, backcross breeding, recurrent selection, and pure-line selection (Lamichhane and Thapa, 2022). The classical approach involves controlled crosses between distinct *Aloe vera* genotypes, with meticulous selection of desirable traits. This method relies on the natural genetic variability within *Aloe vera* populations, requiring thorough phenotypic evaluations to identify and propagate favorable traits. Comparative assessments were often limited to a single GE (Genetic Engineering) line and its closest non-GE counterpart. The transcriptomes, proteomes, and metabolomes of the studied plants all showed certain common differences, according to these data (Liu et al., 2020). Through controlled crosses between distinct genotypes, the efficacy of traditional breeding strategies in enhancing desirable traits, including increased mucilage content, enhanced stress tolerance, and improved yield in *Aloe vera* (*Aloe barbadensis Miller*) varieties, can be evaluated.

Hybridization is the process of producing hybrids by mating genetically distant parents. It may be either natural or artificial and can involve different species, genetically different individuals, or the same species (Lamichhane and Thapa, 2022). Employing hybridization techniques facilitates the amalgamation of the desired traits from diverse parental lines. The methodological pursuit revolves around the crossing of genetically disparate *Aloe vera* varieties to induce novel genetic combinations, enhancing traits such as disease resistance, yield, and secondary metabolite production. See Table 13.2.

In order to produce a hybrid with increased gel content, a more robust growth habit, and disease resistance, the beneficial traits of numerous *Aloe vera* species have been carefully blended. This hybrid is highly attractive for commercial cultivation due to its increased gel yield, which has the potential to significantly increase the efficacy of gel processing techniques. This combination is an excellent option for the health and wellness sector because it has also been shown to have improved medicinal qualities. Widespread hybridization has given non-transgenic methods for crop

TABLE 13.2
Various biotechnological approaches used for genetic improvement of *Aloe vera*

Sl. No.	Crop	Target mechanism for genetic enhancement	Technique used	Purpose	References
1	*Aloe barbadensis*	*Agrobacterium* strains (*EHA105**, *C58C1***); Vector: pBI121; Selective marker: *npt II* gene	Gene overexpression	Generation of antibiotic resistance traits	He et al., 2007
2	*Aloe vera (L.) Burm. f.*	SSR markers are used to identify polymorphisms based on how frequently the base sequence repeats	MAS- SSR marker analysis	Examination of the genetic diversity among *Aloe vera* genotypes	Kumar et al., 2018
3	*Aloe vera*	Root and leaf transcriptome sequencing, assembly, clustering, pathway enrichment, DGE, real-time PCR, TF identification.	Transcriptomics- (NGS using Illumina HiSeq technology)	*Aloe vera*'s metabolically modified genes have been identified; further study on transcription factors and metabolic engineering may be conceivable	Choudhri et al., 2018
4	*Aloe vera* and its relatives in *Alooideae* clade (Asphodelaceae, subfamily Asphodeloideae)	Custom RNA-bait panel for *Alooideae* genes, improving sequencing.	RNA-bait sequencing	Develop tool for *Alooideae* species discrimination, overcome sequencing challenges, aid conservation	Woudstra et al., 2021
5	*Aloe vera*	NGS metagenomics techniques were utilized to analyse the PCR amplicon of 16S rDNA sequences of endophytic bacteria in *Aloe vera* plant, revealing common genera such as *Proteobacteria*, *Firmicutes*, *Actinobacteria*, and *Bacteriodetes*	NGS- Illumina metagenomics technique	Studying microbial diversity in *Aloe vera*	Akinsanya et al., 2015
6	*Aloe vera*	A stress-inducible gene called HVA22 was found in the aleurone layer of Hordeum vulgare. There are homologs of HVA22 in 354 plant species. The goal of the study was to isolate *Aloe vera*'s stress-inducible genes. The cDNA pool was isolated using *Aloe vera* plantlets that were exposed to ABA.	RT-PCR	Isolating stress inducible gene from *Aloe vera* and subjected to ABA were used to isolate cDNA pool to improve stress condition.	Sharon and Suvarna, 2017
7	*Aloe vera*	They aimed to assess the impact of a blend of *Aloe vera* and collagen on the survival, adhesion, and angiogenic capacity of cells by measuring the expression of the genes encoding platelet endothelial cell adhesion molecule (PECAM-1) and integrin $\alpha 1\beta 1$ in human adipose-derived stem cells (hASCs).	MTT Assay, RT-PCR	The study found that *Aloe vera*/ Collagen blended can improves cell proliferation, cell attachment due to high expression of integrin $\alpha 1\beta 1$, and stimulatory effect on PECAM-1.	Sigaroodi et al., 2019

8	*Aloe vera*	In *Aloe vera* plants under water stress, the GMMT gene expression dramatically increased. As endogenous ABA levels rise in correlation with this expression, ABA may regulate the expression of this gene. After 48 hours of ABA treatment, there was a notable rise in GMMT expression in plants that were not under water stress.	RT-qPCR, polysaccharide analysis by carbohydrate gel electrophoresis (PACE)	To study the retention of water in drought conditions.	Salinas et al., 2019
9	*Aloe vera* (both conventionally grown/wild type and tissue cultured)	*Aloe vera* genomic variation detection using RAPD and ISSR	RAPD Marker Assay	Objective is to evaluate genetic stability, identify changes, and comprehend the impacts of regeneration in *Aloe vera*	Badar et al., 2017
10	*Aloe vera*	The study utilized AFLP markers to analyse the genotypic variance of *Aloe vera* in various agroclimatic zones, including population metrics, resolving power, and loci	AFLP markers	To assess genetic variation in *Aloe vera* genotypes across various agro-climatic zones using AFLP markers, evaluate their reproducibility, polymorphism, throughput, and cost-effectiveness for effective utilization	Kumar et al., 2022
11	*Aloe vera*	DNA fingerprinting with RAPD marker	ITS1 and ITS2 of ribosomal DNA sequence analysis	Study assesses *Aloe vera* diversity in West Bengal, Rajasthan using RAPD, ITS, dendrogram, and phylogenetics.	Das et al., 2017

* Rifampicin and Chloramphenicol resistant
**Gentamicin and Rifampicin resistant
SSRs: Simple Sequence Repeats
MAS: Marker-Assisted Selection
HTP: High-Throughput Phenotyping
NGS: Next-Generation Sequencing
DGE: Differential gene expression
TF: Transcription Factor
PACE: Polysaccharide analysis by carbohydrate gel electrophoresis

enhancement programs a platform (Mwangangi et al., 2019). The creation of this high-yielding hybrid signifies a noteworthy breakthrough in plant breeding and presents promising prospects for those working in the *Aloe vera* sector (Banyal, 2023). Hybridization is the process by which conventional breeding produces variability. Interspecific and intraspecific are the two types of hybridization techniques. Intraspecific is one of the most successful methods of hybridization and is widely used in commercial plant breeding (Lamichhane and Thapa, 2022). This technique can widely be used for the betterment of the production and yield of *Aloe vera*.

13.4.2 Mutagenesis in *Aloe vera*

Aloe vera, esteemed for its therapeutic attributes, presents a botanical treasure trove owing to its multifaceted applications across pharmaceutical, cosmetic, and agricultural domains. The pursuit of elevating the inherent genetic attributes, aimed at bolstering yield, potency, and resilience to environmental stressors, has steered the scientific community's interest in employing innovative methodologies for genetic enhancement. Regarding the many facets of *Aloe vera* metabolic engineering and in vitro multiplication, plant tissue culture has become a crucial technology (Jangra et al., 2023). According to previous findings, the shoot tip of *Aloe vera* is the best explant for in vitro micropropagation. This might be explained by the remarkable potential of juvenile tissues for regeneration (Yasmin et al., 2022). Metabolic engineering plays a role when it is necessary to increase the concentration of certain medicinally significant metabolites or regulate the percentage of some harmful metabolites. The goal of metabolic engineering is to modify the genome at the molecular scale, which requires the use of common gene editing techniques. *Agrobacterium tumefaciens*-mediated transformation has been shown to be an extremely effective method for a variety of monocot plant species, from decorative plants to cereal crops (Batra and Kumar, 2003; Cheng et al., 2004). The ability to integrate relatively large T-DNA segments (> 10 kb) with definite ends into the plant genome with only minor rearrangements, resulting in high-quality transgenic plants, is a major benefit of using Agrobacterium for transformation (Ishida et al., 2007; Shivani and Tiwari, 2019). *Aloe vera* was transformed using Agrobacterium-mediated means under several optimized conditions, including explant selection, the degree of explant injury during the infection, the concentration of Agrobacterium, the pH of the co-cultivation, and the length of time and desiccation of plant tissue. These conditions improved the efficiency of the transformation, which may open the door to further functional genomic studies and gene regulation in the aloe plant, ultimately permitting the effective genetic modification of aloe species.

Using mutagens (chemical or physical agents of mutation), plant breeders can generate new genes or non-existent genes in order to provide unique sources of diversity for breeding. In contrast to the more exact methods made possible by contemporary biotechnology, mutation breeding is nevertheless utilized in contemporary traditional plant breeding (Bado et al., 2013; Matijevic et al., 2013; Shu et al., 2012).

13.4.3 Chemical Mutagenesis

Chemical agents, such as ethyl methane sulfonate (EMS) or sodium azide, induce point mutations in the DNA of *Aloe vera*, amplifying genetic variability. This strategy aims to generate phenotypic diversity to uncover potentially valuable traits that may not naturally exist within the genetic pool. In mutation breeding programs, chemical mutagenesis is a useful method for enhancing the essential traits of floriculture crops. The objective of this study was to quantify the effects of two chemical mutagens, sodium azide (SA) and diethyl sulfate (DES), at various concentrations (El-Nashar and Asrar, 2016). Mutagen-treated *Aloe vera* samples underwent screening and selection of desirable traits. In plant biology, induced mutagenesis is a basic method for determining the function of a gene. The process of introducing mutations and the creation of a stable mutagenic population by chemical or physical means (Wani et al., 2021). This process involves growing mutagenized plants under controlled conditions and evaluating them for variations in traits of interest, such as enhanced aloin content, increased polysaccharide levels, improved drought tolerance, or disease resistance. The practice of selecting individuals from a large modified population that satisfy particular selection criteria, such as early flowering or disease resistance relative to the parent, is known as mutant screening (Oladosu et al., 2016). Among these, rare mutants exhibiting favorable characteristics were identified and further evaluated for stability and performance across generations. Considering that chemical mutagenesis can improve the yield of *Aloe vera* production.

13.4.4 Radiation-Induced Mutagenesis

Exposure to ionizing radiation, including gamma rays or X-rays, triggers genetic alterations within the *Aloe vera* genome. These alterations, comprising point mutations or chromosomal rearrangements, offer prospects for the emergence of novel traits, albeit with varied frequencies and modes of genetic modifications. There are three main categories of radiation mutation breeding: classical radiation mutation breeding, particle radiation mutation breeding, and space radiation mutation breeding. X-ray and gamma ray applications are the principal uses for traditional radiation mutation breeding techniques (Ma et al., 2021). Radiation-induced mutagenesis involves exposing *Aloe vera* seeds, tissues, or explants to ionizing radiation, such as gamma rays or X-rays. This exposure induces genetic mutations by causing changes in DNA sequences, including point mutations, chromosomal rearrangements, or deletions. Irradiated *Aloe vera* samples were grown and propagated, generating a population of mutagenized plants. These plants exhibit a wide spectrum of genetic variations that lead to phenotypic differences. The development of new agricultural types has benefited greatly from radiation mutation breeding (Ma et al., 2021). The selection was performed to identify mutants displaying desirable traits such as increased medicinal compound production, enhanced stress tolerance, improved growth characteristics, and disease resistance. Radiation mutation breeding offers exceptional benefits over other breeding techniques like chemical mutagenesis and cross-breeding, due to its high mutation efficiency and broad mutation spectrum (Shirasawa et al., 2016). As per different studies that took place in other plants, it seems radiation-induced mutagenesis can result in a range of phenotypic alterations in the *Aloe vera* population. Among these mutants, rare individuals exhibiting favorable traits were identified and further evaluated for stability and performance across generations.

13.4.5 Modern Approaches for Genetic Enhancement of *Aloe vera*

Plant breeding techniques cover all the processes aimed at improving the genetic characteristics of crops. It is expected that they will contribute to the achievement of desired characteristics, such as disease and pest resistance, susceptibility to climatic stress, increased yield, and improved crop quality (Anand et al., 2023). It is possible to define plant breeding as a process in which, in the form of human intervention, certain heritable changes occur in plants (Orton, 2019). The pursuit of higher yields, resistance to diseases and biotic stresses, and early maturity is being pursued through an ongoing effort to produce plants with superior phenotypes (Bhargava and Srivastava, 2019). The long time period required for conventional breeding methods has created demand for modern breeding procedures. The major problems were incompatibility between sex and sexual barriers that arose before and after conception (Bhargava and Srivastava, 2019). The "Modern Plant-Breeding Triangle" was described as including phenomics, enviromics, and genomics. Furthermore, techniques such as high-throughput phenotyping (HTP), CRISPR-Cas9, marker-assisted selection (MAS), and enviromics have been used to obtain larger genetic gains (Crossa et al., 2017). Some of the most important techniques for the genetic improvement of *Aloe vera*

13.4.6 Genome Editing, CRISPR-Cas9, and NGS Techniques

The application of CRISPR-Cas9 technology presents an unprecedented opportunity for precise and targeted genetic modification in *Aloe vera*. CRISPR/Cas9 vector systems were created to modify plant genomes more effectively. These systems allow for the suppression of virus infections, destruction of cis-regulatory regions, gene knock-in, and gene knockout (Ma et al., 2016). This revolutionary tool allows for the specific editing of genes responsible for desirable traits such as increased medicinal compounds, improved stress tolerance, and enhanced yield. To alter the plant genome, Cas9 and sgRNA expressions are applied. There are numerous plant systems behind the promoters, vectors, such as added genes from the non-profit plasmid repository, that are utilized to express Cas9 and its variations in plant systems (Fateh et al., 2021). Gene expression profiling is improved by next-generation sequencing (NGS), which has several benefits. Researchers can do gene expression profiling to uncover an organism's transcriptional complexity and provide several answers by utilizing the capabilities of next-generation sequencing (NGS) in conjunction with innovative molecular techniques and computational tools for biological inquiries. NGS allows for high-throughput sequencing of DNA, enabling the rapid and cost-effective sequencing of entire plant genomes. This can provide valuable insights into the genetic makeup of *Aloe vera* and identify key genes associated with desirable traits. RNA-Seq, a technique within NGS, enables the study of the transcriptome, providing information on gene expression levels. Transcriptomic analysis can help identify genes involved in specific processes, such as the biosynthesis of bioactive compounds in *Aloe vera*. The single-base resolution provided by RNA-seq has the potential to revise several aspects of gene/ genome annotation in plants. RNA-seq's application in gene investigations of regulation through gene expression will be highly intriguing, as it may offer new perspectives on the gene regulatory network controlling biological processes. Moreover, RNA-seq can be utilized to investigate a number of novel aspects of transcriptional and post-transcriptional gene regulation in plants, including transcriptomes specific to particular cell types, alternative initiation codons, changes in the dominant transcription start site or splice form based on biological context, alternative polyadenylation, antisense loci, RNA secondary structure determination, RNA editing, gene fusions, etc. (Jain, 2012).

With some restrictions, CRISPR/Cas9 is an important genome editing technique in agriculture. Ineffective cargo and non-specific off-targets the fundamental restrictions of this method include vectors and HR as well as the creation of sequences like PAM (Protospacer adjacent motif). Certain advancements are made to the genome editing process to meet the aforementioned problems. CRISPR/CAS9 can be used to check salinity tolerance and to detect any kind of viruses. In the plant genome, foreign DNA is inserted at the target site and screening of mutants helps in the editing process. Modern studies reveal certain innovations regarding genome editing of plants, such as the incorporation of the fluorescent gene in the expression site of CRISPR/Cas9 (Economics et al., 2020). As a result, research on *Aloe vera* has benefited greatly from the combination of CRISPR/Cas9 technology and NGS, which has allowed scientists to better understand the plant's genetic makeup and create superior variants with more advantageous characteristics and increased therapeutic qualities. These technological advancements benefit sectors including pharmaceuticals, cosmetics, and agriculture by facilitating the effective and sustainable cultivation of *Aloe vera*.

13.4.7 Marker-Assisted Selection (MAS)

DNA markers have the potential to significantly increase the accuracy and efficiency of marker-assisted selection (MAS) in traditional plant breeding. An abundance of DNA marker–trait connections has been made possible by the numerous quantitative trait loci (QTLs) mapping studies conducted for a variety of crop species (Collard and Mackill, 2008). Advancements in high-throughput genotyping and phenotyping technologies enable the identification and utilization of molecular markers associated with key traits in *Aloe vera*. It is necessary to improve the qualitative and quantitative characteristics of *Aloe vera*, and this can be done by appreciating the genetic variety within the species. Molecular markers of many kinds can be used to identify polymorphisms in plant nuclear DNA. SSR markers are useful molecular markers for research on genome organization, genetic diversity, marker-assisted selection (MAS), and crop improvement in a variety of plant species (Kumar et al., 2018). MAS expedites the breeding process by allowing the selection of desirable genotypes at early developmental stages and reducing the time required for trait fixation and cultivar development.

13.4.8 Genomics and Transcriptomics

Comprehensive genome sequencing and transcriptomic analyses provide insights into the genetic architecture and regulatory networks governing *Aloe vera* physiology, secondary metabolite production, and stress responses. *Aloe vera* is a drought-tolerant, succulent plant in the genus Aloe of the Asphodelaceae family. The sequence of the *Aloe vera* genome will help to identify the *Asphodelaceae* family's genomic markers and to comprehend the genetic foundation of significant phenotypes in the plant, such as drought tolerance and therapeutic qualities (Jaiswal et al., 2021). Integrating multi-omics data facilitates the identification of candidate genes, regulatory elements, and metabolic pathways crucial for targeted trait improvement. In the current study, full transcriptome sequencing of *Aloe vera* root and leaf tissues was carried out utilizing Illumina Hi-Seq technology and NGS technology. It offers useful sequence data from the tissues of *Aloe vera* roots and leaves, with a focus on genes associated with secondary metabolic processes (Choudhri et al., 2018). The Asphodelaceae family's genomic markers will be helpful in figuring out the genetic basis of significant traits in *Aloe vera*, like its ability to withstand drought and its medical qualities. Two tissues, the root and the leaf, were subjected to transcriptome sequencing and analysis in order to identify the genes with precision and obtain a profound understanding of their expression (Jaiswal et al., 2021).

The development of *Aloe vera* is greatly aided by genomics, the study of an organism's whole gene pool and how each gene operates. Here's a brief explanation of how genomics can help improve the *Aloe vera* plant's growth, health, and general usefulness. *Aloe vera* is a plant belonging to the Asphodelaceae family, which is known for its several therapeutic benefits and tolerance to drought. However, because the genome sequence was not available, the genetic basis of these phenotypes is yet unknown. Therefore, they present the first genome sequencing of *Aloe vera,* containing 86,177 protein-coding genes and measuring 12.93 Gbp. It is the largest angiosperm genome to be sequenced and assembled to date, and the first genome from the Asphodelaceae family (Jaiswal et al., 2021). Researchers can investigate the genetic variation among populations of *Aloe vera* through genomic investigations. Breeders can choose and develop types that are well-adapted to particular settings, assuring optimal growth and health, by having a thorough understanding of natural genetic differences. Rapid investigation of the diversity and makeup of microbial communities across several habitats is made possible by next-generation sequencing (NGS). They utilized the Illumina metagenomics technology to create a total of 5,199,102 reads from the samples, and we applied the high-throughput techniques of NGS to the metagenomics investigation of endophytic bacteria in *Aloe vera* plants by evaluating their PCR amplicon of 16S rDNA sequences (V3–V4 regions). According to the analysis, the most common genera were Actinobacteria, Firmicutes, Proteobacteria, and Bacteroidetes (Akinsanya et al., 2015). The genes and metabolic processes that enable the production of polysaccharides, antioxidants, and other therapeutic components in *Aloe vera* are identified with the use of genomic research. Breeding programs that try to increase the concentration of these important chemicals benefit from this information. Genomic studies help unravel the interactions between *Aloe vera* and its microbiome. Understanding the plant-microbe relationships contributes to strategies for promoting beneficial microbial communities that enhance plant health and nutrient uptake. Genomics is a powerful tool for unravelling the genetic intricacies of *Aloe vera*, guiding breeding efforts, and optimizing its cultivation for various applications. This knowledge not only enhances the plant's medicinal properties but also contributes to the sustainability and commercial viability of *Aloe vera* cultivation.

13.4.9 Metabolomics and Bioinformatics

Measurement and analysis of metabolites in a biological sample using high-throughput technologies such as nuclear magnetic resonance and mass spectrometry is known as metabolomics. The method by which *Aloe vera* extract acts on Raji cells was also discovered through metabolic experiments conducted during their survey. The primary ingredient in *Aloe vera*, a chemical known as aloin, is an anthraquinone that has been shown to have anticancer potential. It inhibits tumor angiogenesis and growth, blocks signal transducers, and activates transcription 3, indicating its potential as a medication candidate for cancer therapy (Noorolahi et al., 2016). Metabolomic studies offer a comprehensive understanding of the intricate metabolic pathways in *Aloe vera*, enabling the characterization of bioactive compounds and their biosynthetic pathways.

The bioinformatics approach provides us with the necessary information about the capabilities of any plant. Leveraging bioinformatics tools for data integration and analysis aids in unraveling complex metabolic networks and facilitating the manipulation of metabolic fluxes for desired compound enhancement. The Asphodelaceae plant family is the first to have its genome sequenced, and the *Aloe vera* genome is the largest angiosperm and the fifth largest genome to date, so its availability is very significant. Additionally, it is the largest genome that has been sequenced with Oxford Nanopore technology to date. The utilization of a hybrid strategy that combined short-read (from Illumina) and long-read (from Nanopore) sequence data proved to be effective in overcoming the difficulty of sequencing one of the biggest plant genomes (Jaiswal et al., 2021). Using NGS technology, transcriptome sequence data for *Aloe vera* root and shoot were produced. With a focus on secondary metabolism, the transcriptome sequences have been assembled, annotated, and examined. De novo transcriptome data assembly in conjunction with DGE analysis was a potent method for identifying the genes responsible for the production of significant secondary metabolites belonging to several chemical classes in *Aloe vera* (Choudhri et al., 2018). Bioinformatics is essential to the progress of our knowledge of *Aloe vera* biology, from functional characterisation to genome analysis, and ultimately to the creation of *Aloe vera*-based products for use in cosmetics, pharmaceuticals, and agriculture. The graphical representation of various modern approaches used for the improvement of *Aloe vera* is presented in Figure 13.3.

13.5 POTENTIAL BENEFITS OF GENETIC ENHANCEMENT

13.5.1 Enhancing *Aloe vera*'s Medicinal Potential

Many ancient cultures, including those of Greece, Egypt, India, Mexico, Japan, and China, have utilized *Aloe vera* as a medicinal plant. Both the pharmacological actions and the therapeutic claims for *Aloe vera* cover a wide spectrum of ailments related to it (Vogler and Ernst, 1999). Researchers have utilized cutting-edge techniques to elucidate the genetic makeup of plants and identify the specific genes responsible for key medicinal compounds, stress tolerance mechanisms, and nutritional components. For example, three commercially viable oils with a linolenic acid (18:3) composition have been produced by genetic modification of soybean oil to fulfill end-user specifications and reduce chemical hydrogenation for food applications. And also, the removal of chemically hydrogenated and trans-hydrogenated fatty acids from the body is crucial for maintaining human health. Lowering palmitic acid (16:0) from 11% to less than 4% results in low amounts of saturated fats and is beneficial for heart health (Fehr, 2007). Rich in protein, soybean meal derived from high-oleic-acid soybeans stops the development of lung, liver, and colon cancer cells (Rayaprolu et al., 2013).

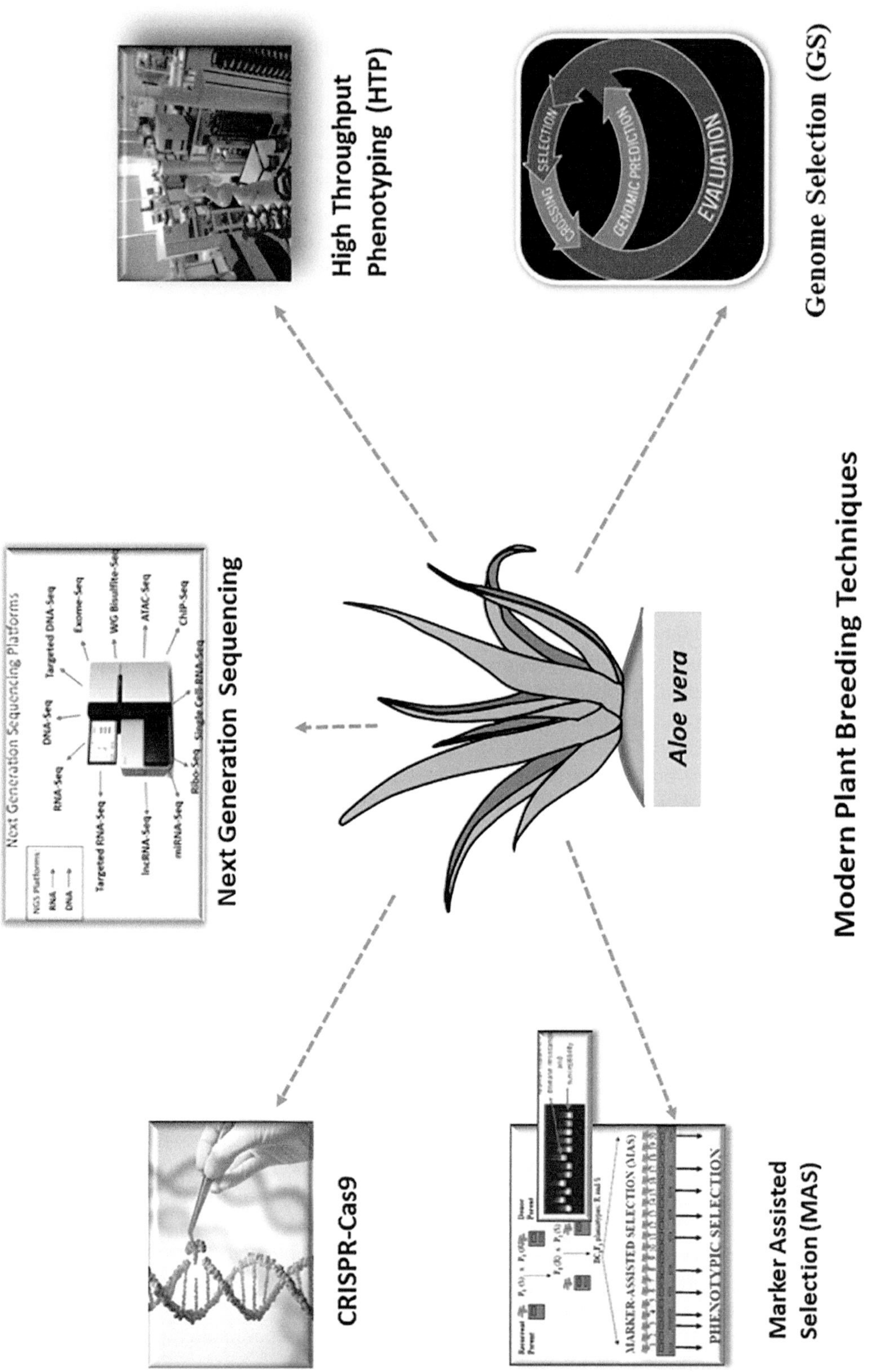

FIGURE 13.3 Modern approaches which can be employed for genetic enhancement in *Aloe vera*.

Another example is the Chinese soybean genome, known as "Zhonghuang 13," which was assembled from scratch in 2018 by researchers (Shen et al., 2018) using cutting-edge sequencing techniques, including optical mapping, Hi-C chromosomal conformation capture, and single-molecule real-time (SMRT) sequencing. When compared to Williams 82, this genome showed more than 250,000 structural changes. The genome of the wild soybean accession W05 was then assembled; it had a contig N50 of 3.3 Mb and a size of 1013.2 Mb. (Xie et al., 2019). Using homology-based gene predictions, about 1.6 million ESTs and 1.5 billion paired-end Illumina RNA-seq data were integrated into the reassembled and re-annotated Williams 82 genome (version Wm82.a4.v1) (Valliyodan et al., 2019). In a similar manner, genetic alteration can be made in *Aloe vera* and its products as well. This scientific groundwork serves as the foundation for targeted genetic modifications aimed at optimizing desired traits within the plant. Genetic enhancement revolutionizes the utility and potency of *Aloe vera*. By selectively manipulating the genetic material of plants, researchers aim to increase the production of bioactive compounds that are critical for therapeutic effects. These enhancements hold the promise of elevating *Aloe vera*'s efficacy in addressing various health conditions, potentially marking a paradigm shift in natural medicine and wellness. The potential to create specialized *Aloe vera* varieties tailored to specific medicinal purposes represents a significant stride in precision agriculture and healthcare. Medicinal plants contain bioactive phytochemicals that are essential for promoting health and preventing illness. These substances contain antispasmodic, antimalarial, analgesic, diuretic, antioxidant, anti-allergenic, antibacterial, antiviral, anticancer, anti-inflammatory, and plant defence properties. They include alkaloids, phenols, flavonoids, terpenoids, glycosides, and saponins. It has long been practiced to employ medicinal herbs and herbal preparations, and new developments in contemporary therapeutics have prompted the application of these natural remedies in the diagnosis, prevention, and treatment of a wide range of illnesses. A wide range of diseases can be effectively treated by a variety of bioactive compounds derived from different plant sources, such as taxol, vinblastine, vincristine, topotecan, irinotecan, curcumin, silymarin, ricinine, tannins, steroidal lactones, limonoids, piperidine, diterpenoid furanolactones, aloin, emodin, campesterol, β-sisosterol from *Aloe vera*, apigenin, taxol, ursolic acid, citral, berberine, digoxin, thymoquinone, quinq quinine, etc. (Alamgir, 2018).

Genetic engineering enables the customization of *Aloe vera* cultivars and optimizes them to address targeted health concerns. This approach not only diversifies the plant's medicinal applications but also presents a path towards personalized medicine by tailoring *Aloe vera* strains to meet specific therapeutic needs. It has been shown that *Aloe vera* gel therapy reduces diabetic foot lesions in rodents while promoting wound healing mechanisms such as contraction, fibroplasia, and collagen synthesis. Its hypoglycemic behavior and fibroblast function might be the cause of this (Daburkar et al., 2014). *Aloe vera* used topically improved the morphological, biochemical, and biomechanical aspects of healing cutaneous lesions. *Aloe vera* and calendula ointment may hasten the healing of episiotomy wounds, according to clinical investigations. A potential natural remedy for wound treatment is *Aloe vera* gel (Haritha Yadav et al., 2012). The *Liliaceae* family has been utilized for about a century as a functional ingredient in many meals and as a traditional medication to treat viral and gastrointestinal illnesses. The use of *Aloe vera* juice in probiotic yoghurt is a trend that shows promise for the use of functional ingredients and herbs in dairy products. *Aloe vera* juice may increase the activity of starter cultures in the product and help explain better textural qualities that result in different syneresis percentages by lowering the amount of liquid separation in gels and causing the low-fat set probiotic yoghurt samples to coagulate more. Because of their well-established health benefits, including their ability to combat cancer, diabetes, diarrhoea, and high cholesterol, probiotics are becoming increasingly popular (Ouwehand et al., 2002; Roberfroid, 2000; Salminen et al., 1999). Genetic enhancement strategy aims to bolster *Aloe vera*'s healing properties, such as burn and wound healing, anti-aging, immune system restoration, anti-inflammatory, antidiabetic, antimutagenic, antioxidant, immunomodulatory, antibacterial, antifungal, and antiviral.

Among all these, the use of *Aloe vera* in the treatment of burns and other ailments is well-known. Applying *Aloe vera* to a wound speeds up the healing process and improves the wound's tensile strength by encouraging cell growth (Kumar and Debjit, 2010). It accomplishes this by quickening the blood flow to the injured region. The greatest wound treatment ever found is aloe (West and Zhu, 2003). The following mechanism explains this acceleration: *Aloe vera* gel promotes increased collagen cross-linking and collagen content in the wound, which improves scar breakup and wound contraction (Chithra et al., 1998; Heggers et al., 1996). It is also documented an increase in dermatan sulfate and hyaluronic acid levels in the granulating tissue of healed wounds. In a human keratinocyte monolayer, a 5.5 kDa glycoprotein that was isolated from *Aloe vera* increased epithelial cell motility and improved the healing process of wounds (Choi et al., 2001). Currently, *Aloe vera* is also used in the production of almost 95% of goods with dermatological value. This is due to its remarkable moisturizing qualities. It aids in the elimination of dead skin cells and enhances the skin's capacity to moisturize itself. By generating collagen and elastin fibers, it makes the skin less wrinkly and more elastic, which reverses the effects of degenerative skin changes. Through the action of amino acids and its cohesive effect on superficially peeling epidermal cells, it softens the skin (Surjushe et al., 2008; West and Zhu, 2003). *Aloe vera* is a perfect element for cosmetics and dermatological procedures because of its amazing qualities. It has been suggested that *Aloe vera* shields the skin from radiation harm. (Roberts and Travis, 1995; Sato et al., 1990). The use of *Aloe vera* gel is thought to cause the production of the antioxidant protein metallothionein, which scavenges hydroxy radicals and shields the skin from oxidative damage. Moreover, it causes the production of the immunosuppressive interleukin IL-10, which stops UV-induced inhibition of delayed hypersensitivity to light (Byeon et al., 1998).

The anti-inflammatory qualities of *Aloe vera* gel reduce pain and hasten recovery. Its acetylated mannan inhibits the cyclooxygenase pathway in a manner similar to that of mannose-6-phosphate (Hamman, 2008). Research indicates that *Aloe vera* extracts may be used to treat H. pylori infection (Prabjone et al., 2006), reduce neutrophil count, and have anti-oedema properties (Vázquez et al., 1996). The polysaccharides found in *Aloe vera* have been linked to immunomodulatory effects by stimulating macrophage cells to produce nitric oxide, release cytokines (such as interleukin-1 or IL-1, interleukin-6 or IL-6, and interferon-γ or INF-γ), and display cell surface markers (Chow et al., 2005; Im et al., 2005; Zhang and Tizard, 1996). Studies have indicated that certain immunomodulatory effects are linked to glycoproteins present in aloe gel, specifically lectins (Hamman, 2008). Because of its many therapeutic benefits, *Aloe vera* has been utilized as a medical plant for centuries. *Aloe vera*'s chemical makeup is especially intriguing since some of its constituents have pharmacological and therapeutic benefits. *Aloe vera* farming has grown significantly in value for the pharmaceutical and cosmetic industries. India's farmers frequently deal with issues including poor soil quality, low groundwater levels, and lack of rain. Therefore, it is imperative that *Aloe vera* be genetically modified in order to enhance its medical and therapeutic qualities for the benefit of humankind. *Aloe vera* may be grown in a variety of ecological settings and yield substantial returns on investment in addition to a consistent income.

13.5.2 Enhancing Therapeutic Compounds of *Aloe vera*

13.5.2.1 Polysaccharides and Glycoproteins

Genetic modification techniques target pathways involved in polysaccharide and glycoprotein synthesis, aiming to enhance their production in *Aloe vera*. Because of their polysaccharides, aloe leaves, especially their inner, transparent pulp, are employed in medicinal, cosmetic, and nutraceutical purposes. As per the information given in (Reynolds, 2004), the most common and extensively researched polysaccharides from aloe leaves were found. Mannans, galactans, arabinogalactans, pectic materials, and sugar combinations are examples of additional polysaccharides. While there are other polysaccharides in the plant, mannan is the most often extracted and researched kind from the pulp. As per the information given in Reynolds (2004), aloes have also been used to isolate glycoproteins, which are frequently discovered to be lectins with enzymatic activity. It is yet unknown, nevertheless, how these glycoproteins' sugar side-chains are structured. While (Okamura et al., 1996) found glucosamine, galactosamine, and N-actylglucosamine in *Aloe* arborescens.

13.5.2.2 Anthraquinones and Phenolic Compounds

Genetic interventions aim to amplify the synthesis of anthraquinones and phenolic compounds, which possess antibacterial, antiviral, and anti-inflammatory properties in *Aloe vera*. Five metabolites were isolated in 2019 (Canche-Escamilla et al., 2019) from the *Aloe vera* root and one from the industrial waste that was created during the aloin extraction process. Both the root and the trash are rich in phenolic components. The antibacterial activities of desoxyerythrolacin and lacaic acid D methyl ester, which are found in its roots, were demonstrated against the phytopathogenic bacterium X. campestris. Furthermore, the primary products, aloesaponarin-I and aloeemodin, exhibit superior dyeing and color fastness characteristics when applied to polyester fabrics. Polyester is brightly colored by aloesaponarin-I, while a deeper yellow color is produced by aloeemodin. It was discovered that aloesaponarin-I uptake was more pH-sensitive than aloe-emodin uptake. The identification of certain phenolic compounds (C-chromones, phenyl pyrones, and anthrones) for chemical profiling and fingerprinting purposes or the straightforward quantitative determination of the major components (aloin A, aloin B, and other aloin derivatives) are the main goals of the various analytical methods that have been published for the analysis of some phenolic compounds in Aloe species. These methods include HPLC-UV (Okamura et al., 1996; Park et al., 1998; Zonta et al., 1995), MEKC (Kuzuya et al., 2001), GC–MS (Saccù et al., 2001), HPLC-NMR (Karagianis et al., 2003), LC–MS (Rebecca et al., 2003; Wang et al., 2012), and nano-LC–MS (Fanali et al., 2010).

13.5.3 Advantages of Genetically Enhanced *Aloe vera*: Market Expansion, Grower Profits, and Industry Impact

The application of genetic enhancement techniques in *Aloe vera* cultivation presents a spectrum of economic opportunities, including expanded market demand, heightened profitability for growers and manufacturers, and potential impacts on related industries. Genetic enhancement endeavors aim to imbue *Aloe vera* with superior qualities, augmenting its nutritional, medicinal, and agronomic attributes, thereby driving increased demand across diverse consumer segments. Genetic modification targets traits that elevate the nutritional content, medicinal potency, and resilience of *Aloe vera* plants. These improved qualities are likely to resonate with health-conscious consumers, pharmaceutical industries, cosmetic manufacturers, and the food and beverage sector, stimulating heightened demand for *Aloe vera*-based products. Products made from *Aloe vera* are being made and sold all over the world. China and Thailand are Asia's two biggest *Aloe vera* producers. South Africa (Steenkamp, 2015) and Colombia (Flórez Martínez, 2014) are two nations that are leading the way in the proliferation of this crop. *Aloe vera* functions as a natural food preservative that may improve food items' texture, boost their nutritional and health-promoting value, and shield them against oxidative and microbiological degradation. At the quantities needed for food preservation and packaging applications, *Aloe vera* poses no threat to humans or the environment, making it a viable and sustainable substitute for artificial chemical-based preservatives, antioxidants, and antibacterial agents (Kumar et al., 2022). The plant *Aloe vera* is home to around 75 bioactive substances, the most important of which are phytosterols, phenolics, and polysaccharides. *Aloe vera* is used to make pressed filleted gel, whole leaf gel, and hand filleted gel in the culinary, cosmetic, and pharmaceutical industries. This product has a million-dollar global market, with sales estimated to be approximately $1.60 billion in 2018, according to the data given by Martínez-Burgos et al. (2022). According to Steenkamp (2015), analysis of the *Aloe vera* profile in South Africa, the presentation and concentration of Aloe intermediate goods on the global market in 2015 influenced their price indicators. Concentrated juice (1:10), for instance, sold for around US$ 12 to 13/kg, or US$ 12,000 to 13,000/ton. Raw aloe fillet cubes were sold for US$ 630 to US$ 1,000/ton internationally, whereas powders (1:100) sold for around US$ 130 to 180/kg or US$ 130,000 to 180,000/ton.

13.6 NAVIGATING CHALLENGES IN GENETIC ENHANCEMENT TOWARDS *ALOE VERA*'S PATH TO IMPROVED TRAITS

The cultivation of crops like *Aloe vera* could be revolutionized by genetically modified organisms (GMOs), but there are drawbacks as well. Plant breeding techniques have been created to produce genetically modified (GM) crops displaying intriguing features, including herbicide tolerance, insect resistance, and abiotic stress resistance, with the goal of improving agricultural operations and nutritional quality. Genetically modified organisms (GMOs) laws have been implemented in several nations to safeguard consumer choice and ensure the traceability of food and feed products on the market. In order to enforce these laws, various GMO identification techniques, mostly based on DNA, have been developed (Kourkoutas et al., 2016). The introduction of Bacillus thuringiensis (Bt) genes into plants has raised issues related to risk assessment and biosafety (Khan and Hakeem, 2015). It has been discovered that certain bacterial and eukaryotic cells are susceptible to mutagenic and genotoxic effects from certain *Aloe vera* anthraquinones. *Aloe vera* is also linked to a number of harmful effects, depending on which anthraquinones, like aloin, are present. Additional research showed that numerous plants and Aloin were not metabolized by microorganisms (Jangra et al., 2022). Thinking carefully and communicating openly are essential when it comes to GMO ethics, which include ownership, equity, and concerns about unintended consequences. The development and application of genetically modified organisms (GMOs) must be done properly, taking sustainability, ethics, and safety into account, or else they will not be able to meet the urgent needs of agriculture. Gene drive methods may be used to eradicate weeds, control or eradicate exotic species that endanger biodiversity, or even modify pathogens that injure crops or spread illness.

Additionally, gene drive technologies could benefit already existing communities and may even be able to revive extinct plant species or conserve endangered ones (Hefferon and Herring, 2017). Thus, there are numerous challenges associated with genetic enhancement in any type of plant similar to *Aloe vera.* Considering this, performing genetic enhancement can lead us to the betterment of *Aloe vera*. This will enhance *Aloe vera* production and potentially benefit the industrial market.

13.7 PUBLIC CONCERNS ABOUT GENETICALLY MODIFIED ORGANISMS (GMOS)

Public concerns about GMOs are widespread and varied. Some people worry about the safety of GMOs, while others are concerned about their potential impact on the environment and human health. Still others have ethical concerns about the creation of modified organisms. For example, the European Commission harmonized national rules in response to the GM food debate that surfaced in the late 1980s. The 1997 Regulation on Novel Foods Ingredients, which recognized consumers' rights to knowledge and labelling for making educated decisions, was the result of community concerns around GMO authorization that have persisted since the 1990s (Bawa and Anilakumar, 2013).

The expression of foreign genes can change an organism's natural state, impacting its growth rate, metabolism, and reaction to external stimuli. This may result in health hazards, including the introduction of novel allergens from genetically modified foods and the emergence of antibiotic-resistant genes in the gut flora. The spread of disease and ecological imbalances can result from the horizontal gene transfer of antibiotics, herbicides, or pesticides. Reduced survivability of progeny can result from vertical gene transfer between GMOs and wild-type organisms, posing a risk to both species (Phillips, 2008). There is a common belief that developing nations stand to gain the most from plant biotechnology while also being the most impacted by its introduction. As a result, ethical concerns about this kind of biotechnology most likely also need to be resolved (Ricroch et al., 2018). The public must see a larger benefit-to-disadvantage ratio for genetically modified (GM) crops to be generally accepted. GM crops are a potent substitute for conventionally cultivated plants in the face of our growing global food demand. A greater degree of openness, as well as improved rules and governing systems, is necessary to ensure that technology is used appropriately. This is seen in the public's serious concerns about modified plants and their products (Munawar et al., 2024).

13.8 FUTURE PERSPECTIVES

Resolving issues concerning disease resistance, stress tolerance, yield optimisation, and environmental adaptability, genetic modification in *Aloe vera* greatly advances sustainable agriculture. *Aloe vera* can be grown more profitably, ecologically, and efficiently by utilizing the potential of genetic technologies. Plant genetic transformation has emerged as a key research technique for the functional characterization of genes and plant development in the field of plant biology. Achieving the given goals requires a transformation efficiency that is both effective and high. In contrast to the tropical cactus plants, the *Aloe vera* family, *Liliaceae*, has gained widespread recognition in the secular world as a medicinal tool and is arguably the most widely used herbal treatment available today. Nonetheless, *Aloe vera*'s genetic transformation is quite minimal and hasn't been documented in the past (Jangra et al., 2023).

Plants, animals, and microorganisms that are utilized as food have been modified to create desired features for around 10,000 years. A common food species now differs genetically from its forebears as a result of these alterations, as well as organic evolutionary changes. Improved flavor and nutritional value; higher food production, yields, and dependability; and a reduction in losses from biotic and abiotic challenges like bacterial and fungal infections are all positive effects of these genetic alterations (National Research Council US, 2004). Introducing desired genetic traits, genetic transformation is a valuable tactic for increasing plant biomass or resistance to harsh environments and population increase. The application of superior features, accurate breeding, and functional analysis of plant genes can all be aided by research on plant genetic transformation technologies (Gidamo, 2023). *Aloe vera* genetic engineering has a wide range of possible advantages that could improve farming methods and other sectors. *Aloe vera* can be genetically modified to display better qualities, like higher output, better disease and insect resistance, and improved nutritional or therapeutic qualities. These improvements may result in increased crop output, more effective growing techniques, and better-quality *Aloe vera* products.

13.9 CONCLUSION

In ancient Egypt, *Aloe vera* was known as the "plant of immortality." Traditional uses of *Aloe vera* involve applying the gel on the skin to soothe burns, wounds, and skin irritations. Its cooling and anti-inflammatory properties make it valuable for topical applications. The latex derived from the plant has been historically used as a laxative, although its use is now regulated due to potential side effects. Ancient civilizations, particularly in Egypt and Greece, incorporated *Aloe vera* into cosmetics for skincare and beauty purposes. Its moisturizing and soothing effects are still highly valued in modern cosmetic formulations. *Aloe vera*'s benefits are becoming increasingly recognized, which could lead to its use being expanded across various industries and geographical areas, ultimately increasing its global significance. *Aloe vera*'s history reflects its enduring significance in traditional practices, and its future importance lies in continued exploration of its therapeutic potential, sustainable cultivation, and versatile applications across diverse industries. The future of *Aloe vera* holds exciting possibilities in areas such as biomedicine, agriculture, cosmetics, and the food industry. Research, technological innovations, and market dynamics will collectively shape the trajectory of *Aloe vera*, making it a multifaceted and valuable resource in various domains.

REFERENCES

Akinsanya, M. A., Goh, J. K., Lim, S. P., and Ting, A. S. Y. (2015). Metagenomics study of endophytic bacteria in *Aloe vera* using next-generation technology. *Genomics Data*, 6, 159–163. https://doi.org/10.1016/j.gdata.2015.09.004

Alamgir, A. N. M. (2018). (ed.) Biotechnology, in vitro production of natural bioactive compounds, herbal preparation, and disease management (treatment and prevention). In: *Therapeutic Use of Medicinal Plants and their Extracts: Phytochemistry and Bioactive Compounds*, Vol. 2 (pp. 585–664). Cham: Springer International Publishing.

Anand, A., Subramanian, M., and Kar, D. (2023). Breeding techniques to dispense higher genetic gains. *Frontiers in Plant Science*, 13, 1076094. https://doi.org/10.3389/fpls.2022.1076094

Badar, Z., Khan, S., and Rao, H. (2017). Analysis of genetic fidelity of wild type and in vitro regenerated *Aloe vera* plants through RAPD and ISSR molecular markers. *International Journal of Biotechnology and Bioengineering*, 3(7), 259–267.

Bado, S., Kozak, K., Sekander, H. et al. (2013). Resurgence of X-rays in mutation breeding. In: *Plant Genetics and Breeding Technologies; Plant Diseases and Resistance Mechanisms: Proceedings*, 18–20 Feb, Vienna, Austria: Medimond-Monduzzi Editore International Proceedings Division, pp. 13–16,

Bai, Y., Niu, Y., Qin, S., and Ma, G. (2023). A new biomaterial derived from *Aloe vera*-Acemannan from basic studies to clinical application. *Pharmaceutics*, 15(7). https://doi.org/10.3390/pharmaceutics15071913

Banyal, S. K. (2023). Aloe ×vera: A new high gel yielding natural hybrid of Aloe vera from CSIR- NBRI botanical garden. 23 August 2023, PREPRINT (Version 1) available at Research Square. https://doi.org/10.21203/rs.3.rs-3274241/v1

Baruah, A., Bordoloi, M., and Baruah, H. P. (2016). *Aloe vera*: A multipurpose industrial crop. *Industrial Crops and Products*, 94, 951–963.

Batra, S., and Kumar, S. (2003). Agrobacterium-mediated transient GUS gene expression in buffel grass (Cenchrus ciliaris L.). *Journal of Applied Genetics*, 44(4), 449–458.

Bawa, A. S., and Anilakumar, K. R. (2013). Genetically modified foods: Safety, risks and public concerns-a review. *Journal of Food Science and Technology*, 50(6), 1035–1046. https://doi.org/10.1007/s13197-012-0899-1

Bhargava, A., and Srivastava, S. (2019). Participatory plant breeding: Concept and applications. *Springer Singapore*, 10, 978–981.

Byeon, S. W., Pelley, R. P., Ullrich, S. E., Waller, T. A., Bucana, C. D., and Strickland, F. M. (1998). Aloe barbadensis extracts reduce the production of interleukin-10 after exposure to ultraviolet radiation. *Journal of Investigative Dermatology*, 110(5), 811–817.

Canche-Escamilla, G., Colli-Acevedo, P., Borges-Argaez, R., Quintana-Owen, P., May-Crespo, J. F., Cáceres-Farfan, M., Yam Puc, J. A., Sansores-Peraza, P., and Vera-Ku, B. M., Sansores-Peraza, P., and Vera-Ku, B. M. (2019). Extraction of phenolic components from an *Aloe vera* (Aloe barbadensis Miller) crop and their potential as antimicrobials and textile dyes. *Sustainable Chemistry and Pharmacy*, 14, 100168.

Cheng, M., Lowe, B. A., Spencer, T. M., Ye, X., and Armstrong, C. L. (2004). Factors influencing Agrobacterium-mediated transformation of monocotyledonous species. *In Vitro Cellular and Developmental Biology-Plant*, 40, 31–45.

Chithra, P., Sajithlal, G. B., and Chandrakasan, G. (1998). Influence of *Aloe vera* on the glycosaminoglycans in the matrix of healing dermal wounds in rats. *Journal of Ethnopharmacology*, 59(3), 179–186. https://doi.org/10.1016/S0378-8741(97)00112-8.

Choi, S. W., Son, B. W., Son, Y. S., Park, Y. I., Lee, S. K., and Chung, M. H. (2001). The wound-healing effect of a glycoprotein fraction isolated from *Aloe vera*. *British Journal of Dermatology*, 145(4), 535–545.

Choudhri, P., Rani, M., Sangwan, R. S., Kumar, R., Kumar, A., and Chhokar, V. (2018). De novo sequencing, assembly and characterisation of *Aloe vera* transcriptome and analysis of expression profiles of genes related to saponin and anthraquinone metabolism. *BMC Genomics*, 19(1), 1–21. https://doi.org/10.1186/s12864-018-4819-2

Chow, J. T. N., Williamson, D. A., Yates, K. M., and Goux, W. J. (2005). Chemical characterization of the immunomodulating polysaccharide of *Aloe vera* L. *Carbohydrate Research*, 340(6), 1131–1142.

Collard, B. C. Y., and Mackill, D. J. (2008). Marker-assisted selection: An approach for precision plant breeding in the twenty-first century. *Philosophical Transactions of the Royal Society B: Biological Sciences*, 363(1491), 557–572. https://doi.org/10.1098/rstb.2007.2170

Crossa, J., Pérez-Rodríguez, P., Cuevas, J., Montesinos-López, O., Jarquín, D., de los Campos, G., Burgueño, J., González-Camacho, J. M., Pérez-Elizalde, S., Beyene, Y., Dreisigacker, S., Singh, R., Zhang, X., Gowda, M., Roorkiwal, M., Rutkoski, J., and Varshney, R. K. (2017). Genomic selection in plant breeding: Methods, models, and perspectives. *Trends in Plant Science,* 22(11), 961–975. https://doi.org/10.1016/j.tplants.2017.08.011.

Daburkar, M., Lohar, V., Rathore, A. S., Bhutada, P., and Tangadpaliwar, S. (2014). An in vivo and in vitro investigation of the effect of *Aloe vera* gel ethanolic extract using animal model with diabetic foot ulcer. *Journal of Pharmacy and Bioallied Sciences*, 6(3), 205.

Danish, P., Ali, Q., Hafeez, M., and Malik, A. (2020). Antifungal and antibacterial activity of *Aloe vera* plant extract. *Biological and Clinical Sciences Research Journal*, 2020(1), 1–8. https://doi.org/10.54112/bcsrj.v2020i1.4

Das, A., Nandagopal, K., and Jha, T. B. (2017). Molecular characterization of some Indian *Aloe vera* populations through RAPD and ITS markers. *Plant Biosystems-An International Journal Dealing with all Aspects of Plant Biology*, 151(4), 695–703.

Economics, P., Khaldoon, A., Ahmad, A., Wei, H., Yousaf, I., Ali, S. S., Naveed, M., Latif, A. S., Abdullah, F., Ab Razak, N. H., Palahuddin, S. H., Sajjad, T., Abbas, N., Hussain, S., SabeehUllah, A. W., Gulzar, M. A., Zongjun, W., Gunderson, M., Gloy, B., Rodgers, C., Orazalin, N., Mahmood, M., … Ishak, R. B. (2020). No 主観的健康感を中心とした在宅高齢者における 健康関連指標に関する共分散構造分析Title. *Corporate Governance (Bingley)*, 10(1), 54–75. https://doi.org/10.7537/marslsj170720.13.

El-Nashar, Y. I., and Asrar, A. A. (2016). Phenotypic and biochemical profile changes in calendula (Calendula officinalis L.) plants treated with two chemical mutagenesis. *Genetics and Molecular Research*, 15(2). https://doi.org/10.4238/gmr.15028071

Fanali, S., Aturki, Z., D'Orazio, G., Rocco, A., Ferranti, A., Mercolini, L., and Raggi, M. A. (2010). Analysis of Aloe-based phytotherapeutic products by using nano-LC-MS. *Journal of Separation Science*, 33(17–18), 2663–2670.

Fateh, A., Idrees, N., Siddique, A., and Arif, A. (2021). Crispr/Cas9 for crop improvement. *Plant Cell Biotechnology and Molecular Biology*, 22 (13), 166–178.

Fehr, W. R. (2007). Breeding for modified fatty acid composition in soybean. *Crop Science*, 47, S–72.

Flórez Martínez, D. H. (2014). Estudio de inteligencia competitiva para la cadena productiva de la sábila. http://agris.fao.og.

Frébortová, J., and Frébort, I. (2021). Biochemical and structural aspects of cytokinin biosynthesis and degradation in bacteria. *Microorganisms*, 9(6), 1314.

Frutos, R., Rang, C., and Royer, M. (1999). Managing insect resistance to plants producing Bacillus thuringiensis toxins. *Critical Reviews in Biotechnology*, 19(3), 227–276. https://doi.org/10.1080/0738-859991229251

Gidamo, G. H. (2023). *Genomic Selection: A Faster Strategy for Plant Breeding*. IntechOpen. https://doi.org/10.5772/intechopen.105398

Gupta, A., and Rawat, S. (2017). Clinical importance of *Aloe vera*: Review. *Research Journal of Topical and Cosmetic Sciences*, 8(1), 30. https://doi.org/10.5958/2321-5844.2017.00004.8

Gupta, V., and Malhotra, S. (2012). Pharmacological attribute of *Aloe vera*: Revalidation through experimental and clinical studies. *AYU (An International Quarterly Journal of Research in Ayurveda)*, 33(2), 193. https://doi.org/10.4103/0974-8520.105237

Hamman, J. H. (2008). Composition and applications of *Aloe vera* leaf gel. *Molecules*, 13(8), 1599–1616.

Haritha Yadav, K. C., Ravi Kumar, J., Basha, S. I., Deshmukh, G. R., Gujjula, R., and Santhamma, B. (2012). Wound healing activity of topical application of *Aloe vera* gel in experimental animal models. *International Journal of Pharma and Bio Sciences*, 3(2), P63–P72.

He, C., Zhang, J., Chen, J., Ye, X., Du, L., Dong, Y., and Zhao, H. (2007). Genetic transformation of Aloe barbadensis miller by Agrobacterium tumefaciens. *Journal of Genetics and Genomics*, 34(12), 1053–1060. https://doi.org/10.1016/S1673-8527(07)60120-9

Hefferon, K. L., and Herring, R. J. (2017). The end of the GMO? Genome editing, gene drives and new frontiers of plant technology. *Review of Agrarian Studies*, 7(1), 1–32.

Heggers, J. P., Kucukcelebi, A., Listengarten, D., Stabenau, J., Ko, F., Broemeling, L. D., ... and Winters, W. D. (1996). Beneficial effect of Aloe on wound healing in an excisional wound model. *The Journal of Alternative and Complementary Medicine*, 2(2), 271–277.

Im, S. A., Oh, S. T., Song, S., Kim, M. R., Kim, D. S., Woo, S. S., ... and Lee, C. K. (2005). Identification of optimal molecular size of modified Aloe polysaccharides with maximum immunomodulatory activity. *International Immunopharmacology*, 5(2), 271–279.

Ishida, Y., Hiei, Y., and Komari, T. (2007). Agrobacterium-mediated transformation of maize. *Nature Protocols*, 2(7), 1614–1621. https://doi.org/10.1038/nprot.2007.241

Jain, M. (2012). Next-generation sequencing technologies for gene expression profiling in plants. *Briefings in Functional Genomics*, 11(1), 63–70. https://doi.org/10.1093/bfgp/elr038

Jaiswal, S. K., Mahajan, S., Chakraborty, A., Kumar, S., and Sharma, V. K. (2021). The genome sequence of *Aloe vera* reveals adaptive evolution of drought tolerance mechanisms. *IScience*, 24(2), 102079. https://doi.org/10.1016/j.isci.2021.102079

Jangra, A., Chaturvedi, S., Kumar, N., Singh, H., Sharma, V., Thakur, M., Tiwari, S., and Chhokar, V. (2023). Polyamines: The gleam of next-generation Plant growth regulators for growth, development, stress mitigation, and hormonal crosstalk in plants—A systematic review. *Journal of Plant Growth Regulation*, 42(8), 5167–5191.

Jangra, A., Chaturvedi, S., Sharma, G., Sihag, S., Tiwari, S., and Chhokar, V. (2023). Efficient Agrobacterium tumefaciens-mediated genetic transformation of *Aloe vera*. *Plant Cell, Tissue and Organ Culture (PCTOC)*, 154, 189–200.

Jangra, A., Sharma, G., Sihag, S., and Chhokar, V. (2022). The dark side of miracle plant-*Aloe vera*: A review. *Molecular Biology Reports*, 49(6), 5029–5040.

Kahramanoğlu, I., Chen, C., Chen, J., and Wan, C. (2019). Chemical constituents, antimicrobial activity and food preservative characteristics of *Aloe vera* gel. *Agronomy*, 9(12). https://doi.org/10.3390/agronomy9120831

Kapoor, D., Bhardwaj, S., Landi, M., Sharma, A., Ramakrishnan, M., and Sharma, A. (2020). The impact of drought in plant metabolism: How to exploit tolerance mechanisms to increase crop production. *Applied Sciences*, 10(16), 5692.

Karagianis, G., Viljoen, A., and Waterman, P. G. (2003). Identification of major metabolites in Aloe littoralis by high-performance liquid chromatography–nuclear magnetic resonance spectroscopy. *Phytochemical Analysis: An International Journal of Plant Chemical and Biochemical Techniques*, 14(5), 275–280.

Khan, F., and Hakeem, K. R. (2015). Genetic modification of crop plants: Issues and challenges. In: Khalid R. Hakeem (ed.) *Crop Production and Global Environmental Issues* (pp. 369–384). Cham: Springer.

Kourkoutas, Y., Chorianopoulos, N., Nisiotou, A., Valdramidis, V. P., and Karatzas, K. A. (2016). Application of innovative technologies for improved food quality and safety. *BioMed Research International*, 2016, 9160375.

Kumar, K. S., and Debjit, B. (2010). *Aloe vera*: A potential herb and its medicinal importance. *Journal of Chemical and Pharmaceutical Research*, 2(1), 21–29.

Kumar, R., Salar, R. K., Kumar, A., Kumar, A., and Chhokar, V. (2018). A low cost, high throughput gel electrophoresis method for separation of SSR markers in *Aloe vera*. *The Pharma Innovation Journal*, 7(3), 622–627.

Kumar, R., Salar, R. K., Naik, P. K., Yadav, M., Kumar, A., Kumar, A., ... and Chhokar, V. (2022). Elucidation of genetic diversity and population structure of sixty genotypes of *Aloe vera* using AFLP markers. *South African Journal of Botany*, 147, 1146–1155.

Kumar, S., Kalita, S., Das, A., Kumar, P., Singh, S., Katiyar, V., and Mukherjee, A. (2022). *Aloe vera*: A contemporary overview on scope and prospects in food preservation and packaging. *Progress in Organic Coatings*, 166, 106799.

Kuzuya, H., Tamai, I., Beppu, H., Shimpo, K., and Chihara, T. (2001). Determination of aloenin, barbaloin and isobarbaloin in Aloe species by micellar electrokinctic chromatography. *Journal of Chromatography B: Biomedical Sciences and Applications*, 752(1), 91–97.

Lamichhane, S., and Thapa, S. (2022). Advances from conventional to modern plant breeding methodologies. *Plant Breeding and Biotechnology*, 10(1), 1–14. https://doi.org/10.9787/PBB.2022.10.1.1

Liu, Q., Yang, X., Tzin, V., Peng, Y., Romeis, J., and Li, Y. (2020). Plant breeding involving genetic engineering does not result in unacceptable unintended effects in rice relative to conventional cross-breeding. *Plant Journal*, 103(6), 2236–2249. https://doi.org/10.1111/tpj.14895

Ma, L., Kong, F., Sun, K., Wang, T., and Guo, T. (2021). from classical radiation to modern radiation: Past, present, and future of radiation mutation breeding. *Frontiers in Public Health*, 9(December), 1–11. https://doi.org/10.3389/fpubh.2021.768071

Ma, X., Zhu, Q., Chen, Y., and Liu, Y. G. (2016). CRISPR/Cas9 platforms for genome editing in plants: Developments and applications. *Molecular Plant*, 9(7), 961–974. https://doi.org/10.1016/j.molp.2016.04.009

Martínez-Burgos, W. J., Serra, J. L., Marsiglia F. R. M., Montoya, P., Sarmiento-Vásquez, Z., Marin, O., ... and Paternina-Arboleda, C. D. (2022). *Aloe vera*: From ancient knowledge to the patent and innovation landscape–A review. *South African Journal of Botany*, 147, 993–1006.

Matijevic, M., Bado, S., Lagoda, P. J. L., and Forster, B. P. (2013). Impact of induced mutations in plant breeding. In: *Plant Genetics and Breeding Technologies; Plant Diseases And Resistance Mechanisms: Proceedings*, 18–20 Feb, Vienna, Austria: Medimond-Monduzzi Editore international Proceedings Division, pp 45–47.

Chen, J. Q., Meng, X. P., Zhang, Y., Xia, M., and Wang, X. P. (2008). Over-expression of OsDREB genes lead to enhanced drought tolerance in rice. *Biotechnology Letters*, 30(12), 2191–2198.

Munawar, N., Ahsan, K., and Ahmad, A. (2024). CRISPR-edited plants' social, ethical, policy, and governance issues. In: Kamel A. Abd-Elsalam and Aftab Ahmad (eds.) *Global Regulatory Outlook for CRISPRized Plants* (pp. 367–396). United Kingdom: Academic Press.

Mwangangi, I. M., Kiilu Muli, J., and Neondo, J. O. (2019). Plant hybridization as an alternative technique in plant breeding improvement. *Asian Journal of Research in Crop Science*, 4(1), 1–11. https://doi.org/10.9734/ajrcs/2019/v4i130059

Noorolahi, S. M., Sadeghi, S., Mohammadi, M., Azadi, M., Rahimi, N. A., Vahabi, F., Arjmand, M., Hosseini, H., Mosallatpur, S., and Zamani, Z. (2016). Metabolomic profiling of cancer cells to *Aloe vera* extract by 1HNMR spectroscopy. *Journal of Metabolomics*, 2(1), 1. https://doi.org/10.7243/2059-0008-2-1

Okamura, N., Asai, M., Hine, N., and Yagi, A. (1996). High-performance liquid chromatographic determination of phenolic compounds in Aloe species. *Journal of Chromatography A*, 746(2), 225–231.

Oladosu, Y., Rafii, M. Y., Abdullah, N., Hussin, G., Ramli, A., Rahim, H. A., Miah, G., and Usman, M. (2016). Principle and application of plant mutagenesis in crop improvement: A review. *Biotechnology and Biotechnological Equipment*, 30(1), 1–16. https://doi.org/10.1080/13102818.2015.1087333

Orton, T. J. (2019). *Horticultural Plant Breeding*. United Kingdom: Elsevier Science. https://books.google.co.in/books?id=rte_DwAAQBAJ

Ouwehand, A. C., Salminen, S., and Isolauri, E. (2002). Probiotics: An overview of beneficial effects. In: *Lactic Acid Bacteria: Genetics, Metabolism and Applications: Proceedings of the Seventh Symposium on Lactic Acid Bacteria: Genetics, Metabolism and Applications*, 1–5 September, (pp. 279–289). Netherlands: Springer.

Park, M. K., Park, J. H., Kim, N. Y., Shin, Y. G., Choi, Y. S., Lee, J. G., ... and Lee, S. K. (1998). Analysis of 13 phenolic compounds in Aloe species by high performance liquid chromatography. *Phytochemical Analysis: An International Journal of Plant Chemical and Biochemical Techniques*, 9(4), 186–191.

Phillips, T. (2008). Genetically modified organisms (GMOs): Transgenic crops and recombinant DNA technology. *Nature Education,* 1(1), 213. www.nature.com/scitable/topicpage/genetically-modified-organisms-gmos-transgenic-crops-nbsp-732/

Prabjone, R., Thong-Ngam, D., Wisedopas, N., Chatsuwan, T., and Patumraj, S. (2006). Anti-inflammatory effects of *Aloe vera* on leukocyte–endothelium interaction in the gastric microcirculation of Helicobacter pylori-infected rats. *Clinical Hemorheology and Microcirculation*, 35(3), 359–366.

Radha, M. H., and Laxmipriya, N. P. (2015). Evaluation of biological properties and clinical effectiveness of *Aloe vera*: A systematic review. *Journal of Traditional and Complementary Medicine*, 5, 21–26.

Raksha, B., Pooja, S., and Babu, S. (2014). Bioactive compounds and medicinal properties of *Aloe vera* L.: An update. *Journal of Plant Sciences*, 2(3), 102–107.

Rayaprolu, S. J., Hettiarachchy, N. S., Chen, P., Kannan, A., and Mauromostakos, A. (2013). Peptides derived from high oleic acid soybean meals inhibit colon, liver and lung cancer cell growth. *Food Research International*, 50(1), 282–288.

Rebecca, W., Kayser, O., Hagels, H., Zessin, K. H., Madundo, M., and Gamba, N. (2003). The phytochemical profile and identification of main phenolic compounds from the leaf exudate of Aloe secundiflora by high-performance liquid chromatography–mass spectroscopy. *Phytochemical Analysis: An International Journal of Plant Chemical and Biochemical Techniques*, 14(2), 83–86.

Reynolds, T. (Ed.). (2004). *Aloes: The Genus Aloe*. Florida: CRC Press.

Reynolds, T., and Dweck, A. C. (1999). *Aloe vera* leaf gel: A review update. *Journal of Ethnopharmacology*, 68, 3–37.

Ricroch, A. E., Guillaume-Hofnung, M., and Kuntz, M. (2018). The ethical concerns about transgenic crops. *Biochemical Journal*, 475(4), 803–811.

Roberfroid, M. B. (2000). Concepts and strategy of functional food science: The European perspective. *The American Journal of Clinical Nutrition*, 71(6), 1660S–1664S.

Roberts, D. B., and Travis, E. L. (1995). Acemannan-containing wound dressing gel reduces radiation-induced skin reactions in C3H mice. *International Journal of Radiation Oncology* Biology* Physics*, 32(4), 1047–1052.

Saccù, D., Bogoni, P., and Procida, G. (2001). Aloe exudate: Characterization by reversed phase HPLC and headspace GC-MS. *Journal of Agricultural and Food Chemistry*, 49(10), 4526–4530.

Salminen, S., Ouwehand, A., Benno, Y., and Lee, Y. K. (1999). Probiotics: How should they be defined? *Trends in Food Science and Technology*, 10(3), 107–110.

Salinas, P., Salinas, C., Contreras, R. A., Zuñiga, G. E., Dupree, P., and Cardemil, L. (2019). Water deficit and abscisic acid treatments increase the expression of a glucomannan mannosyltransferase gene (GMMT) in Aloe vera Burm. F. *Phytochemistry*, 159, 90-101.

Sánchez, M., González-Burgos, E., Iglesias, I., and Gómez-Serranillos, M. P. (2020). Pharmacological update properties of *Aloe vera* and its major active constituents. *Molecules*, 25(6), 1–37. https://doi.org/10.3390/molecules25061324

Sato, Y., Ohta, S., and Shinoda, M. (1990). Studies on chemical protectors against radiation. XXXI. Protection effects of Aloe arborescens on skin injury induced by X-irradiation. *Yakugaku Zasshi: Journal of the Pharmaceutical Society of Japan*, 110(11), 876–884.

Sharon, K., and Suvarna, S. (2017). Cloning of HVA22 homolog from Aloe vera and preliminary study of transgenic plant development. *International Journal of Pure & Applied Bioscience*, 5(6), 1113–1121.

Shen, Y., Liu, J., Geng, H., Zhang, J., Liu, Y., Zhang, H., ... and Tian, Z. (2018). De novo assembly of a Chinese soybean genome. *Science China Life Sciences*, 61, 871–884.

Shirasawa, K., Hirakawa, H., Nunome, T., Tabata, S., and Isobe, S. (2016). Genome-wide survey of artificial mutations induced by ethyl methanesulfonate and gamma rays in tomato. *Plant Biotechnology Journal*, 14(1), 51–60. https://doi.org/10.1111/pbi.12348

Shivani, and Tiwari, S. (2019). Enhanced Agrobacterium-mediated transformation efficiency of banana cultivar Grand Naine by reducing oxidative stress. *Scientia Horticulturae*, 246(October 2018), 675–685. https://doi.org/10.1016/j.scienta.2018.11.024

Shu, Q. Y., Forster, B. P., and Nakagawa, H. (2012). *Plant Mutation Breeding and Biotechnology*. Wallingford/Cambridge: CABI International.

Sierra-García, G. D., Castro-Ríos, R., González-Horta, A., Lara-Arias, J., and Chávez-Montes, A. (2014). Acemannan, an extracted polysaccharide from *Aloe vera*: A literature review. *Natural Product Communications*, 9(8), 1217–1221. https://doi.org/10.1177/1934578x1400900836

Sigaroodi, F., Shafaei, H., Karimipour, M., Dolatkhah, M. A., and Delazar, A. (2019). Aloe vera/collagen mixture induces integrin α1β1 and PECAM-1 genes expression in human adipose-derived stem cells. *Advanced Pharmaceutical Bulletin*, 9(4), 662.

Silva, H., Sagardia, S., Seguel, O., Torres, C., Tapia, C., Franck, N., and Cardemil, L. (2010). Effect of water availability on growth and water use efficiency for biomass and gel production in *Aloe vera* (Aloe barbadensis M.). *Industrial Crops and Products*, 31(1), 20–27. https://doi.org/10.1016/j.indcrop.2009.08.001

Steenkamp, E. J. (2015). A profile on the aloe industry for export-A focus on South Africa. Directorate International Trade, Department of Agriculture. Forestry and Fisheries, Pretoria.

Stintzi, A., Heitz, T., Prasad, V., Wiedemann-Merdinoglu, S., Kauffmann, S., Geoffroy, P., Legrand, M., and Fritig, B. (1993). Plant "pathogenesis-related" proteins and their role in defense against pathogens. *Biochimie*, 75(8), 687–706. https://doi.org/10.1016/0300-9084(93)90100-7

Surjushe, A., Vasani, R., and Saple, D. G. (2008). *Aloe vera*: A short review. *Indian Journal of Dermatology*, 53(4), 163–166. https://doi.org/10.4103/0019-5154.44785. PMID: 19882025; PMCID: PMC2763764.

Valliyodan, B., Cannon, S. B., Bayer, P. E., Shu, S., Brown, A. V., Ren, L., ... and Nguyen, H. T. (2019). Construction and comparison of three reference-quality genome assemblies for soybean. *The Plant Journal*, 100(5), 1066–1082.

Vázquez, B., Avila, G., Segura, D., and Escalante, B. (1996). Antiinflammatory activity of extracts from *Aloe vera* gel. *Journal of Ethnopharmacology*, 55(1), 69–75.

Vogler, B. K., and Ernst, E. (1999). *Aloe vera*: A systematic review of its clinical effectiveness. *British Journal of General Practice*, 49(447), 823–828.

Wang, P. G., Zhou, W., Wamer, W. G., Krynitsky, A. J., and Rader, J. I. (2012). Simultaneous determination of aloin A and aloe emodin in products containing *Aloe vera* by ultra-performance liquid chromatography with tandem mass spectrometry. *Analytical Methods*, 4(11), 3612–3619.

Wani, M. R., Laskar, R. A., Raina, A., Khan, S., and Khan, T. U. (2021). Application of chemical mutagenesis for improvement of productivity traits in Lentil (Lens culinaris medik). *Annals of Biology*, 37(1), 69–75.

West, D. P., and Zhu, Y. F. (2003). Evaluation of *Aloe vera* gel gloves in the treatment of dry skin associated with occupational exposure. *American Journal of Infection Control*, 31(1), 40–42.

Woudstra, Y., Viruel, J., Fritzsche, M., Bleazard, T., Mate, R., Howard, C., ... and Grace, O. M. (2021). A customised target capture sequencing tool for molecular identification of Aloe vera and relatives. *Scientific Reports*, 11(1), 24347.

Xie, M., Chung, C. Y. L., Li, M. W., Wong, F. L., Wang, X., Liu, A., ... and Lam, H. M. (2019). A reference-grade wild soybean genome. *Nature Communications*, 10(1), 1216.

Yasmin, S., Hasan, M. J., Hossain, M. S., Saha, S., and Khatun, F. (2022). Auxin and cytokinin synergism in micropropagation for mass production of *Aloe vera*. *BioTechnologia*, 103(3), 301.

Zhang, L., and Tizard, I. R. (1996). Activation of a mouse macrophage cell line by acemannan: The major carbohydrate fraction from *Aloe vera* gel. *Immunopharmacology*, 35(2), 119–128.

Zonta, F., Bogoni, P., Masotti, P., and Micali, G. (1995). High-performance liquid chromatographic profiles of aloe constituents and determination of aloin in beverages, with reference to the EEC regulation for flavouring substances. *Journal of Chromatography A*, 718(1), 99–106.

Index

For Product Safety Concerns and Information please contact our EU representative GPSR@taylorandfrancis.com Taylor & Francis Verlag GmbH, Kaufingerstraße 24, 80331 München, Germany